T0358307

Bounded Symmetric Domains in Banach Spaces

Bounded Symmetric Domains in Banach Spaces

Cho-Ho Chu
Queen Mary, University of London, UK

World Scientific

NEW JERSEY · LONDON · SINGAPORE · BEIJING · SHANGHAI · HONG KONG · TAIPEI · CHENNAI · TOKYO

Published by

World Scientific Publishing Co. Pte. Ltd.

5 Toh Tuck Link, Singapore 596224

USA office: 27 Warren Street, Suite 401-402, Hackensack, NJ 07601

UK office: 57 Shelton Street, Covent Garden, London WC2H 9HE

Library of Congress Cataloging-in-Publication Data

Names: Chu, Cho-Ho, author.

Title: Bounded symmetric domains in Banach spaces /
  Cho-Ho Chu, Queen Mary, University of London, UK.

Description: New Jersey : World Scientific, [2020] | Includes bibliographical references and index.

Identifiers: LCCN 2020025418 | ISBN 9789811214103 (hardcover) |
  ISBN 9789811214110 (ebook for institutions) | ISBN 9789811214127 (ebook for individuals)

Subjects: LCSH: Banach spaces. | Symmetric domains.

Classification: LCC QA322.2 .C48 2020 | DDC 515/.732--dc23

LC record available at https://lccn.loc.gov/2020025418

British Library Cataloguing-in-Publication Data

A catalogue record for this book is available from the British Library.

For any available supplementary material, please visit
https://www.worldscientific.com/worldscibooks/10.1142/11659#t=suppl

Printed in Singapore

Contents

Preface

This book discusses aspects of bounded symmetric domains and their algebraic structures. The focus is on infinite dimensional domains and some recent advances in the geometric and analytic theory of these domains, of which the dissemination has been confined to research papers so far and a self-contained monographic exposition seems timely. With a concise bibliography, it is intended as a convenient reference for a broad readership including research students.

Since the seminal work of É. Cartan, symmetric spaces in finite dimensions have been intensively studied and constitute an important area of research in geometry and literature abounds. Lie theory has been an important tool in the investigation of these manifolds and their classification. In finite dimensions, bounded symmetric domains are exactly the class of Hermitian symmetric spaces of non-compact type, via the Harish-Chandra realisation.

The finite dimensional concept of a symmetric domain can be extended naturally to infinite dimension and in recent decades, Jordan algebras and Jordan triple systems have gradually become a significant part of the theory of bounded symmetric domains due to the successful application of them in providing a unified and fruitful treatment of both finite and infinite dimensional symmetric domains. Nevertheless, this remarkable success also hinges on the close relationship between Jordan and Lie algebras. A far-reaching accomplishment is the discovery that every bounded symmetric domain can be realised as the open unit ball of a complex Banach space equipped with a Jordan structure. This led to fertile interactions with infinite dimensional analysis and operator theory.

In this book, we present an introduction to some basic theory of infinite dimensional Jordan and Lie algebras (but not excluding the finite dimensional ones) and explain in detail how they are used to show that a bounded symmetric domain is biholomorphic to the open unit ball of a Banach space with a Jordan structure. We discuss various applications of this realisation of a bounded symmetric domain. The most important ones concern the classification and geometric function theory of these domains.

The book begins with some basic concepts and notations in complex analysis in Banach spaces and a brief review of Banach manifolds. Symmetric Banach manifolds, which generalise finite dimensional Hermitian symmetric spaces, are introduced, together with the associated Lie structures.

As alluded to earlier, Jordan and Lie algebras play an important role in the theory of bounded symmetric domains. We discuss Jordan and Lie structures in Chapter 2, with a view for later applications, which requires a dimension-free setting: infinite dimensional algebras are not excluded. We first discuss the structures of Jordan algebras and their generalisation, Jordan triple systems. The Tits-Kantor-Koecher construction is shown in Section 2.3, which establishes the one-one correspondence between Jordan triple systems and a subclass of Lie algebras. This will be vital later for constructing a holomorphic embedding of a bounded symmetric domain in a complex Banach space equipped with a Jordan triple structure, alias JB*-triple. The concept of JB*-triples is introduced in Section 2.4, where a number of examples are given. These include complex spaces of matrices, operators, Hilbert spaces and interestingly, C*-algebras. This hints at a close connection with functional analysis. The fundamental examples of JB*-triples, known as *Cartan* *factors*, are discussed in Section 2.5. They are the building blocks of JB*-triples, but crucially, the classifying spaces of bounded symmetric domains.

Chapter 3 is the highlight of the book, where a full discussion of bounded symmetric domains takes place. We begin by discussing the Jordan and Lie structures of symmetric Banach manifolds, which are of fundamental importance. This paves the way for the major task in Section 3.2 of proving one of the most important theorems in the book, due to Kaup [99], which asserts, in Theorem 3.2.18 and Theorem 3.2.20, that a domain in a complex Banach space is a bounded symmetric domain if and only if it is biholomorphic to the open unit ball of a JB*-triple. A finite dimensional precursor has been shown by Loos [125]. Kaup's result can be viewed as a version of the Riemann mapping theorem for all dimensions since the complex plane is a one-dimensional JB*-triple. It is hard to exaggerate the consequences of this theorem. It offers us, in addition to other methods, a Jordan approach to geometry of symmetric manifolds and we do just that in ensuing sections, by identifying a bounded symmetric domain as the open unit ball of a JB*-triple. We show in Section 3.3 that the rank of a bounded symmetric domain can be defined by the ambient JB*-triple structure. This leads to a classification of finite-rank bounded symmetric domains in terms of JB*-triples. In Section 3.4, we study the boundary structures of a bounded symmetric domain and describe its boundary components in terms of Jordan structures. The success of the Jordan approach relies on two fundamental bounded linear operators on a JB*-triple, namely, the Bergman operator and the left multiplication, called a *box operator*. They can be used to describe the Möbius transformations and automorphisms of a bounded symmetric domain. Using this device, we study the Carathéodory metric and Kobayashi metric on bounded symmetric domains in Section 3.5 and inevitably, the Schwarz lemma comes to the fore. We discuss variants of the Schwarz lemma, leading to the Denjoy-Wolff theorem and the investigation of iteration of holomorphic self-maps on a bounded symmetric domain. The holomorphic equivalence of the open unit disc and the upper half-plane in $\mathbb{C}$, via the Cayley transform, is fundamental in complex analysis. In Section 3.6, we discuss the finite and infinite dimensional generalisations of the upper half-plane. These are called the *Siegel domains*, which are defined over a positive cone. The main question is, regarding bounded symmetric domains as a generalisation of the open unit disc in $\mathbb{C}$, when are they holomorphically equivalent to Siegel domains? We provide an answer and generalise the Cayley transform in this section, where Jordan algebras play an important role. Siegel domains of the first kind over a cone C in a Hilbert space is biholomorphic to a symmetric domain exactly when C is a linearly homogeneous self-dual cone. In finite dimensions, Koecher [109] and Vinberg [170] have shown that these cones are in one-one correspondence with the formally real Jordan algebras. One can find in Section 2.4 an infinite dimensional extension of this correspondence. In Section 3.7, as a further application of Jordan theory, we determine completely which bounded symmetric domains are holomorphic homogeneous regular manifolds. The notion of these manifolds in finite dimensions has been introduced by Liu, Sun and Yau [120] in connection with the estimation of several canonical metrics on the moduli and Teichmüller spaces of Riemann surfaces. We extend this concept to infinite dimension and show its connection with the rank of a bounded symmetric domain. The last section of this chapter is devoted to the classification of bounded symmetric domains. We discuss a Jordan approach to the classification of these domains and show that É. Cartan's seminal classification of finite dimensional bounded symmetric domains can be viewed as a special case of this Jordan classification.

In the last chapter, the realisation of bounded symmetric domains in JB*-triples is used again to render an effective and unified treatment of various topics in function theory on these domains, in all dimensions. The main discussions include distortion theorems, Bloch functions and the Bloch constant in the first three sections, as well as composition operators on Banach spaces of Bloch functions in the last section. These are familiar topics in several complex variables, but have only been studied recently in infinite dimension. However, the content of the chapter is by no means exhaustive and needless to say, the geometric function theory of infinite dimensional bounded symmetric domains has yet to be fully developed. It is hope that this 'handbook' may help to facilitate the enterprise.

To conclude, I wish to thank my wife for her constant support and encouragement, without which this book would not have materialised.

Acknowledgement

This work was partly supported by a research grant (EP/R044228/1) from the Engineering and Physical Sciences Research Council in UK.

Chapter 1

Introduction

1.1 Holomorphic maps in Banach spaces

To facilitate ensuing discussions, we begin with an introduction to some basic concepts in complex analysis and geometry as well as notations. In this first section, we discuss holomorphic maps in finite and infinite dimensional settings.

Throughout the book, the open unit disc in the complex plane $\mathbb{C}$ is denoted by

$$\mathbb{D} = \{z \in \mathbb{C} : |z| < 1\}$$

and a *domain* in a Banach space is meant to be a non-empty open connected set. The open unit ball of a complex Hilbert space will be called a *Hilbert ball*. A domain D in a complex Banach space is called *circular* if $z \in D$ implies $e^{i\theta}z \in D$ for all $\theta \in \mathbb{R}$; it is called *balanced* if $\lambda D \subset D$ for all $\lambda \in \mathbb{D}$. A balanced domain contains the origin 0.

As usual, we denote the *closure* of a set E in a Banach space by $\overline{E}$, the *interior* of E by E^0 or int E, and the topological boundary of E by ∂E. It would be impossible for the latter notation to be confused with partial differentiation.

Let V and W be (real or complex) Banach spaces, and let $\mathcal{U}$ be an

1

open subset of V. A function $f : \mathcal{U} \longrightarrow W$ is said to be *differentiable* at a point $a \in \mathcal{U}$ if there is a continuous linear map $f'(a) : V \longrightarrow W$ satisfying

$$\lim_{h \to 0} \frac{\|f(a+h) - f(a) - f'(a)(h)\|}{\|h\|} = 0.$$

The map $f'(a)$ is unique and called the (*Fréchet*) *derivative of f at a*, which is sometimes denoted by $D_a f$ or df_a. The function f is called *differentiable* in $\mathcal{U}$ if it is differentiable at every point in $\mathcal{U}$, in which case the *derivative f'* is a mapping

$$f' : \mathcal{U} \longrightarrow L(V, W)$$

where $L(V, W)$ is the Banach space of continuous linear operators from V to W. We say that f is *continuously differentiable* in $\mathcal{U}$, or of *class C^1*, if the derivative f' is continuous on $\mathcal{U}$. One defines k-times continuously differentiable functions, or C^k-functions, for $k \in \mathbb{N}$, by iteration. A *smooth function* on $\mathcal{U}$ is one that is infinitely differentiable, that is, it is in the class C^k for all $k \in \mathbb{N}$.

A basic rule in differentiation is the chain rule which states that the composite $f \circ g$ of two differentiable functions f and g, whenever well-defined, is differentiable and the derivative is given by

$$f'(g(a)) \circ g'(a)$$

for a in an open set $\mathcal{U}$ where g is defined. A very useful theorem on differentiation is the following mean value theorem (cf. [57, 8.5.4]).

Theorem 1.1.1. *Let $f : \mathcal{U} \longrightarrow W$ be a differentiable function on an open set $\mathcal{U}$ which contains the segment $\{a + sh : a, h \in \mathcal{U}, 0 \le s \le 1\}$. Then we have*

$$\|f(a+h) - f(h)\| \le \|h\| \sup_{0 \le s \le 1} \|f'(a+sh)\|.$$

In finite dimensions, the complex d-dimensional Euclidean space $\mathbb{C}^d$ is often identified with the real $2d$-dimensional Euclidean space $\mathbb{R}^{2d}$ via the real linear isomorphism

$$(z_1, \ldots, z_j, \ldots, z_d) \in \mathbb{C}^d \mapsto (x_1, y_1, \ldots, x_j, y_j, \ldots, x_d, y_d) \in \mathbb{R}^{2d}$$

where we usually write $z_j = x_j + iy_j \in \mathbb{R} + i\mathbb{R}$, and the real linear isomorphism

$$z = (z_1, \ldots, z_j, \ldots, z_d) \in \mathbb{C}^d \mapsto iz = (iz_1, \ldots, iz_j, \ldots, iz_d) \in \mathbb{C}^d$$

gives rise to a real linear isomorphism

$$\begin{aligned} J \ : \ &(x_1, y_1, \ldots, x_j, y_j, \ldots, x_d, y_d) \in \mathbb{R}^{2d} \\ &\mapsto (-y_1, x_1, \ldots, -y_j, x_j, \ldots, -y_d, x_d) \in \mathbb{R}^{2d} \end{aligned} \tag{1.1}$$

such that $-J^2$ is the identity map on $\mathbb{R}^{2d}$. We call J the *canonical complex structure* on $\mathbb{R}^{2d}$.

Differentiability of a complex function in $\mathbb{C}^d$ can be characterized by the Cauchy-Riemann equations, which can be described in terms of the differential operators

$$\frac{\partial}{\partial z_j} = \frac{1}{2} \left(\frac{\partial}{\partial x_j} - i \frac{\partial}{\partial y_j} \right) \quad \text{and} \quad \frac{\partial}{\partial \bar{z}_j} = \frac{1}{2} \left(\frac{\partial}{\partial x_j} + i \frac{\partial}{\partial y_j} \right).$$

By a well-known theorem of Hartogs, a function $f : \mathcal{U} \longrightarrow \mathbb{C}$ having continuous partial derivatives at each point of an open set $\mathcal{U} \subset \mathbb{C}^d$ is differentiable if and only if it satisfies the Cauchy-Riemann equations

$$\frac{\partial f}{\partial \bar{z}_j} = 0 \qquad (j = 1, \ldots, d).$$

If V and W are complex Banach spaces, a differentiable function $f : \mathcal{U} \longrightarrow W$ is smooth and is usually called *holomorphic*. In addition, it has a local power series expansion which is made precise below.

First, for $n \in \mathbb{N}$, the vector space $L^n(V, W)$ of all continuous n-linear maps

$$F : \underbrace{V \times \cdots \times V}_{n\text{-times}} \longrightarrow W$$

is a Banach space in the norm

$$\|F\| = \sup_{v_k \neq 0} \frac{\|F(v_1, \ldots, v_n)\|}{\|v_1\| \cdots \|v_n\|}.$$

We define $L^0(V, W) = W$. If W is the underlying scalar field of the Banach space V, then $L^1(V, W)$ is just the dual space V^* of V.

Definition 1.1.2. A continuous map $p : V \longrightarrow W$ between Banach spaces is called a *homogeneous polynomial of degree n* if there exists $P \in L^n(V, W)$ such that

$$p(v) = P(v, \ldots, v) \qquad (v \in V).$$

If P is chosen to be *symmetric*, that is, invariant under permutation of variables, then P is uniquely determined by p and is called the *polar form* of p (cf. [84]).

We denote by $\mathcal{P}^n(V, W)$ the vector space of all homogeneous polynomials of degree n from V to W, and equip it with the norm

$$\|p\| = \sup_{v \neq 0} \frac{\|p(v)\|}{\|v\|} \qquad (p \in \mathcal{P}^n(V, W)).$$

We note that $\mathcal{P}^n(V, W)$ is a closed subspace of the space $C(V, W)$ of continuous maps from V to W, in the pointwise topology.

Definition 1.1.3. Let V and W be Banach spaces. A *power series* from V to W is a formal sum

$$\sum_{n=0}^{\infty} p_n$$

where $p_n \in \mathcal{P}^n(V, W)$. Its *radius of convergence* is defined to be the largest non-negative number $R \leq \infty$ such that the series

$$\sum_{n=0}^{\infty} p_n(v) \qquad (v \in V)$$

converges uniformly for $\|v\| \leq r$ and $r < R$.

As in the scalar case, the radius of convergence R can be obtained by

$$R = \frac{1}{\limsup_n \|p_n\|^{1/n}}.$$

Likewise, one can define the *radius of convergence* R for the series $\sum_{n=0}^{\infty} F_n$ where $F_n \in L^n(V, W)$, for which we also have

$$R = \frac{1}{\limsup_n \|F_n\|^{1/n}},$$

so that

$$\sum_{n=0}^{\infty} F_n(v_1, \ldots, v_n)$$

converges uniformly whenever $\max(\|v_1\|, \ldots, \|v_n\|) \leq r$ and $r < R$.

A function $f : \mathcal{U} \longrightarrow W$ from an open set $\mathcal{U}$ in a Banach space V to another one W is said to be *analytic* at a point $a \in \mathcal{U}$ if it can be expressed as a convergent power series about a which means that there is a power series $\sum_n p_n$ with positive radius of convergence such that

$$f(v) = \sum_{n=0}^{\infty} p_n(v - a)$$

for each v in some neighbourhood of a. An *analytic* function $f : \mathcal{U} \longrightarrow W$ is one that is analytic at every point in $\mathcal{U}$. If an analytic function $f : \mathcal{U} \longrightarrow W$ is bijective and the inverse f^{-1} is analytic, then f is called *bianalytic*.

Analytic functions are smooth but not vice versa. For instance, the function

$$f(x) = \begin{cases} \exp(-x^{-1}) & (x > 0) \\ 0 & (x \leq 0) \end{cases}$$

is smooth on $\mathbb{R}$ but not analytic at the origin 0. (We note that the same example has been given in [37, p. 64], but a minus sign is missing there.) For complex Banach spaces, however, holomorphic functions are analytic and the term *"biholomorphic"* is a synonym of "bianalytic".

If an analytic function f has a power series representation $\sum_n p_n$ about $a \in \mathcal{U}$ with $p_n(v) = P_n(v, \dots, v)$, then its n-th derivative at a is given by

$$f^{(n)}(a) = n! P_n \in L^n(V, W).$$

Given a bounded holomorphic function $f : \mathcal{U} \longrightarrow W$ on a bounded open set $\mathcal{U}$ with $a \in \mathcal{U}$, we have the Cauchy inequality

$$\|f'(a)\| \leq \frac{1}{R} \sup\{\|f(z)\| : z \in \mathcal{U}\} \tag{1.2}$$

where R is the distance between a and the topological boundary of $\mathcal{U}$.

Example 1.1.4. For any Banach spaces V, W and Z over $\mathbb{F} = \mathbb{R}, \mathbb{C}$, a bilinear map $f : V \times W \longrightarrow Z$ is analytic and the derivative $f'(a, b) : V \times W \longrightarrow Z$ at any point $(a, b) \in V \times W$ is given by

$$f'(a, b)(x, y) = f(a, y) + f(x, b).$$

The following two fundamental properties of analytic functions are frequently used. We refer to [168, Theorem 1.11, Theorem 1.23] for the proofs.

Theorem 1.1.5. (Principle of analytic continuation) *Let $f : \mathcal{U} \longrightarrow W$ be an analytic function on an open connected set $\mathcal{U}$ such that $f(x) = 0$ for all x in some non-empty open set $\mathcal{S} \subset \mathcal{U}$. Then f is identically 0.*

Theorem 1.1.6. (Inverse function theorem) *Let* $f : \mathcal{U} \longrightarrow W$ *be an analytic function on an open set* $\mathcal{U}$ *such that the derivative* $f'(a) : V \longrightarrow W$ *has a continuous inverse for some* $a \in \mathcal{U}$. *Then* f *is bianalytic from a neighbourhood of* a *onto a neighbourhood of* $f(a)$.

Another useful fact is Cartan's uniqueness theorem which states that two holomorphic maps f and g, one of which is biholomorphic, from a bounded domain $\mathcal{U}$ to another domain in a complex Banach space must coincide if $f(p) = g(p)$ and $f'(p) = g'(p)$ for some $p \in \mathcal{U}$. In [33], H. Cartan's proof of this result is for two-dimensional domains, but it can be extended readily to domains in Banach spaces. We will make use of the following two useful consequences.

Lemma 1.1.7. *Let* D *be a bounded domain in a complex Banach space* V. *Given a holomorphic map* $f : D \longrightarrow V$ *such that* $f(p) = 0$ *and* $f'(p) = 0$ *for some* $p \in D$, *then* f *is identically zero on* D.

Proof. Apply Cartan's uniqueness theorem to the identity map $id : D \longrightarrow D$ and the holomorphic map $id + f$. $\qquad\square$

Lemma 1.1.8. *Let* D *be a circular bounded domain in a complex Banach space* V *containing the origin* 0. *A biholomorphic map* $f : D \longrightarrow D$ *satisfying* $f(0) = 0$ *is (the restriction of) a linear map (on* V *).*

Proof. Pick $\theta \in \mathbb{R} \backslash \{0\}$. Define a holomorphic map $F : D \longrightarrow D$ by

$$F(z) = f^{-1}(e^{-i\theta} f(e^{i\theta} z)) \qquad (z \in D).$$

Then we have $F(0) = 0$ and $F'(0) = (f^{-1})'(0) f'(0)$ which is the identity map on V. Hence Cartan's uniqueness theorem gives

$$f(e^{i\theta} z) = e^{i\theta} f(z) \qquad (z \in D).$$

Since $f(0) = 0$, the power series of f about 0 has the form

$$f = \sum_{k=1}^{\infty} p_k$$

and we have

$$\sum_{k=1}^{\infty} e^{i\theta} p_k(z) = e^{i\theta} f(z) = f(e^{i\theta} z) = \sum_{k=1}^{\infty} p_k(e^{i\theta} z) = \sum_{k=1}^{\infty} e^{ik\theta} p_k(z).$$

Hence $p_k = 0$ for $k \neq 1$ and f is linear. $\square$

We end this section with a review of different modes of convergence of sequences of holomorphic maps. We are concerned with three kinds of convergence, namely, pointwise convergence, uniform convergence on compact sets and locally uniform convergence.

Let $\mathcal{U}$ be a non-empty open subset in a complex normed vector space V and let $C(\mathcal{U}, W)$ denote the complex vector space of all continuous maps from $\mathcal{U}$ to a complex normed vector space W. Given two (non-empty) sets A and B in V, we write

$$d(A, B) = \inf\{\|x - y\| : x \in A, y \in B\}$$

and for $p \in V$, simply write $d(p, B)$ for $d(\{p\}, B)$ which is the distance from p to B. A subset K of $\mathcal{U}$ is said to be *strictly contained* in $\mathcal{U}$ if $d(K, V \backslash \mathcal{U}) > 0$. Consider the following three families of subsets of $\mathcal{U}$:

$\mathcal{K}_1 = \{F : F \text{ is a finite subset of } \mathcal{U}\}$,

$\mathcal{K}_2 = \{K : K \text{ is a compact subset of } \mathcal{U}\}$,

$\mathcal{K}_3 = \{B : B \text{ is a finite union of closed balls strictly contained in } \mathcal{U}\}$

where a *closed ball* is a closed set of the form

$$\overline{B}(a, r) := \{v \in V : \|v - a\| \leq r\}$$

for some $a \in V$ and $r > 0$, and as usual,

$$B(a, r) := \{v \in V : \|v - a\| < r\}$$

is called an *open ball* (centred at a, with radius r). These three families determine the locally convex topologies for the kinds of convergence just mentioned. The topology $\mathcal{T}_j$ determined by the family $\mathcal{K}_j$ ($j = 1, 2, 3$) can be defined by a local base. A local base for the topology $\mathcal{T}_j$, is given by

$$\{U(K, \varepsilon) : K \in \mathcal{K}_j, \varepsilon > 0\}$$

where

$$U(K, \varepsilon) = \{h \in C(\mathcal{U}, W) : \sup\{\|h(x)\| : x \in K\} < \varepsilon\}.$$

A sequence (f_n) in $C(\mathcal{U}, W)$ converges to a function $f \in C(\mathcal{U}, W)$ in the topology $\mathcal{T}_1$ means that the convergence is pointwise. The topology $\mathcal{T}_2$ is the topology of uniform convergence on compact sets in $\mathcal{U}$, and $\mathcal{T}_3$ is the topology of locally uniform convergence. Evidently, we have $\mathcal{T}_1 \subset \mathcal{T}_2 \subset \mathcal{T}_3$. The topology $\mathcal{T}_2$ is also called the *compact-open topology*. We refer to [63, Proposition IV.3.1, Lemma IV.3.2, Lemma IV.3.3] for a proof of the following fundamental result.

Proposition 1.1.9. *Let V, W be complex normed vector spaces and $\mathcal{U}$ a non-empty open subset of V. In the space $C(\mathcal{U}, W)$ of complex continuous functions from $\mathcal{U}$ to W, the topology $\mathcal{T}_3$ of locally uniform convergence is equivalent to the topology $\mathcal{T}_2$ of uniform convergence on compact sets if and only if $\dim V < \infty$. The subspace $H(\mathcal{U}, W)$ of holomorphic maps in $C(\mathcal{U}, W)$ is closed in both topologies $\mathcal{T}_2$ and $\mathcal{T}_3$.*

Definition 1.1.10. A family $\mathcal{F}$ of functions in $H(\mathcal{U}, W)$ is called a *normal family* if every sequence (f_n) in $\mathcal{F}$ admits a locally uniformly convergent subsequence.

This notion of normality is a natural extension of the finite dimensional one (cf. [7]). The well-known Montel's theorem asserts that, for an open set $\mathcal{U} \subset \mathbb{C}^m$, a family $\mathcal{F} \subset H(\mathcal{U}, \mathbb{C})$ is normal if (and only if) it is uniformly bounded on compact subsets of $\mathcal{U}$ (see, for example, [134, p. 8]). This condition for normality is still true for a family $\mathcal{F} \subset H(\mathcal{U}, \mathbb{C}^d)$ of $\mathbb{C}^d$-valued functions.

Lemma 1.1.11. *For $\mathcal{U} \subset \mathbb{C}^m$, a family $\mathcal{F} \subset H(\mathcal{U}, \mathbb{C}^d)$ is normal if it is uniformly bounded on compact subsets of $\mathcal{U}$.*

Proof. Let $(f_n)_{n=1}^{\infty}$ be a sequence in $\mathcal{F}$. We can write

$$f_n(z) = (f_n^1(z), \ldots, f_n^d(z)) \in \mathbb{C}^d$$

for each $z \in \mathcal{U}$, where the $\mathbb{C}$-valued holomorphic functions $\{f_n^j : n = 1, 2, \ldots ; j = 1, \ldots, d\}$ are uniformly bounded on compact subsets of $\mathcal{U}$. Applying Motel's theorem repeatedly, one can find a locally uniformly convergent subsequence $(f_{n_1}^1)$ of (f_n^1), then there is a locally uniformly convergent subsequence $(f_{n_2}^2)$ of $(f_{n_1}^2)$, and so on, finally one finds a locally uniformly convergent subsequence $(f_{n_d}^d)$ of $(f_{n_{d-1}}^d)$. It can be seen readily that the subsequence $f_{n_d} = (f_{n_d}^1, \ldots, f_{n_d}^d)$ of (f_n) is locally uniformly convergent. $\square$

An infinite dimensional version of the Vitali theorem states that a sequence $f_n : D_1 \longrightarrow D_2$ of holomorphic maps between bounded domains in Banach spaces converges locally uniformly to a holomorphic map $f : D_1 \longrightarrow D_2$ whenever it converges uniformly to f on an open ball strictly contained in D_1 (cf. [146, Theorem 2.13]).

Notes. There are several books devoted to complex analysis in Banach spaces, the ones we will refer to occasionally in the sequel are [63] and [133]. One can find a proof in [63, Theorem II.3.4] of the maximum principle for holomorphic maps f in $H(\mathcal{U}, W)$, which states that the function $x \in \mathcal{U} \mapsto \|f(x)\|$ cannot have a maximum at $x_0 \in \mathcal{U}$ unless $\|f(\cdot)\|$ is constant on a neighbourhood of x_0. Further, if $\mathcal{U}$ is a domain and W is strictly convex, then the map $x \in \mathcal{U} \mapsto \|f(x)\|$ cannot achieve its maximum in $\mathcal{U}$ unless f is constant [63, Corollary III.1.5].

1.2 Banach manifolds

Infinite dimensional symmetric domains are Banach manifolds. We discuss some basic properties of Banach manifolds in this section.

Definition 1.2.1. Let $\mathbb{F} = \mathbb{R}$ or $\mathbb{C}$. A *Banach manifold M over $\mathbb{F}$* is a Hausdorff topological space with a family $\mathcal{A} = \{(\mathcal{U}_\varphi, \varphi, V_\varphi)\}$ of *local charts* $(\mathcal{U}_\varphi, \varphi, V_\varphi)$ satisfying the following conditions:

(i) $\mathcal{U}_\alpha$ is an open subset of M and $M = \bigcup_\varphi \mathcal{U}_\varphi$;

(ii) $\varphi : \mathcal{U}_\varphi \longrightarrow V_\varphi$ is a homeomorphism onto an open subset of a Banach space V_φ over $\mathbb{F}$;

(iii) the local charts are *compatible*, that is, the change of charts

$$\psi\varphi^{-1} : \varphi(\mathcal{U}_\varphi \cap \mathcal{U}_\psi) \longrightarrow \psi(\mathcal{U}_\varphi \cap \mathcal{U}_\psi)$$

is bianalytic;

(iv) the family $\mathcal{A}$ is maximal relative to conditions (i), (ii) and (iii), that is, if $(\mathcal{U}, \varphi, V)$ is a local chart compatible with all the local charts in $\mathcal{A}$, then $(\mathcal{U}, \varphi, V) \in \mathcal{A}$.

A family $\{(\mathcal{U}_\varphi, \varphi, V_\varphi)\}$ satisfying conditions (i), (ii) and (iii) above is called an *atlas* of M or an *analytic structure* on M. Since an atlas can always be extended to a maximal one satisfying condition (iv), it is often sufficient to exhibit an atlas for a topological space to be a

Banach manifold. A *local chart* (or *system of local coordinates*) at a point $p \in M$ is a chart $(\mathcal{U}_\varphi, \varphi, V_\varphi)$ with $p \in \mathcal{U}_\varphi$. Note that the Banach spaces V_φ in an atlas need not be the same space, but V_φ and V_ψ are isomorphic if $\mathcal{U}_\varphi \cap \mathcal{U}_\psi$ contains a point p, in which case the derivative

$$(\psi\varphi^{-1})'(\varphi(p)) : V_\varphi \longrightarrow V_\psi$$

is an isomorphism. If all Banach spaces V_φ in the atlas are isomorphic, we can always find an equivalent atlas in which they are all equal to some Banach space V, in which case, we say that the manifold M is *modelled on the Banach space* V and that V is a *model space* for M. We define the *dimension* $\dim M$ of M to be that of V if $\dim V < \infty$. If V is infinite dimensional, we define $\dim M = \infty$.

For a local chart $(\mathcal{U}, \varphi, V)$ of a Banach manifold M, the set

$$\{p \in M : \exists \text{ a local chart } (\mathcal{U}_\psi, \psi, V_\psi) \text{ at } p \text{ and } V_\psi \simeq V\}$$

is open and closed in M. Consequently, on a connected component of M, we can choose an atlas in which all V_φ are the same space. In particular, if M is connected, then we can find a model space for M.

We call a manifold *real* or *complex* according to the underlying scalar field of the spaces V_φ. Since a complex Banach space can be viewed as a real Banach space when the scalar multiplication is restricted to the real field, a complex Banach manifold can be viewed as a real Banach manifold with the underlying real analytic structure. In particular, a complex manifold modelled on a d-dimensional complex Euclidean space $\mathbb{C}^d$ can be viewed as a real analytic manifold modelled on the $2d$-dimensional Euclidean space $\mathbb{R}^{2d}$.

If, in Definition 1.2.1, the Banach spaces V_φ are real and the coordinate transformations

$$\psi\varphi^{-1} : \varphi(\mathcal{U}_\varphi \cap \mathcal{U}_\psi) \longrightarrow \psi(\mathcal{U}_\varphi \cap \mathcal{U}_\psi)$$

are only smooth, then we call M a (real) *smooth Banach manifold* and, its atlas a *differentiable structure*. Assuming analytic structures in our definition of Banach manifolds has the advantage of unifying both real and complex cases. Of course, we can (and will) always regard a Banach manifold (according to our definition) as a smooth Banach manifold. Sometimes we speak of *analytic* Banach manifolds to highlight the underlying analytic structure.

Example 1.2.2. Every Banach space over $\mathbb{F}$ is itself a Banach manifold over $\mathbb{F}$, with the analytic structure given by the identity map. Also, an open subset U of a Banach manifold M is a Banach manifold, endowed with the atlas $\{(U \cap \mathcal{U}_\varphi, \varphi|_{U \cap \mathcal{U}_\varphi}, V_\varphi)\}$ derived from the atlas $\{(\mathcal{U}_\varphi, \varphi, V_\varphi)\}$ of M.

In particular, a domain in a Banach space V is a Banach manifold modelled on V.

Example 1.2.3. Given two Banach manifolds M and N over the same field with analytic structures $\{(\mathcal{U}_\varphi, \varphi, V_\varphi)\}$ and $\{(\mathcal{V}_\psi, \psi, W_\psi)\}$ respectively, the Cartesian product $M \times N$ is a Banach manifold with the natural analytic structure $\{(\mathcal{U}_\varphi \times \mathcal{V}_\psi, \varphi \times \psi, V_\varphi \times W_\psi)\}$.

Example 1.2.4. Let V be a real Banach space and let

$$S(V) = \{(\lambda, v) \in \mathbb{R} \times V : \lambda^2 + \|v\|^2 = 1\}$$

be the unit sphere in the product manifold $\mathbb{R} \times V$. Let $p = (1,0)$ and $q = (-1,0)$. On $S(V)$, define two local charts $(S(V)\setminus\{p\}, \varphi, V)$ and $(S(V)\setminus\{q\}, \psi, V)$ by

$$\varphi(\lambda, v) = \frac{v}{1 - \lambda}; \quad \psi(\lambda, v) = \frac{v}{1 + \lambda}.$$

We have $\varphi(-1, 0) = 0 = \psi(1, 0)$. For $v \in \varphi(S(V)\setminus\{p\})\setminus\{0\}$, we have $\varphi^{-1}(v) = \left(\frac{\|v\|^2 - 1}{\|v\|^2 + 1}, \frac{2v}{\|v\|^2 + 1}\right)$. Therefore

$$\psi \circ \varphi^{-1}(v) = \frac{v}{\|v\|^2} \qquad (v \in \varphi(S(V)\setminus\{p\} \cap S(V)\setminus\{q\})).$$

Hence the above charts define an analytic structure on $S(V)$. If $V = \mathbb{R}^n$, we adopt the usual notation S^n for the unit sphere $S(\mathbb{R}^n)$ in $\mathbb{R}^{n+1}$.

A mapping $f : M \longrightarrow N$ between Banach manifolds (over the same field $\mathbb{F}$) is called *analytic* if, for each $x \in M$, there are charts $(\mathcal{U}, \varphi, V)$ of M and $(\mathcal{V}, \psi, W)$ of N such that $x \in \mathcal{U}$, $f(\mathcal{U}) \subset \mathcal{V}$ and the composed map

$$\psi \circ f \circ \varphi^{-1} : \varphi(\mathcal{U}) \longrightarrow \psi(\mathcal{V}) \subset W$$

is analytic. A bijection $f : M \longrightarrow N$ is called *bianalytic* if both f and the inverse f^{-1} are analytic, in which case M is said to be *bianalytic to N*. The coordinate map $\varphi : \mathcal{U} \longrightarrow \varphi(\mathcal{U})$ of a chart $(\mathcal{U}, \varphi, V)$ is bianalytic.

Holomorphic and *biholomorphic* maps on complex Banach manifolds are defined likewise. Holomorphic maps on complex Banach manifolds are analytic and in the sequel, the terms "*holomorphic*" (respectively, *biholomorphic*) and "*analytic*" (respectively, *bianalytic*) are interchangeable for complex manifolds. Occasionally, a biholomorphic map is also called a biholomorphism. One can also define smooth maps between real smooth Banach manifolds M and N in the same manner, and a smooth map $f : M \longrightarrow N$ is called a *diffeomorphism* if it is bijective and the inverse f^{-1} is also smooth, in which case we say that M is *diffeomorphic to N*.

A bianalytic map $f : M \longrightarrow M$ is called an *automorphism* of M. The automorphisms of M form a group, with composition product, called the *automorphism group* of M and is denoted by $\mathrm{Aut}\, M$.

We define tangent vectors on a manifold as directional derivatives along smooth curves. A *smooth curve* in a Banach manifold M over $\mathbb{F}$ is a smooth map $\gamma : (-c, c) \longrightarrow M$ on some open interval in $\mathbb{R}$, where $c > 0$.

Let $p \in M$ and let $\gamma : (-c, c) \longrightarrow M$ be a smooth curve such that

$\gamma(0) = p$. Let $(\mathcal{U}, \varphi, V)$ be a local chart at p. Then $\gamma(t) \in \mathcal{U}$ for t near 0 and the derivative $(\varphi \circ \gamma)'(0) : \mathbb{R} \longrightarrow V$ is (identified with) a vector in V. If $\gamma_1 : (-c_1, c_1) \longrightarrow M$ is another smooth curve such that $\gamma_1(0) = p$ and $(\varphi \circ \gamma)'(0) = (\varphi \circ \gamma_1)'(0)$, then for any chart $(\mathcal{U}_\psi, \psi, V_\psi)$ at p, we have $(\psi \circ \gamma)'(0) = (\psi \circ \varphi^{-1})'(\varphi(p))(\varphi \circ \gamma)'(0) = (\psi \circ \gamma_1)'(0)$. Hence we can define an equivalence relation $\sim$ on smooth curves γ in M with $\gamma(0) = p$ by

$$\gamma \sim \gamma_1 \text{ if } (\varphi \circ \gamma)'(0) = (\varphi \circ \gamma_1)'(0)$$

for a local chart $(\mathcal{U}, \varphi, V)$ at p. An equivalence class $[\gamma]_p$ is called a *tangent vector to M at p*. We define the *tangent space of M at p* to be the set $T_p M$ of all tangent vectors $[\gamma]_p$ at p.

For the local chart $(\mathcal{U}, \varphi, V)$ at p, the map

$$[\gamma]_p \in T_p M \mapsto (\varphi \circ \gamma)'(0) \in V$$

is a bijection. Indeed, given a vector $v \in V$, there is a smooth curve $\gamma_v(t) = \varphi^{-1}(\varphi(p) + tv)$ in M from some interval in $\mathbb{R}$ such that $(\varphi \circ \gamma_v)'(0) = v$. Hence $T_p M$ can be identified with V via the above bijection and is equipped with a Banach space structure. In particular, if M is an open subset of a Banach space W and $p \in M$, then $T_p M = W$ via the identification $[\gamma]_p \in T_p M \mapsto \gamma'(0) \in W$.

In the above chart, if $\gamma(t) \in \mathcal{U}$ is defined, then $(\varphi \circ \gamma)'(t) \in V$ is a tangent vector at $\gamma(t)$. Indeed, $(\varphi \circ \gamma)'(t) = (\varphi \circ \alpha)'(0)$ with $[\alpha]_{\gamma(t)} \in T_{\gamma(t)} M$, where $\alpha(s)$ is the curve

$$\alpha(s) = \varphi^{-1}(\varphi(\gamma(t)) + s(\varphi \circ \gamma)'(t))$$

defined on some interval in $\mathbb{R}$.

Remark 1.2.5. In view of the above identification of tangent vectors, we often denote by $\gamma'(t)$ the tangent vector to a curve γ at $\gamma(t)$.

Let $f : M \longrightarrow N$ be an analytic map between Banach manifolds and let $p \in M$. The map

$$df_p : [\gamma]_p \in T_p M \mapsto [f \circ \gamma]_{f(p)} \in T_{f(p)} N$$

is a continuous linear map and is called the *differential of f* at p. The differential of a smooth map is defined in the same manner.

Example 1.2.6. Let M and N be open subsets of Banach spaces V and W respectively. Let $f : M \longrightarrow N$ be an analytic map and let $p \in M$. Identifying the tangent spaces $T_p M$ and $T_{f(p)} N$ with V and W as above, the differential $df_p : T_p M \longrightarrow T_{f(p)} N$ is the derivative $f'(p) : V \longrightarrow W$. Indeed, given $v \in V$ with $v = \gamma'(0)$ for some $[\gamma]_p \in T_p M$, we have

$$df_p([\gamma]_p) = [f \circ \gamma]_{f(p)} = (f \circ \gamma)'(0) = f'(\gamma(0))\gamma'(0) = f'(p)v.$$

Example 1.2.7. Let $(\mathcal{U}, \varphi, V)$ be a local chart of a Banach manifold M. For $p \in \mathcal{U}$, the differential $d\varphi_p : T_p \mathcal{U} \longrightarrow T_{\varphi(p)} V$ is given by $d\varphi_p([\gamma]_p) = [\varphi \circ \gamma]_{\varphi(p)}$. Under the identification

$$[\varphi \circ \gamma]_{\varphi(p)} \in T_{\varphi(p)} V \mapsto (\varphi \circ \gamma)'(0) \in V \mapsto [\gamma]_p \in T_p M,$$

we have $d\varphi_p([\gamma]_p) = [\gamma]_p$.

The following result is an important consequence of the inverse function theorem for Banach spaces.

Theorem 1.2.8. *Let $f : M \longrightarrow N$ be an analytic (a smooth) map between (smooth) Banach manifolds such that the differential $df_p : T_p M \longrightarrow T_{f(p)} N$ is bijective at some point $p \in M$. Then f is locally bianalytic (diffeomorphic) at p, that is, f is bianalytic (diffeomorphic) from a neighbourhood of p onto a neighbourhood of $f(p)$.*

A closed subspace E of a Banach space V is called a *complemented subspace* if there is a closed subspace E^c of V such that $V = E \oplus E^c$

is the direct sum of E and E^c, or equivalently, if there is a continuous linear projection from V onto E. If $\dim V < \infty$, then every subspace of V is complemented.

Definition 1.2.9. Let $f : M \longrightarrow N$ be an analytic (a smooth) map between (smooth) Banach manifolds M and N. Let $p \in M$ and let $df_p : T_pM \longrightarrow T_{f(p)}N$ be the differential of f at p. The map f is called an *immersion* at p if df_p is injective and its image is a complemented subspace of $T_{f(p)}N$. We call f a *submersion* at p if df_p is surjective and its kernel is a complemented subspace of T_pM.

Compared with the case of bijective differential df_p where the inverse function theorem implies that f is locally invertible, the weaker condition of injectivity or surjectivity of df_p alone entails the existence of one-sided inverse of f locally. More precisely, we have the following characterizations of immersion and submersion.

Lemma 1.2.10. *An analytic (a smooth) map $f : M \longrightarrow N$ between (smooth) Banach manifolds is a submersion at $p \in M$ if, and only if, there is an open neighbourhood $\mathcal{V}$ of $f(p) \in N$ and an analytic (a smooth) map $g : \mathcal{V} \longrightarrow M$ such that $g(f(p)) = p$ and $f \circ g$ is the identity map on $\mathcal{V}$.*

Proof. We consider the analytic case. The proof for the smooth case is verbatim. Let f be a submersion at $p \in M$. Then the differential $df_p : T_pM \longrightarrow T_{f(p)}N$ is surjective with complemented kernel K. Therefore $T_pM = K \oplus K^c$ and df_p is bijective on K^c.

Let $(\mathcal{U}, \varphi, T_pM)$ be a chart at p with $\varphi(p) = 0$ and $(\mathcal{V}_\psi, \psi, T_{f(p)}N)$ a chart at $f(p)$, with $\psi(f(p)) = 0$ and $f(\mathcal{U}) \subset \mathcal{V}_\psi$, so that $\psi \circ f \circ \varphi^{-1}$ is analytic on $\varphi(\mathcal{U})$.

On an open neighbourhood B of $(0, 0) \in K \times K^c \approx T_pM$, we can define an analytic map $F : B \longrightarrow K \times T_{f(p)}N$ by

$$F(u, v) = (u, \psi f \varphi^{-1}(v)) \qquad ((u, v) \in B)$$

where $v \in K^c \cap \varphi(\mathcal{U})$. The derivative

$$F'(0, 0) : K \times K^c \longrightarrow K \times T_{f(p)}N$$

is bijective since

$$F'(0, 0)(u, v) = (u, (\psi f \varphi^{-1})'(0)(v))$$

where $(\psi f \varphi^{-1})'(0) = df_p : K^c \longrightarrow T_{f(p)}N$ is an isomorphism. By the inverse function theorem, F is bianalytic from a neighbourhood of $(0, 0) \in K \times K^c$ onto a neighbourhood $S \times O$ of $(0, \psi f(p)) \in K^c \times T_{f(p)}N$, where S and O are open. The inverse F^{-1} induces a well-defined analytic map $h : \psi f \varphi^{-1}(v) \in O \mapsto v \in K^c \cap \varphi(\mathcal{U})$. Let

$$\mathcal{V} = \psi^{-1}(O \cap \psi(\mathcal{V}_\psi))$$

and define $g : \mathcal{V} \longrightarrow \mathcal{U} \subset M$ by

$$g(y) = \varphi^{-1} \circ h \circ \psi(y) \qquad (y \in \mathcal{V})$$

which satisfies $g(f(p)) = p$. For each $y \in \mathcal{V}$, we have $\psi(y) = \psi f \varphi^{-1}(v)$ for some $v \in K^c \varphi(\mathcal{U})$ and

$$f(g(y)) = f(\varphi^{-1}(h(\psi(y)))) = f(\varphi^{-1}(v)) = y \qquad (y \in \mathcal{V}).$$

Conversely, let f have local right inverse $g : \mathcal{V} \longrightarrow M$. Then the differential $df_p \circ dg_{f(p)}$ is the identity map on $T_{f(p)}N$. Hence the differential df_p is surjective and $dg_{f(p)}$ is injective. It follows that

$$dg_{f(p)} \circ df_p : T_p M \longrightarrow T_p M$$

is a continuous projection with kernel $(df_p)^{-1}(0)$ which is complemented in $T_p M$. $\square$

Remark 1.2.11. The above lemma implies that a submersion $f : M \longrightarrow N$ is an open map.

If $f : M \longrightarrow N$ is an immersion at $p \in M$, then we have $T_{f(p)}N = V \oplus V^c$ where $V = df_p(T_pM)$ is the image of the differential df_p. With local charts $(\mathcal{U}, \varphi, T_pM)$ at p and $(\mathcal{V}_\psi, \psi, T_{f(p)}M)$ at $f(p)$, as in the above proof, we can define an analytic map $F : T_pM \times V^c \longrightarrow V \times V^c \approx T_{f(p)}N$ *locally* by

$$F(u, v) = ((\psi f \varphi^{-1})(u), v)$$

for (u, v) in a neighbourhood of $(0, 0) \in T_pM \times V^c$. Again, with similar arguments as before, the derivative $F'(0, 0)$ is bijective and the local inverse F^{-1} induces an analytic map $g : \mathcal{V} \longrightarrow M$ from a neighbourhood $\mathcal{V}$ of $f(p)$ such that $g \circ f$ is the identity map on a neighbourhood of p. This gives the following characterization of an immersion.

Lemma 1.2.12. *An analytic (a smooth) map* $f : M \longrightarrow N$ *between (smooth) Banach manifolds is an immersion at* $p \in M$ *if, and only if, there are open neighbourhoods* $\mathcal{U}$ *and* $\mathcal{V}$ *of* p *and* $f(p) \in N$ *respectively, and an analytic (a smooth) map* $g : \mathcal{V} \longrightarrow M$ *such that* $f(\mathcal{U}) \subset \mathcal{V}$ *and* $g \circ f$ *is the identity map on* $\mathcal{U}$.

A topological subspace N of a Banach manifold M is called a *submanifold of M* if N is itself a Banach manifold and the inclusion map $\iota : N \hookrightarrow M$ is an immersion. In this case, the tangent space T_pN of N at a point $p \in N$ identifies as a complemented subspace of the tangent space T_pM of M at p via the differential $d\iota_p : T_pN \longrightarrow T_pM$. An equivalent way of saying that N is a submanifold of M is that at each point $p \in N$, there is a chart $(\mathcal{U}_\varphi, \varphi, V_\varphi)$ of M and a complemented subspace $W_\varphi \subset V_\varphi$ such that

$$\varphi(N \cap \mathcal{U}_\varphi) = W_\varphi \cap \varphi(\mathcal{U}_\varphi)$$

in which case, $\{(N \cap \mathcal{U}_\varphi), \varphi|_{(N \cap \mathcal{U}_\varphi)}, W_\varphi\}$ forms an atlas of N.

Every open topological subspace of a Banach manifold M is a submanifold of M.

Let M be a Banach manifold over $\mathbb{F}$ and let

$$T(M) = \bigcup_{p \in M} T_p M$$

be the union of tangent spaces of M. Then $T(M)$ is a Banach manifold over $\mathbb{F}$ with an analytic structure induced from that of M. Indeed, let $\{(\mathcal{U}_\varphi, \varphi, V_\varphi)\}$ be an atlas of M and write $T\mathcal{U}_\varphi = \bigcup_{p \in \mathcal{U}_\varphi} T_p M$. Define a combine map

$$T(\varphi) : T\mathcal{U}_\varphi \longrightarrow V_\varphi \times V_\varphi$$

by

$$T(\varphi)[\gamma]_p = (\varphi(p), (\varphi \circ \gamma)'(0)) \qquad ([\gamma]_p \in T_p M).$$

The image of $T\mathcal{U}_\varphi$ is $\varphi(\mathcal{U}_\varphi) \times V_\varphi$. Define a topology on $T(M)$ in which the open sets $\mathcal{O}$ are those such that $T(\varphi)(\mathcal{O} \cap T\mathcal{U}_\varphi)$ is open in $V_\varphi \times V_\varphi$. Then $\{(T\mathcal{U}_\varphi, T(\varphi), V_\varphi \times V_\varphi)\}$ is an analytic structure on $T(M)$.

The Banach manifold $T(M)$ is a vector bundle on M, with bundle projection

$$\pi : [\gamma]_p \in TM \mapsto p \in M,$$

and is called the *tangent bundle of M*.

An *analytic vector field* on M is an analytic map $X : M \longrightarrow TM$ such that $X(p) \in T_p M$, that is, X is a section of the tangent bundle $T(M)$. We often write X_p for $X(p)$. The analytic vector fields on M form a vector space $\mathfrak{X}M$ under pointwise addition and scalar multiplication.

Given an analytic map $f : M \longrightarrow N$ between Banach manifolds M and N. We can define the *tangent map $df : T(M) \longrightarrow T(N)$* as the combine map

$$df(X_p) = df_p(X_p)$$

which in turn induces the map $df \circ X : M \longrightarrow T(N)$ for each $X \in \mathfrak{X}M$.

Analytic vector fields can be regarded as differential operators in the following way. Given an analytic map $f : \mathcal{U} \longrightarrow W$ from an open subset $\mathcal{U}$ of M to a Banach space W (over the same field of M), and given $X \in \mathfrak{X}M$, one can define an analytic map $Xf : \mathcal{U} \longrightarrow W$ by

$$Xf(p) = df_p(X_p) \qquad (p \in M).$$

The above mapping is written as $Xf = df \circ X$. We can define the n-th iterate X^n of X inductively by

$$X^0 f = f, \quad X^n f = X(X^{n-1} f) \qquad (n = 1, 2, \ldots). \tag{1.3}$$

Example 1.2.13. Let $\mathcal{U}$ be an open set in the n-dimensional Euclidean space $\mathbb{C}^n$, considered as a Banach manifold, with local chart given by the inclusion map $\varphi : \mathcal{U} \hookrightarrow \mathbb{C}^n$. Let $X : \mathcal{U} \longrightarrow T\mathcal{U}$ be a holomorphic vector field. For each $z = (z_1, \ldots, z_n) \in \mathcal{U}$, we have

$$X(z) = [\gamma_z]_z \in T_z\mathcal{U}$$

for some smooth curve $\gamma_z = ((\gamma_z)_1, \ldots, (\gamma_z)_n) : (-c, c) \longrightarrow \mathcal{U}$ with $\gamma_z(0) = z$.

Identify $T_z\mathcal{U}$ with $\mathbb{C}^n$ via $[\gamma_z]_z \mapsto \gamma_z'(0)$ and define a holomorphic map $h : \mathcal{U} \longrightarrow \mathbb{C}^n$ by

$$h(z) = \gamma_z'(0) = ((\gamma_z)_1'(0), \ldots, (\gamma_z)_n'(0))$$

and write $h(z) = (h_1(z), \ldots, h_n(z))$. Then for each holomorphic function $f : \mathcal{U} \longrightarrow \mathbb{C}$, we have

$$\begin{aligned} Xf(z) &= df_z(X(z)) = \left(\frac{\partial f}{\partial z_1}(z), \ldots, \frac{\partial f}{\partial z_n}(z) \right) (\gamma_z'(0)) \\ &= \sum_{j=1}^n (\gamma_z)_j'(0) \frac{\partial f}{\partial z_j}(z) = \sum_{j=1}^n h_j(z) \frac{\partial f}{\partial z_j}(z). \end{aligned}$$

Therefore we can write, as a differential operator,

$$X = \sum_{j=1}^{n} h_j \frac{\partial}{\partial z_j}$$

and write for short $X = h\frac{\partial}{\partial z}$.

In view of this example, we shall represent a holomorphic vector field X, in a local chart $\mathcal{U}_\varphi$ of a manifold M modelled on a complex Banach space V, by a holomorphic function $h : \mathcal{U}_\varphi \longrightarrow V$ and denote this by $X = h\frac{\partial}{\partial z}$, and sometimes write, by abuse of notation, $X(z) = h(z) \in V$. One can consider X as the vector field $\underline{X} = \underline{h}\frac{\partial}{\partial z}$ on $\varphi(\mathcal{U}_\varphi)$, where $\underline{h}$ is the holomorphic map $h \circ \varphi^{-1} : \varphi(\mathcal{U}_\varphi) \longrightarrow V$.

Given two analytic vector fields X, Y on a Banach manifold over $\mathbb{F}$, there is a unique analytic vector field $[X, Y]$ on M, called the *commutator* (or *brackets*) of X and Y, such that for each open set $\mathcal{U} \subset M$, we have

$$[X, Y]f = X(Yf) - Y(Xf)$$

for all analytic maps f from $\mathcal{U}$ to a Banach space W (cf. [168, p. 60]). Using the above formula, one can verify easily that the brackets

$$[\cdot, \cdot] : \mathfrak{X}M \times \mathfrak{X}M \longrightarrow \mathfrak{X}M$$

is antisymmetric, bilinear and satisfies the Jacobi identity

$$[X, [Y, Z]] + [Y, [Z, X]] + [Z, [X, Y]] = 0 \qquad (X, Y, Z \in \mathfrak{X}M).$$

It follows that $\mathfrak{X}M$ is a Lie algebra in the above brackets.

Let $(\mathcal{U}_\varphi, \varphi, V_\varphi)$ be a local chart at $p \in M$. A tangent vector $[\gamma]_q$ at a point $q \in \mathcal{U}_\varphi$ identifies with the vector $(\varphi \circ \gamma)'(0) \in V_\varphi$. On this chart, an analytic vector field X can be viewed as an analytic function $X : \mathcal{U}_\varphi \longrightarrow V_\varphi$ and $X\varphi = d\varphi \circ X = X$ by Example 1.2.7. It follows that

$$[X, Y]\varphi = X(Y\varphi) - Y(X\varphi) = dY\varphi \circ X - dX\varphi \circ Y = dY \circ X - dX \circ Y$$

and

$$[X, Y](p) = dY_p(X_p) - dX_p(Y_p) \qquad (p \in \mathcal{U}).$$

Example 1.2.14. Given holomorphic vector fields $X = h\frac{\partial}{\partial z}$ and $Y = k\frac{\partial}{\partial z}$ on an open set D in a complex Banach space V, we have $[X, Y] = f\frac{\partial}{\partial z}$ where

$$f(z) = k'(z)h(z) - h'(z)k(z) \qquad (z \in D).$$

A bianalytic map $f : M \longrightarrow N$ between Banach manifolds induces a map

$$f_* : \mathfrak{X}M \longrightarrow \mathfrak{X}N \tag{1.4}$$

defined by

$$(f_*X)_{f(p)} = df_p(X_p) \qquad (X \in \mathfrak{X}M, p \in M).$$

If g is another bianalytic map on N, then we have the tangent map $d(g \circ f) = dg \circ df$ and it follows that $(g \circ f)_* = g_* \circ f_*$. In particular, taking $g = f^{-1}$ implies that f_* is bijective. Further, it is straightforward to verify that f_* is in fact a Lie algebra isomorphism (cf. [168, 4.5]).

Given an analytic vector field X on a Banach manifold M, using the existence theorem for differential equations in Banach spaces, one can find an open interval I_p in $\mathbb{R}$ containing 0 and an analytic curve

$$\gamma_p : I_p \longrightarrow \mathcal{U}_\varphi$$

satisfying the differential equation

$$\frac{d\gamma_p(t)}{dt} = X(\gamma_p(t)) \in V_\varphi$$

and the initial condition $\gamma_p(0) = p$ (cf. Remark 1.2.5). Since the solution also depends analytically on the initial point, the theory of differential equations gives further the following result (cf. [140, §I.7] and [118, Chapter IV]).

Theorem 1.2.15. *Given an analytic vector field X on a Banach manifold M and $p \in M$, there is an open neighbourhood $\mathcal{U}$ of p and, an open interval I in $\mathbb{R}$ containing 0, such that for all $q \in \mathcal{U}$, the curve γ_q satisfying $\gamma_q'(t) = X(\gamma_q(t))$ and $\gamma_q(0) = q$ is defined on I. The map $\alpha : (t, q) \in I \times \mathcal{U} \mapsto \gamma_q(t) \in M$ is analytic and satisfies $\alpha(s + t, q) = \alpha(s, \alpha(t, q))$ for $s, t, s + t \in I$ and $\alpha(t, q) \in \mathcal{U}$.*

For $t \in I$, let $\alpha_t : \mathcal{U} \longrightarrow M$ be the analytic map

$$\alpha_t(q) = \alpha(t, q) = \gamma_q(t) \qquad (q \in \mathcal{U}).$$

Since $\alpha_0(p) = p \in \mathcal{U}$, we have $\alpha_t(p) \in \mathcal{U}$ for t near 0 and hence the differential $(d\alpha_{-t})_{\alpha_t(p)}$ sends a tangent vector $Y_{\alpha_t(p)}$ to a tangent vector in T_pM. Noting that $X(p) = [\gamma_p] \in T_pM$ and for an analytic map f from $\mathcal{U}$ to a Banach space, we have $Xf(p) = [f \circ \gamma_p] = (f \circ \gamma_p)'(0) = \lim_{t \to 0} \frac{1}{t}(f(\alpha_t(p)) - f(p))$, a direct computation gives

$$[X, Y](p) = \lim_{t \to 0} \frac{1}{t}(d\alpha_{-t}(Y_{\alpha_t(p)}) - Y_p) \tag{1.5}$$

which expresses in some sense the rate of change of Y in the direction of X, known as the *Lie derivative* of Y in the direction of X (cf. [118, p. 121]).

Definition 1.2.16. The map $\alpha : (t, q) \in I \times \mathcal{U} \mapsto \gamma_q(t) \in M$ in Theorem 1.2.15 is called a *local flow* or *an integral curve* of the vector field X. If the flow α is defined on $\mathbb{R} \times M$, then X is called a *complete analytic vector field*.

In a local chart $(\mathcal{U}_\varphi, p, V)$ of a Banach manifold modelled on a complex Banach space V, with $\varphi(p) = 0$, a local flow $\alpha(t, q)$ of an analytic vector field $X = h\frac{\partial}{\partial z}$ on M can be written more explicitly in the following way (cf. [168, Theorem 5.11]). One can find an open neighbourhood B of 0 in $\varphi(\mathcal{U}_\varphi)$ and some $\delta > 0$ such that the local flow $\underline{\alpha}(t, z)$ of the vector field $\underline{X} = (h\varphi^{-1})\frac{\partial}{\partial z}$ on $\varphi(\mathcal{U}_\varphi)$ is given by

$$\underline{\alpha}(t, z) = \sum_{n=0}^{\infty} \frac{t^n}{n!} \underline{X}^n \iota(z), \qquad (t, z) \in (-\delta, \delta) \times B \qquad (1.6)$$

where $\iota : \varphi(\mathcal{U}_\varphi) \longrightarrow \varphi(\mathcal{U}_\varphi)$ is the identity map and $\underline{X}^n \iota$ is as defined in (1.3).

By the uniqueness theorem in differential equations (cf. [118, p. 88]), there is only one flow $\alpha : \mathbb{R} \times M \longrightarrow M$ of a complete analytic vector field X, satisfying $\alpha(0, p) = p$. For each $t \in \mathbb{R}$, we define an analytic map

$$\exp tX : M \longrightarrow M$$

by $\exp tX(p) = \alpha(t, p)$ for each $p \in M$. The notation is suggested by the property

$$\exp (s + t)X = \exp sX \circ \exp tX \qquad (s, t \in \mathbb{R})$$

which also implies that each map $\exp tX$ is bianalytic since $\exp tX \circ \exp -tX = \exp 0X$ is the identity map on M. We call the homomorphism

$$t \in \mathbb{R} \mapsto \exp tX \in \operatorname{Aut} M$$

the *one-parameter group* of X. In this notation, we have

$$X(p) = \frac{d}{dt}\bigg|_{t=0} \exp tX(p) \qquad (p \in M).$$

Denote by $\operatorname{aut} M$ the set of all complete analytic vector fields on a Banach manifold M. The map

$$X \in \operatorname{aut} M \mapsto \exp 1X \in \operatorname{Aut} M$$

is denoted by $\exp$.

Let $g : M \longrightarrow N$ be a bianalytic map between Banach manifolds. Then the induced Lie algebra isomorphism $g_* : \mathfrak{X}M \longrightarrow \mathfrak{X}N$, as defined in (1.4), maps $\operatorname{aut} M$ to $\operatorname{aut} N$ and for $X \in \operatorname{aut} M$, we have the commutative diagram

$$
\begin{array}{ccc}
(t,p) \in \mathbb{R} \times M & \longrightarrow & \exp tX(p) \in M \\
\downarrow \iota \times g & & \downarrow g \\
(t,g(p)) \in \mathbb{R} \times N & \longrightarrow & \exp tg_*X(g(p)) \in N
\end{array}
$$

where $g(\exp tX(p)) = \exp tg_*X(g(p))$ gives

$$
g \circ \exp tX \circ g^{-1} = \exp tg_*X. \tag{1.7}
$$

Remark 1.2.17. For a smooth Banach manifold M, the tangent bundle $T(M)$ is a smooth Banach manifold in which case one considers smooth vector fields $X : M \longrightarrow T(M)$, smooth flows and so on. A parallel theory for smooth Banach manifolds can be developed along with analytic manifolds, we suppress the repetition but will make use of the results without more ado.

We now discuss two important classes of Banach manifolds, namely, the Riemannian manifolds and the Lie groups.

First, Riemannian manifolds. In what follows, by a *Riemannian manifold*, we shall mean a real smooth Banach manifold M, modelled on a real Hilbert space and equipped with a Riemannian metric. A Riemannian metric g on M is a *'smooth'* choice of an inner product $g(p)$ on the tangent space T_pM for each $p \in M$, where *smoothness* refers to g as a function of p. Let us make this precise below.

Recall that $L^2(V, \mathbb{F})$ denotes the Banach space of continuous bilinear forms on a Banach space V over $\mathbb{F} = \mathbb{R}, \mathbb{C}$. If V is finite dimensional, we have the identification $L^2(V, \mathbb{F}) = V^* \otimes V^*$ where the latter is equipped with the injective tensor norm. Hence bilinear forms on V are the so-called $(0, 2)$-*type tensors* or *covariant tensors of order* 2. If V is infinite dimensional, we have the identification $L^2(V, \mathbb{F}) = (V \hat{\otimes} V)^*$ which only contains the algebraic tensor $V^* \otimes V^*$ as a subspace, where $\hat{\otimes}$ denotes the projective tensor product. Let $L_0^2(V, \mathbb{F})$ be the closed subspace of $L^2(V, \mathbb{F})$, consisting of all *symmetric* bilinear forms on V. One can view $L_0^2(\,\cdot\,, \mathbb{F})$ as a functor on the category of Banach spaces over $\mathbb{F}$.

Let M be a smooth manifold modelled on a real Banach space V. A symmetric bilinear form $g \in L_0^2(V, \mathbb{R})$ is called *positive definite* if

$$g(x, x) > 0 \quad \text{for all} \quad x \in V \backslash \{0\}$$

in which case, g is often called an *inner product*. We call g *completely positive definite* if g is a complete inner product on V. By the open mapping theorem for Banach spaces, g is completely positive definite on V if, and only if, there is a constant $c_g > 0$ such that

$$g(x, x) \geq c_g \|x\|^2 \qquad (x \in V).$$

Of course, if $\dim V < \infty$, then complete positive definiteness is the same as positive definiteness.

Now let V be a real Hilbert space with inner product $\langle \cdot, \cdot \rangle$. Then by the Riesz representation theorem, $L_0^2(V, \mathbb{R})$ is linearly isometric to the closed subspace $L(V)_s$ of $L(V)$, consisting of all symmetric operators in $L(V)$. The linear isometry $g \in L_0^2(V, \mathbb{R}) \mapsto L_g \in L(V)_s$ is implemented by

$$g(x, y) = \langle L_g x, y \rangle \qquad (x, y \in V)$$

where $g(x, x) \geq 0$ for all $x \in V$ if, and only if, $L_g \geq 0$. In this case, positive definiteness of g is equivalent to the symmetric operator L_g being positive and injective. Complete positive definiteness of g is equivalent to the existence of a constant $c_g > 0$ such that

$$g(x, x) \geq c_g \langle x, x \rangle \qquad (x \in V)$$

which is also equivalent to L_g being positive and invertible.

We can apply the functor $L_0^2(\,\cdot\,, \mathbb{R})$ to the tangent bundle

$$\pi : \bigcup_{p \in M} T_p M \longrightarrow M$$

and form a vector bundle

$$L_0^2(\pi) : \bigcup_{p \in M} L_0^2(T_p M, \mathbb{R}) \longrightarrow M.$$

A *Riemannian metric* on M is a smooth section

$$g : M \longrightarrow \bigcup_{p \in M} L_0^2(T_p M, \mathbb{R})$$

of this bundle such that each $g(p) : T_p M \times T_p M \longrightarrow \mathbb{R}$ is completely positive definite. The differential structure on $\bigcup_{p \in M} L_0^2(T_p M, \mathbb{R})$ can be described as follows (cf. [118, p. 170]).

Let $\{(\mathcal{U}_\varphi, \varphi, V)\}$ be an atlas of M where each tangent space $T_p M$ identifies with the model space $V = \{(\varphi \circ \gamma)'(0) : [\gamma]_p \in T_p M\}$, and $g(p) \in L_0^2(T_p M, \mathbb{R})$ identifies with $g_\varphi(p) \in L_0^2(V, \mathbb{R})$, defined by

$$g_\varphi(p)(\,(\varphi \circ \alpha)'(0), (\varphi \circ \beta)'(0)\,) = g(p)([\alpha]_p, [\beta]_p).$$

Write $L_0^2(M) = \bigcup_{p \in M} L_0^2(T_p M, \mathbb{R})$ and $L_0^2 \mathcal{U}_\varphi = \bigcup_{p \in \mathcal{U}_\varphi} L_0^2(T_p M, \mathbb{R})$. Define an injective map

$$L(\varphi) : L_0^2 \mathcal{U}_\varphi \longrightarrow V \times L_0^2(V, \mathbb{R})$$

$$L(\varphi)(g(p)) = (\varphi(p), g_\varphi(p)) \qquad (g(p) \in L_0^2(T_p M, \mathbb{R}))$$

The image of $L(\varphi)$ is $\varphi(\mathcal{U}_\varphi) \times L_0^2(V, \mathbb{R})$. Then

$$\{(L_0^2 \mathcal{U}_\varphi, L(\varphi), V \times L_0^2(V, \mathbb{R}))\}$$

is an atlas on $L_0^2(M)$ whose open sets are the sets $\mathcal{O}$ for which

$$L(\varphi)(\mathcal{O} \cap L_0^2 \mathcal{U}_\varphi)$$

is open in $V \times L_0^2(V, \mathbb{R})$.

Smoothness of the Riemannian metric $g : M \longrightarrow L_0^2(M)$ means that in local charts $(\mathcal{U}_\varphi, \varphi, V)$, the map

$$g_\varphi : \mathcal{U}_\varphi \longrightarrow L_0^2(V, \mathbb{R}) = L(V)_s \qquad (1.8)$$

is smooth.

If a smooth manifold M modelled on a real Hilbert space V admits a Riemannian metric g, we call (M, g), or M (if g is understood), a *Riemannian manifold*.

We usually denote the Riemannian metric $g(p)$ on $T_p M$ by $\langle \cdot, \cdot \rangle_p$, and if confusion is unlikely, the symbol g is also often used to denote a diffeomorphism between manifolds.

It is well known that every finite dimensional paracompact smooth manifold admits a Riemannian metric. We recall that a topological space M is *paracompact* if every open covering of M has a locally finite refinement. Metric spaces are paracompact. We refer to [118, p. 36, 171] for a proof of the following existence theorem.

Theorem 1.2.18. *Let M be a paracompact smooth manifold modelled on a separable Hilbert space. Then M admits a Riemannian metric.*

Definition 1.2.19. Let M and N be Riemannian manifolds. A diffeomorphism $g : M \longrightarrow N$ is called an *isometry* if it satisfies the condition

$$\langle X_p, Y_p \rangle_p = \langle dg_p(X_p), dg_p(Y_p) \rangle_{g(p)} \qquad (p \in M; X_p, Y_p \in T_p M).$$

By polarization, the preceding condition is equivalent to

$$\langle X_p, X_p \rangle_p = \langle dg_p(X_p), dg_p(X_p) \rangle_{g(p)} \qquad (p \in M; \, X_p \in T_p M).$$

The isometries of a Riemannian manifold M form a group $G(M)$ under composition. We call $G(M)$ the *isometry group* of M. We say that $G(M)$ *acts transitively on* M if given two points $p, q \in M$, there exists $g \in G(M)$ such that $g(p) = q$.

We end this section with a review of Lie groups. A (real or complex) Banach manifold G is called a (resp. *real* or *complex*) *Banach Lie group* if G is a group and the group operations are analytic, that is, the multiplication $(x, y) \in G \times G \mapsto xy \in G$ and the inverse map $x \in G \mapsto x^{-1} \in G$ are analytic. In practice, to verify analyticity of the group operations, it suffices to verify the following three conditions (cf. [21, III.1.1]):

(i) the left translation $\ell_a : g \in G \mapsto ag \in G$ is analytic, for all $a \in G$;

(ii) the mapping $(x, y) \in G \times G \mapsto xy^{-1} \in G$ is analytic in a neighbourhood of (e, e) where e is the identity of G;

(iii) the conjugation $g \in G \mapsto aga^{-1} \in G$ is analytic in a neighbourhood of e, for all $a \in G$.

Let G be a Banach Lie group with identity e. For each $a \in G$, the right translation $r_a : g \in G \mapsto ga \in G$ is an analytic map. An analytic vector field $X : G \longrightarrow T(G)$ is called *left invariant* (respectively, *right invariant*) if $(d\ell_a)_g(X_g) = X_{ag}$ (respectively, $(dr_a)_g(X_g) = X_{ga}$) for all $a, g \in G$. This can be written as $d\ell_a \circ X = X \circ \ell_a$ and $dr_a \circ X = X \circ r_a$ in terms of the tangent maps $d\ell_a, dr_a : T(G) \longrightarrow T(G)$. An invariant vector field X is determined entirely by its value at the identity e since $X_a = (d\ell_a)_e(X_e)$ for all $a \in G$ if X is left invariant.

Given two left invariant analytic vector fields X and Y on G, it can be verified that the commutator $[X, Y]$ is also left invariant. Therefore the vector space $\mathfrak{L}$ of all left invariant analytic vector fields on G forms a Lie algebra in the product $[X, Y]$. Given a tangent vector $[\gamma]_e$ in the tangent space $T_e G$ at the identity e, the vector field X defined by $X(g) = (d\ell_g)_e([\gamma]_e)$ is left invariant and $X(e) = [\gamma]_e$. It follows that the map $X \in \mathfrak{L} \mapsto X_e \in T_e G$ is a linear isomorphism since two left invariant vector fields are identical if they have the same value at e. Therefore the tangent space $T_e G$ is a Lie algebra in the brackets

$$[X_e, Y_e] := [X, Y]_e \qquad (X, Y \in \mathfrak{L}).$$

We call $\mathfrak{g} = T_e G = \{X_e : X \in \mathfrak{L}\}$ the *Lie algebra of G*.

Lemma 1.2.20. *A left invariant analytic vector field X on a Banach Lie group G is complete.*

Proof. By left invariance of X, given $X(e) = [\gamma]_e \in T_e G$, we have $X(a) = [a\gamma]_a \in T_a G$. If $\gamma_e : I \longrightarrow G$ satisfies

$$\frac{d\gamma_e(t)}{dt} = X(\gamma_e(t)) \quad \text{and} \quad \gamma_e(0) = e,$$

then the curve $\gamma_a = a\gamma_e$ solves

$$\frac{d\gamma_a(t)}{dt} = X(\gamma_a(t)) \quad \text{and} \quad \gamma_a(0) = a.$$

Let $\alpha_e : I \times \mathcal{U} \longrightarrow G$ be a local flow of X satisfying $\alpha(0, p) = p$ for p in a neighbourhood $\mathcal{U}$ of e. Then the map $\alpha_a : (t, ap) \in I \times a\mathcal{U} \mapsto a\alpha_e(t, p) \in G$ is a local flow of X satisfying $\alpha_a(0, ap) = ap$ for all $ap \in a\mathcal{U}$. Taking the union $\bigcup_{a \in G} I \times a\mathcal{U}$ and applying the uniqueness theorem, we arrive at a local flow $\alpha : I \times G \longrightarrow G$ of X satisfying $\alpha(0, a) = a$ for $a \in G$. Since

$$\alpha(s + t, \cdot) = \alpha(s, \alpha(t, \cdot))$$

for $s, t, s+t$ in I, one can extend α to a map $\alpha_X : \mathbb{R} \times G \longrightarrow G$ by defining

$$\alpha_X(t, \cdot) = \overbrace{\alpha\left(\frac{t}{n}, \alpha\left(\frac{t}{n}, \left(\cdots \alpha\left(\frac{t}{n}, \cdot\right)\right)\right)\right)}^{n\text{-times}}$$

for sufficiently large n, which can be seen to be well-defined. Since $\alpha_X(s+t, a) = \alpha_X(s, \alpha_X(t, a))$ and α_X is analytic on $I \times G$, it follows that α_X is analytic on $\mathbb{R} \times G$ and is the flow of X. $\qquad\square$

A *one-parameter subgroup* of a Banach Lie group G is an analytic homomorphism $\theta : (\mathbb{R}, +) \longrightarrow G$ such that $\theta(0) = e$. Given such a homomorphism θ, we have $d\theta_0(1) \in T_e G$. Conversely, for each left invariant vector field X on G, the homomorphism

$$t \in \mathbb{R} \mapsto \exp tX(e) \in G$$

is a one-parameter subgroup of G satisfying $d(\exp tX(e))_0(1) = X_e$. The *exponential map*

$$\exp : T_e G \longrightarrow G$$

is defined by

$$\exp(X_e) = \exp X(e) \qquad (X \in \mathfrak{L}).$$

The exponential map need not be surjective, but if it is, then G is called *exponential*.

Definition 1.2.21. Let G be a Banach Lie group with Lie algebra $\mathfrak{g} = T_e G$. The *adjoint representation of G* is the homomorphism

$$Ad : G \longrightarrow \operatorname{Aut} \mathfrak{g}$$

defined by the differential of the inner automorphism:

$$Ad(g) = d(r_{g^{-1}} \ell_g)_e : \mathfrak{g} \longrightarrow \mathfrak{g}.$$

We note that Ad is a homomorphism since $r_{(gh)^{-1}}\ell_{gh} = r_{g^{-1}}\ell_g \circ r_{h^{-1}}\ell_h$. In the notation of (1.4), we have $Ad(g) = (r_{g^{-1}}\ell_g)_*$ and by (1.7),

$$\exp Ad(g)tX_e = \exp tAd(g)X_e = g(\exp tX_e)g^{-1} \qquad (X \in \mathfrak{g}). \qquad (1.9)$$

The automorphism group Aut $\mathfrak{g}$ is contained in the Banach space $L(T_eG)$ of continuous linear operators on the tangent space $\mathfrak{g} = T_eG$. Considering Ad as the map $Ad : G \longrightarrow L(T_eG)$, we can take its differential which defines the *adjoint representation* $ad : \mathfrak{g} \longrightarrow L(T_eG)$ of $\mathfrak{g}$:

$$ad(X_e) = d(Ad)_e(X_e).$$

Lemma 1.2.22. *Given two left invariant analytic vector fields X and Y on a Banach Lie group G, we have $[X,Y]_e = ad(X_e)(Y_e)$.*

Proof. We have $Ad(g)(Y_e) = (dr_{g^{-1}})_g(d\ell_g)_e(Y_e) = (dr_{g^{-1}})_g(Y_g)$ by left invariance. Let $\alpha_t(\cdot) = \alpha_X(t, \cdot)$ be the flow of X in Lemma 1.2.20. Since X is left invariant, we have $\ell_a \circ \alpha_t = \alpha_t \circ \ell_a$ for $a \in G$. Hence we have

$$\alpha_{-t}(a) = \alpha_{-t}(\ell_a(e)) = \ell_a(\alpha_{-t}(e) = r_{\alpha_{-t}(e)}(a).$$

It follows from (1.5) that

$$
\begin{aligned}
[X,Y](e) &= \lim_{t\to 0} \frac{1}{t}\left((d\alpha_{-t})_{\alpha_t(e)}Y_{\alpha_t(e)} - Y_e\right) \\
&= \lim_{t\to 0} \frac{1}{t}\left((dr_{\alpha_{-t}(e)})_{\alpha_t(e)}Y_{\alpha_t(e)} - Y_e\right) \\
&= \lim_{t\to 0} \frac{1}{t}\left(Ad(\alpha_t(e))Y_e - Y_e\right) \\
&= ad(X_e)(Y_e). \qquad\qquad\qquad\qquad \square
\end{aligned}
$$

Let G be a Banach Lie group with identity e and Lie algebra $\mathfrak{g}$. A subgroup K of G is called a *Banach Lie subgroup* if it is a submanifold of G, in which case K is closed and a Banach Lie group in the induced topology of G [168, p. 128] and it can be shown that the subalgebra

$$\mathfrak{k} = \{X \in \mathfrak{g} : \exp tX \in K, \forall t \in \mathbb{R}\} \qquad (1.10)$$

of $\mathfrak{g}$ identifies with the Lie algebra of K. Further, the left coset space

$$G/K = \{gK : g \in G\} \qquad (1.11)$$

carries the structure of a Banach manifold and the quotient map $\pi : G \longrightarrow G/K$ is a submersion [168, Theorem 8.19]. Manifolds of the form G/K are called *homogeneous spaces*. Let $p = \pi(e) = K$. Then the differential $d\pi_e : \mathfrak{g} \longrightarrow T_p(G/K)$ has kernel $\ker d\pi_e = \mathfrak{k}$ and gives the canonical isomorphism $\mathfrak{g}/\mathfrak{k} \approx T_p(G/K)$.

Notes. The basic material of Banach manifolds in this section is extracted from the author's book [37]. There is a substantial literature on infinite dimensional manifolds, Lie groups and Lie algebras including, for example, [21, 106, 118, 140, 168].

1.3   Symmetric Banach manifolds

Symmetric Banach manifolds are an infinite dimensional generalisation of Hermitian symmetric spaces in $\mathbb{C}^d$. Bounded symmetric domains form a special class of these manifolds, which need not possess a Riemannian structure.

Let M be a (smooth or analytic) Banach manifold, with tangent bundle $TM = \{(p, v) : p \in M, v \in T_pM\}$. A mapping

$$\nu : TM \longrightarrow [0, \infty)$$

is called a *tangent norm* if $\nu(p, \cdot)$ is a norm on the tangent space T_pM for each $p \in M$. We call ν a *compatible tangent norm* if it satisfies the following two conditions.

(i) ν is continuous.

(ii) For each $p \in M$, there is a local chart $(\mathcal{U}, \varphi, V)$ at p, and constants $0 < c < C$ such that

$$c\|d\varphi_a(v)\|_V \leq \nu(a,v) \leq C\|d\varphi_a(v)\|_V \qquad (a \in \mathcal{U}, v \in T_a M).$$

Remark 1.3.1. A compatible tangent norm satisfying certain smoothness and convexity conditions is known as a *Finsler metric* or *Finsler function* (cf. [3, 13]).

Given an analytic Banach manifold M with a tangent norm ν, a bianalytic map $f : M \longrightarrow M$ is called ν-*isometric* or a ν-*isometry* if it satisfies

$$\nu(f(p), df_p(\cdot)) = \nu(p, \cdot) \quad \text{for all} \quad (p, \cdot) \in TM. \tag{1.12}$$

For a real smooth Banach manifold M, ν-*isometries* of M are defined to be diffeomorphisms satisfying condition (1.12).

Example 1.3.2. A Riemannian manifold (M, g) modelled on a real Hilbert space V, with inner product $\langle \cdot, \cdot \rangle$, admits a compatible tangent norm $\nu : TM \longrightarrow [0, \infty)$ defined by

$$\nu(p, v) := g(p)(v, v)^{1/2} \qquad (p \in M, v \in T_p M \approx V)$$

where, as noted before, $g(p)(v, v) = \langle L_{g(p)} v, v \rangle$ for some symmetric operator $L_{g(p)}$ in $L(V)_s$ and, complete positivity of $g(p)$ implies

$$g(p)(v, v) \geq c_{g(p)} \langle v, v \rangle$$

for some $c_{g(p)} > 0$. Smoothness of the metric g implies that ν is continuous and for each $p \in M$, there is a local chart $(\mathcal{U}_\varphi, \varphi, V)$ at p such that the smooth map $g_\varphi : \mathcal{U}_\varphi \longrightarrow L(V)_s$ in (1.8) satisfies $\|L_{g_\varphi(a)} - L_{g_\varphi(p)}\| \leq c_{g_\varphi(p)}/2$ for all $a, p \in \mathcal{U}_\varphi$, by continuity.

It follows from $\langle g_\varphi(p)v, v\rangle \geq c_{g_\varphi(p)}\|v\|^2$ and

$$|\langle g_\varphi(p)v, v\rangle - \langle g_\varphi(a)v, v\rangle| = |\langle L_{g_\varphi(p)}v, v\rangle - \langle L_{\langle g_\varphi(a)}v, v\rangle|$$

$$\leq \|L_{g_\varphi(a)} - L_{g_\varphi(p)}\|\|v\|^2 < \frac{c_{g_\varphi(p)}}{2}\|v\|^2$$

that

$$\frac{c_{g_\varphi(p)}}{2}\|v\|^2 \leq \langle g_\varphi(a)v, v\rangle \leq \left(L_{\|g_\varphi(p)\|} + \frac{c_{g_\varphi(p)}}{2}\right)\|v\|^2 \qquad (v \in V)$$

where $\nu(a, v) = \langle g_\varphi(a)v, v\rangle^{1/2}$.

The ν-isometries of M are exactly the isometries of M with respect to the Riemannian metric g.

Example 1.3.3. Let M be a finite dimensional complex manifold modelled on $\mathbb{C}^d$. Then there is a canonical *almost complex structure J* on M, viewed as a $2d$-dimensional real manifold (cf. [80, Chapter VIII, §1]). The almost complex structure J is a map $p \in M \mapsto J_p \in L(T_pM)$ such that $J_p : T_pM \longrightarrow T_pM$ is a complex structure on the tangent space T_pM, that is, J_p is a real linear isomorphism such that $-J_p^2$ is the identity map, and for each smooth vector field X on M, the vector field JX is smooth, where $JX(p) = J_p(X_p)$ for $p \in M$. If we identify T_pM with $\mathbb{C}^d = \mathbb{R}^{2d}$, then J_p is the canonical complex structure given in (1.1). We call J a *Hermitian structure* on M if M admits a Riemannian metric g satisfying

$$g_p(J_pu, J_pv) = g_p(u, v) \qquad (u, v \in T_pM, p \in M)$$

in which case, g is called *Hermitian* with respect to J. A finite dimensional complex manifold with a Hermitian structure is called a *Hermitian manifold*. In view of Example 1.3.2, a Hermitian manifold admits a compatible tangent norm defined by the Riemannian metric g.

We now show that a finite dimensional bounded domain D in $\mathbb{C}^d$ carries the structure of a Hermitian manifold. For this, we first introduce the *Bergman kernel* of D. Let λ be the Lebesgue measure on $\mathbb{C}^d$ and let $H_2(D)$ be the complex vector space of square λ-integrable holomorphic functions on D. Then $H_2(D)$ is a separable Hilbert space with inner product

$$\langle f, g \rangle = \int_D f(z)\overline{g(z)}d\lambda(z)$$

which is called the *Bergman space* of D. For each $w \in D$, the map $f \in H_2(D) \mapsto f(w) \in \mathbb{C}$ is a continuous linear functional on $H_2(D)$. By the Riesz representation theorem, there is a unique function $k(\cdot, w) \in H(D)$ such that

$$f(w) = \langle f, k(\cdot, w) \rangle = \int_D f(z)\overline{k(z, w)}d\lambda(z) \qquad (f \in H_2(D)).$$

The function $k(z, w)$ on $D \times D$ is called the *Bergman kernel* of D. Let $\{\varphi_n\}_{n=1}^{\infty}$ be an orthonormal basis in $H_2(D)$. Then we have

$$k(\cdot, w) = \sum_n \langle k(\cdot, w), \varphi_n \rangle \varphi_n = \sum_n \overline{\varphi_n(w)}\varphi_n.$$

From this we deduce that $\overline{k(z, w)} = k(w, z)$ and that $\overline{k(z, \cdot)}$ is holomorphic on D. By considering the holomorphic polynomials on D, it can be seen that $k(w, w) > 0$ for all $w \in D$.

The domain $D \subset \mathbb{C}^d$ is a complex manifold modelled on $\mathbb{C}^d$. Using the Bergman kernel $k(z, w)$, one can define a Hermitian (Riemannian) metric on D as follows. For each $z \in D$, define

$$h_z(u, v) := \sum_{i,j=1}^{d} \frac{\partial^2}{\partial z_i \partial \overline{z}_j} \log k(z, z) u_i \overline{v}_j$$

for $u = (u_1, \ldots, u_d), v = (v_1, \ldots, v_d) \in \mathbb{C}^d$. Then $h_z(\cdot, \cdot)$ is a positive definite Hermitian bilinear form on the tangent space $T_z(D) \approx \mathbb{C}^d$ and its real part

$$g_z(u, v) := \operatorname{Re} \sum_{i,j=1}^{d} \frac{\partial^2}{\partial z_i \partial \overline{z}_j} \log k(z, z) u_i \overline{v}_j$$

is an inner product on $T_z(D) \approx \mathbb{R}^{2d}$ and g is a Riemannian metric on D, which is Hermitian with respect to the canonical almost complex structure J on D. With this Hermitian structure, a bounded domain in $\mathbb{C}^d$ is therefore endowed with a compatible tangent norm. The metric g (or h) is known as the *Bergman metric* on D. The biholomorphic maps between bounded domains are invariant under the Bergman metric. We refer to [80, Chapter VIII, Proposition 3.5] for a proof of the following basic result.

Lemma 1.3.4. *Let $f : D_1 \longrightarrow D_2$ be a biholomorphic map between two bounded domains D_1 and D_2 in $\mathbb{C}^d$. Then f is an isometry with respect to the Bergmann metric, that is,*

$$h_z(u, v) = h_{f(z)}(f'(z)u, f'(z)v) \qquad (z \in D, u, v \in \mathbb{C}^d).$$

Besides the tangent norm defined by the Bergman metric on a bounded domain D in $\mathbb{C}^d$, one can define another tangent norm on any bounded domain in a complex Banach space. This is given in Example 1.3.6 below.

Example 1.3.5. Given a complex Banach manifold M, define a mapping

$$\mathcal{C} : TM \longrightarrow [0, \infty)$$

by

$$\mathcal{C}(p, v) = \sup\{|f'(p)(v)| : f \in H(M, \mathbb{D}) \text{ and } f(p) = 0\}$$

for $(p, v) \in TM$, where $H(M, \mathbb{D})$ denotes the set of all holomorphic maps from M to $\mathbb{D}$. Then clearly $\mathcal{C}(p, \cdot)$ is a semi-norm on the tangent space T_pM. Given a holomorphic map $h : M \longrightarrow M$ and $(p, v) \in TM$, we have

$$
\begin{aligned}
&\mathcal{C}(h(p), dh_p(v)) \\
={} &\sup\{|df_{h(p)}(dh_p(v))| : f \in H(M, \mathbb{D}) \text{ and } f(h(p)) = 0\} \\
={} &\sup\{|d(f \circ h)_p(v)| : f \in H(M, \mathbb{D}) \text{ and } (f \circ h)(p) = 0\} \\
\leq{} &\sup\{|df_p(v)| : f \in H(M, \mathbb{D}) \text{ and } f(p) = 0\} = \mathcal{C}(p, v).
\end{aligned}
$$

In particular, if $h : M \longrightarrow M$ biholomorphic, we have $\mathcal{C}(h(p), dh_p(v)) = \mathcal{C}(p, v)$. The mapping $\mathcal{C} : TM \longrightarrow [0, \infty)$ is called the *Carathéodory tangent semi-norm* on M.

The Carathéodory tangent semi-norm on M need not be a compatible tangent norm. However, it is indeed such if M is a bounded domain in a Banach space.

Example 1.3.6. Let D be a bounded domain in a complex Banach space V. Then the Carathéodory tangent semi-norm $\mathcal{C} : TD \longrightarrow [0, \infty)$ defined in the preceding example is a compatible tangent norm on D. Indeed, pick $p \in D$ and a chart $(\mathcal{U}, \varphi, V)$ at p with $r = d(\mathcal{U}, \partial D) > 0$. Then for each holomorphic map $f : D \longrightarrow \mathbb{D}$, the Cauchy inequality (1.2) implies

$$
r|f'(a)v| \leq r\|f'(a)\|\|v\| \leq \|v\| \qquad (a \in \mathcal{U}, v \in V)
$$

and hence $r\mathcal{C}(a, v) \leq \|v\|$ for $(a, v) \in \mathcal{U} \times V$. On the other hand, let D be contained in some open ball $B(0, s)$. For each linear functional $h \in V^*$ with $\|h\| < 1$, the holomorphic function $f = \frac{1}{2s}(h - h(a)) : D \longrightarrow \mathbb{D}$ satisfies $f(a) = 0$ and hence $|h(v)| = |2sf'(a)v| \leq 2s\mathcal{C}(a, v)$. It follows that $\|v\| \leq 2s\mathcal{C}(a, v)$ for $a \in \mathcal{U}$ and $v \in V$. In particular, $\mathcal{C}(a, \cdot)$ is a norm on V, which is equivalent to the norm of V. Moreover, it can be seen that $\mathcal{C}$ is continuous on TD (cf. [168, 12.23]).

The mapping $\mathcal{C}$ is called the *Carathéodory tangent norm* on D and we sometimes write $\|\cdot\|_a$ for the norm $\mathcal{C}(a, \cdot)$ on V.

Definition 1.3.7. Let M be a complex Banach manifold equipped with a compatible tangent norm ν. We denote by

$$\mathrm{Aut}(M, \nu) = \{f \in \mathrm{Aut}\, M : f \text{ is a } \nu\text{-isometry}\}$$

the subgroup of all ν-isometries in the automorphism group $\mathrm{Aut}\, M$ of M.

Lemma 1.3.8. *Let D be a bounded domain in a complex Banach space V, with the Carathéodory tangent norm ν. Then we have $\mathrm{Aut}(D, \nu) = \mathrm{Aut}\, D$.*

Proof. This follows from Example 1.3.5.   □

Given a Banach manifold M with a compatible tangent norm ν, one can define a distance function on M by the integrated form of ν. Let $\gamma : [0, 1] \longrightarrow M$ be a piecewise smooth curve. The length $\ell(\gamma)$ of γ is given by

$$\ell(\gamma) = \int_0^1 \nu(\gamma(t), \gamma'(t))dt.$$

We define the integrated form $d_\nu : M \times M \longrightarrow [0, \infty)$ of ν by

$$d_\nu(a, b) = \inf_\gamma\{\ell(\gamma) : \gamma(0) = a, \gamma(1) = b\} \tag{1.13}$$

where $\gamma : [0, 1] \to M$ is a piecewise smooth curve. It can be verified that (M, d_ν) is a metric space.

Example 1.3.9. Consider the Carathéodory tangent norm

$$\mathcal{C}(p, v) = \sup\{|f'(p)(v)| : f \in H(\mathbb{D}, \mathbb{D}) \text{ and } f(p) = 0\} \qquad (p \in \mathbb{D}, v \in \mathbb{C})$$

on the complex open unit disc $\mathbb{D}$, where we have

$$|f'(p)(v)| = |f'(p)||v| \le \frac{1 - |f(p)|^2}{1 - |p|^2}|v| \le \frac{|v|}{1 - |p|^2}$$

by the Schwarz-Pick lemma. On the other hand, the Möbius transformation

$$g_{-p}(z) = \frac{z - p}{1 - \bar{p}z} \qquad (z \in \mathbb{D})$$

maps the point p to the origin 0, with derivative $g'_{-p}(p) = \frac{1}{1-|p|^2}$ and it follows that

$$C(p, v) = \frac{|v|}{1 - |p|^2}$$

which is the infinitesimal *Poincaré metric* on $\mathbb{D}$.

In this case, the integrated distance $d_C(a, b)$, the *Poincaré distance*, is given by

$$d_C(a, b) = \tanh^{-1}\left|\frac{a - b}{1 - \bar{a}b}\right| = \frac{1}{2}\log\frac{1 + \left|\frac{a-b}{1-\bar{a}b}\right|}{1 - \left|\frac{a-b}{1-\bar{a}b}\right|} \qquad (a, b \in \mathbb{D}). \quad (1.14)$$

Remark 1.3.10. On a bounded domain D in a complex Banach space, the *Carathéodory distance* c_D is defined by

$$c_D(a, b) = \sup\{d_C(f(a), f(b)) : f \in H(D, \mathbb{D})\} \qquad (a, b \in D)$$

which is invariant under biholomorphic maps on D and will be studied in Section 3.5. Although $c_{\mathbb{D}} = d_C$ is the integrated distance of the Carathéodory tangent norm on $\mathbb{D}$, the Carathéodory distance c_D need not coincide with the integrated distance of the Carathéodory tangent norm on other domains D.

Let M be a connected manifold modelled on a Banach space V, with a compatible tangent norm ν. One can define the topology of locally uniform convergence on $\text{Aut}(M, \nu)$ using the metric d_ν. Choose

a chart $(\mathcal{U}, \varphi, V)$ at $p \in M$ such that $\varphi(\mathcal{U})$ is bounded in V and $\varphi(p) = 0$.
Let $r > 0$ so that

$$B_{d_\nu}(p, 4r) := \{x \in M : d_\nu(x, p) < 4r\} \subset \mathcal{U}.$$

Let $B_p = B_{d_\nu}(p, r/2) := \{x \in M : d_\nu(x, p) < r/2\}$ and define

$$\rho_{B_p}(f, h) = \sup\{d_\nu(f(x), h(x)) : x \in B_p\} \qquad (f, h \in \operatorname{Aut}(M, \nu)).$$

Then ρ_{B_p} is a well-defined metric on $\operatorname{Aut}(M, \nu)$ by the principle of analytic continuation and moreover, the metric ρ_{B_q} defined by another
chart $(\mathcal{V}, \psi, V)$ at $q \in M$ induces the same topology as the one induced by ρ_{B_p}. This topology is called the *topology of locally uniform
convergence on* $\operatorname{Aut}(M, \nu)$.

The following fundamental result is due to Upmeier [167].

Theorem 1.3.11. *Let M be a connected complex Banach manifold
equipped with a compatible tangent norm ν. A subgroup G of $\operatorname{Aut}(M, \nu)$,
which is closed in the topology of locally uniform convergence, can be
topologised to a real Banach Lie group of which the Lie algebra is given
by $\mathfrak{g} = \{X \in \operatorname{aut} M : \exp tX \in G, \forall t \in \mathbb{R}\}$.*

In particular, the group $\operatorname{Aut}(M, \nu)$ of ν-isometric automorphisms
of M carries the structure of a real Banach Lie group with Lie algebra

$$\operatorname{aut}(M, \nu) = \{X \in \operatorname{aut} M : \exp tX \in \operatorname{Aut}(M, \nu), \forall t \in \mathbb{R}\},$$

which is a real Banach Lie algebra where the norm is determined by
the preceding chart $(\mathcal{U}, \varphi, V)$ at some $p \in M$. The norm of $X = h\frac{\partial}{\partial z} \in \operatorname{aut}(M, \nu)$ is defined by

$$\|X\| = \sup\{\|h(x)\| : x \in B_p\}. \tag{1.15}$$

We refer to [168, 13.14, 13.17] for a detailed proof of this theorem. In the special case where M is a bounded domain in a Banach

space, equipped with the Carathéodory tangent norm ν on TM, we have $\text{Aut}(M, \nu) = \text{Aut}\, M$ from Lemma 1.3.8 and hence the following useful result.

Corollary 1.3.12. *Let M be a bounded domain in a complex Banach space. Then its automorphism group $\text{Aut}\, M$ admits a real Banach Lie group structure and its Lie algebra is the Lie algebra $\text{aut}\, M$ of complete holomorphic vector fields on M.*

Remark 1.3.13. With the topology of locally uniform convergence, the automorphism group $\text{Aut}\, D$ of a bounded domain D in an infinite dimensional Banach space need not be a Lie group, as shown in [172, §2.4]. The topology making $\text{Aut}\, D$ into a Lie group is finer than the topology of locally uniform convergence (cf. [168, 13.6]), but they coincide if D is a *symmetric* domain as defined in Definition 1.3.15 (cf. [168, p. 228]), in which case the preceding corollary has also been proved by Vigué [172]. For finite dimensional bounded domains, it is a well-known result of H. Cartan [34] that their automorphism groups are Lie groups in the compact-open topology. It seems appropriate here to amend a slip in the Notes of the author's book [37, p. 169] where it should have been said that '$\text{Aut}\, M$ is a Lie group in a topology finer than the topology of locally uniformly convergence'.

We now introduce the fundamental concept of a symmetric Banach manifold. It is convenient to begin with the notion of a *symmetry* of a manifold. Let M be a Banach manifold endowed with a compatible tangent norm ν and let $p \in M$. A *symmetry* at p is a ν-isometry

$$s : M \longrightarrow M$$

satisfying the following two conditions:

(i) s is involutive, that is, s^2 is the identity map on M,

(ii) p is an isolated fixed-point of s, in other words, p is the only point in some neighbourhood of p satisfying $s(p) = p$.

Lemma 1.3.14. *Let M be a Banach manifold and let $s : M \longrightarrow M$ be a symmetry at $p \in M$. Then the differential $(ds)_p : T_pM \longrightarrow T_pM$ is the map $-id$, where id is the identity map of T_pM.*

Proof. By [168, 17.1], on can find a local chart $(\mathcal{U}, \varphi, T_pM)$ at p with $\varphi(p) = 0$ and a continuous linear map $\ell : T_pM \longrightarrow T_pM$ such that the following diagram commutes.

$$
\begin{array}{ccc}
\mathcal{U} \cap s^{-1}(\mathcal{U}) & \xrightarrow{s} & \mathcal{U} \cap s^{-1}(\mathcal{U}) \\
\downarrow{\varphi} & & \downarrow{\varphi} \\
T_pM & \xrightarrow{\ell} & T_pM
\end{array}
$$

Since s^2 is the identity map, so is ℓ^2 and hence there is a direct decomposition

$$
T_pM = \{X_p : \ell(X_p) = X_p\} \oplus \{X_p : \ell(X_p) = -X_p\}.
$$

Moreover, $\{X_p : \ell(X_p) = X_p\} = \{0\}$ because p is an isolated fixed-point of s. It follows that $-\ell$ is the identity map on T_pM and so is $(-ds)_p$. $\square$

It follows from Lemma 1.3.14 and Cartan's uniqueness theorem that there can only be *one* symmetry at a point p in a bounded domain in a complex Banach space, viewed as a Banach manifold equipped with the Carathéodory tangent norm. For a connected Riemannian manifold, a symmetry is also unique if it exists (cf. [37, p. 101]).

Definition 1.3.15. By a *symmetric Banach manifold*, we mean a connected Banach manifold M, equipped with a compatible tangent norm ν, such that there is a symmetry $s_p : M \longrightarrow M$ at each $p \in M$.

Given a symmetric Banach manifold M modelled on a complex Banach space V, it has been shown in [168, Theorem 17.16] that the map $\epsilon : h \in \mathrm{Aut}(M, \nu) \mapsto h(a) \in M$ is a submersion at the identity e of $\mathrm{Aut}(M, \nu)$, for each $a \in M$. In particular, the differential $d\epsilon$ of ϵ at e, which is the evaluation map

$$X \in \mathrm{aut}(M, \nu) \mapsto X(a) \in V, \tag{1.16}$$

is surjective, and we have the following equivalent definition of a complex symmetric Banach manifold.

Lemma 1.3.16. *Let M be a connected Banach manifold with a compatible tangent norm ν, modelled on a complex Banach space V. The following conditions are equivalent.*

(i) *M is a symmetric Banach manifold.*

(ii) *There is a symmetry s_p at some point $p \in M$ and the map $h \in \mathrm{Aut}(M, \nu) \mapsto h(a) \in M$ is a submersion at e for each $a \in M$.*

Proof. The preceding remark establishes (i) $\Rightarrow$ (ii). Conversely, condition (ii) implies that $\mathrm{Aut}(M, \nu)$ acts transitively on M, that is, given $p, q \in M$, there exists $h \in \mathrm{Aut}(M, \nu)$ with $h(p) = q$. Indeed, the submersion $h \in \mathrm{Aut}(M, \nu) \mapsto h(a) \in M$ is an open map by Remark 1.2.11 and hence the orbit

$$O(a) = \{h(a) : h \in \mathrm{Aut}(M, \nu)\}$$

is open and closed in M. Since M is connected and a disjoint union of orbits, we must have $M = O(a)$, which implies transitivity of the action. It follows that there is a symmetry at every point $a \in M$, given by $h^{-1} \circ s_p \circ h$, where $h \in \mathrm{Aut}(M, \nu)$ satisfies $h(a) = p$. $\square$

The preceding proof shows that a complex symmetric Banach manifold (M, ν) is a homogeneous space of $\mathrm{Aut}(M, \nu)$. Let $K = \{h \in \mathrm{Aut}(M, \nu) : h(a) = a\}$ be the isotropy subgroup at a. Then K is closed in the topology of locally uniform convergence and hence, by Theorem 1.3.11, a real Banach Lie group with Lie algebra

$$\mathfrak{k} = \{X \in \mathrm{aut}(M, \nu) : \exp tX(a) = a, \forall t \in \mathbb{R}\}.$$

Further, the inclusion map $\iota : K \hookrightarrow \mathrm{Aut}(M, \nu)$ is an immersion as its differential at the identity is the inclusion map $\mathfrak{k} \hookrightarrow \mathrm{aut}(M, \nu)$ and $\mathfrak{k}$ is a direct summand of $\mathrm{aut}(M, \nu)$, by Lemma 3.1.1. Hence K is a Banach Lie subgroup of $\mathrm{Aut}(M, \nu)$ and as in (1.11), the left coset space $\mathrm{Aut}(M, \nu)/K$ is a Banach manifold and the quotient map $\pi : \mathrm{Aut}(M, \nu) \longrightarrow \mathrm{Aut}(M, \nu)/K$ is a submersion.

By transitivity, the map $\rho : hK \in \mathrm{Aut}(M, \nu)/K \mapsto h(a) \in M$ is a bijection and since submersions are open maps, the following commutative diagram implies that ρ is a homeomorphism.

$$
\begin{array}{ccc}
\mathrm{Aut}(M, \nu)/K & \xrightarrow{\ \rho\ } & M \\
{\scriptstyle \pi} \nwarrow & & \nearrow {\scriptstyle \epsilon} \\
& \mathrm{Aut}(M, \nu) &
\end{array}
\qquad (1.17)
$$

Further, ρ is bianalytic since the submersions π and ϵ have local left inverses by Lemma 1.2.10. For instance, the submersion π has a local left inverse f_π on some neighbourhood $\mathcal{V}$ of each point in $\mathrm{Aut}(M, \nu)/K$ which implies $\rho|_{\mathcal{V}} = \epsilon \circ f_\pi$ is analytic. Likewise, ρ^{-1} is analytic. Therefore, we can identity M with the homogeneous space $\mathrm{Aut}(M, \nu)/K$ via ρ.

However, we note that a homogeneous space of $\mathrm{Aut}(M, \nu)$ need not be a symmetric manifold in general [143], but if it admits one symmetry s_p, then it is a symmetric manifold since s_p can be moved to a symmetry at any point by the transitive action of $\text{Aut}\,(M, \nu)$ (cf. Example 1.3.22 below).

Example 1.3.17. A (finite dimensional) Hermitian manifold M with a Hermitian metric g is called a *Hermitian symmetric space* if each point $p \in M$ is an isolated fixed-point of an involutive biholomorphic map $s_p : M \longrightarrow M$ which is isometric with respect to g (cf. [80, p. 372]). In view of Examples 1.3.2 and 1.3.3, Hermitian symmetric spaces are complex symmetric Banach manifolds and we can regard the latter as an infinite dimensional generalisation of the former. In finite dimensions, Riemannian symmetric spaces are real symmetric Banach manifolds modelled on $\mathbb{R}^d$.

Definition 1.3.18. A connected Riemannian manifold (M, g) modelled on a real Hilbert space is called a *Riemannian symmetric space* if it is a symmetric Banach manifold with respect to the tangent norm defined by g.

By definition, a Hermitian symmetric space is a finite dimensional Riemannian symmetric space.

Example 1.3.19. A real Hilbert space V is a Riemannian symmetric space. The symmetry s_p at $p \in V$ is given by $s_p(x) = 2p - x$.

An important class of complex symmetric Banach manifolds are the bounded symmetric domains.

Definition 1.3.20. A *bounded symmetric domain* is defined to be a bounded domain D in a complex Banach space such that each point $p \in D$ is an isolated fixed-point of an involutive biholomorphic map $s_p : D \longrightarrow D$. We call s_p a *symmetry* of D.

In Section 3.6, we will discuss an interesting class of real symmetric Banach manifolds associated to bounded symmetric domains.

By considering the Carathéodory tangent norm and Lemma 1.3.8, we see that bounded symmetric domains are complex symmetric Banach manifolds. In view of Lemma 1.3.16, they are (strictly) homogeneous domains according to the following definition.

Definition 1.3.21. A domain D in a complex Banach space is called *homogeneous* if the automorphism group $\operatorname{Aut} D$ acts transitively on D. If D satisfies the stronger condition that the map $h \in \operatorname{Aut} D \mapsto h(a) \in D$ is a submersion (at the identity) for some $a \in D$, then it is called *strictly homogeneous*.

When a finite dimensional bounded symmetric domain is viewed as a Hermitian manifold with the Bergman metric, it is a Hermitian symmetric space by Lemma 1.3.4. In fact, finite dimensional bounded symmetric domains are Hermitian symmetric spaces of the so-called *non-compact type*, this says they have non-positive sectional curvature [32]. The seminal works of É. Cartan [32] and Harish-Chandra [78] show that a Hermitian symmetric space of the non-compact type is biholomorphic to a bounded symmetric domain in $\mathbb{C}^d$ (cf. [80, Chapter VIII, Theorem 7.1]). In this way, one can identify the class of Hermitian symmetric spaces of the non-compact type with the class of finite dimensional bounded symmetric domains, and regard bounded symmetric domains in Banach spaces as an infinite dimensional generalisation of Hermitian symmetric spaces of the non-compact type.

Example 1.3.22. The open unit disc $\mathbb{D} \subset \mathbb{C}$ is the simplest example of a bounded symmetric domain. Evidently, the reflection $\ell(z) = -z$ is a symmetry at 0. Given $a \in \mathbb{D}$, the Möbius transform

$$g_{-a}(z) = \frac{z - a}{1 - \bar{a}z} \qquad (z \in \mathbb{D})$$

(induced by $-a$) is biholomorphic on $\mathbb{D}$, satisfying $g_{-a}(a) = 0$, and the biholomorphic map $g_{-a}^{-1} \circ \ell \circ g_{-a}$ is a symmetry at a.

One can extend the previous example to higher dimensions. Indeed, all finite and infinite dimensional (complex) Euclidean unit balls, that is, open unit balls of complex Hilbert spaces, are bounded symmetric domains. Let $D = \{v \in H : \|v\| < 1\}$ be the open unit ball of a Hilbert space H with inner product $\langle \cdot, \cdot \rangle$. Then, as before, $\ell(v) = -v$ is a symmetry at 0 and for $a \in H$, the Möbius transform induced by $-a$ is defined by

$$g_{-a}(v) = -a + \frac{\sqrt{1 - \|a\|^2}}{1 - \langle v, a \rangle} \left(v + (\sqrt{1 - \|a\|^2} - 1)\langle v, a \rangle \frac{a}{\|a\|^2} \right) \quad (v \in H)$$

and $g_{-a}^{-1} \circ \ell \circ g_{-a}$ is a symmetry at a. We will explain in Chapter 3 how one arrives at the above Möbius transform g_{-a} in the general setting of bounded symmetric domains in Banach spaces.

Finite dimensional bounded symmetric domains were classified by É. Cartan [32] using Lie algebras and Lie groups. This classification can be expressed in terms of Jordan algebraic structures and thereby extended to a larger class of infinite dimensional bounded symmetric domains. To achieve this, we first need to develop the relevant theory of Jordan algebras and Jordan triples, which is the task of the next chapter. Details of the classification will be discussed in Section 3.8.

Notes. The books by Helgason [80] and Satake [155] are classics on finite dimensional symmetric spaces. Satake's book also discusses the related Jordan triple structures, so does Loo's book [125], which focuses on finite dimensional bounded symmetric domains. These references are relevant to our later discussions.

Chapter 2

Jordan and Lie algebraic structures

2.1 Jordan algebras

Since É. Cartan's seminal work, Lie theory has played a pivotal role in the study of finite dimensional bounded symmetric domains. It has been found relatively recently that the closely related Jordan theory offers a transparent algebraic description of these domains as well as their geometric structures, which is also accessible in infinite dimension. Indeed, bounded symmetric domains are biholomorphic to the open unit balls of JB*-triples, which are complex Banach spaces equipped with a Jordan triple structure. This fundamental result and many of its applications will be discussed later, including Cartan's classification of finite dimensional bounded symmetric domains in terms of JB*-triples. To prepare for this endeavour, we commence a chapter on the relevant basics of Jordan and Lie algebras.

We begin with Jordan algebras, which are closely related to Lie algebras. One important feature of these algebras is that the multiplication need not be associative.

51

By an *algebra* we mean a vector space $\mathcal{A}$ over a field, equipped with a bilinear product $(a, b) \in \mathcal{A}^2 \mapsto ab \in \mathcal{A}$. We do not assume associativity of the product. If the product is associative, we call $\mathcal{A}$ *associative.*

Homomorphisms and isomorphisms between two algebras are defined as in the case of associative algebras. An *antiautomorphism* of an algebra $\mathcal{A}$ is a linear bijection $\varphi : \mathcal{A} \longrightarrow \mathcal{A}$ such that $\varphi(ab) = \varphi(b)\varphi(a)$ for all $a, b \in \mathcal{A}$. If $\mathcal{A}$ is over the field $\mathbb{F} = \mathbb{R}$ or $\mathbb{C}$, an antiautomorphism φ is called an *involution* if $\varphi(\varphi(a)) = a$ and $\varphi(\lambda a) = \overline{\lambda}\varphi(a)$, where $\overline{\lambda}$ denotes the usual conjugate of $\lambda \in \mathbb{F}$.

We call an algebra $\mathcal{A}$ *unital* if it contains an identity which will always be denoted by $\mathbf{1}$, unless stated otherwise. As usual, one can adjoint an identity $\mathbf{1}$ to a non-unital algebra $\mathcal{A}$ to form a unital algebra $\mathcal{A}_1$, called the *unit extension* of $\mathcal{A}$.

A *Jordan algebra* is a commutative algebra over a field $\mathbb{F}$ and satisfies the Jordan identity

$$(ab)a^2 = a(ba^2) \qquad (a, b \in \mathcal{A}).$$

We always assume that $\mathbb{F}$ is not of characteristic two, but in later sections, $\mathbb{F}$ is usually either $\mathbb{R}$ or $\mathbb{C}$.

To avoid confusion, homomorphisms and isomorphisms between Jordan algebras are sometimes called *Jordan homomorphisms* and *Jordan isomorphisms* to distinguish them from the ones for other algebraic structures.

The concept of a Jordan algebra was introduced by P. Jordan, J. von Neumann and E. Wigner [94], with the aim to formulate an algebraic model for quantum mechanics. They introduced the notion of an *r-number system* which is, in modern terminology, a finite-dimensional, formally real Jordan algebra. In fact, the term *"Jordan algebra"* first appeared in a paper by Albert [9]. It denotes an algebra of linear transformations closed in the product

$$A \cdot B = \frac{1}{2}(AB + BA).$$

Although Jordan algebras were motivated by quantum formalism, un-expected and important applications in algebra, geometry and analysis have been discovered. In particular, the applications of Jordan theory to symmetric manifolds are the subject of our discussions.

On any associative algebra $\mathcal{A}$, one can define a product $\circ$ by

$$a \circ b = \frac{1}{2}(ab + ba) \qquad (a, b \in \mathcal{A})$$

where the product on the right-hand side is the original product of $\mathcal{A}$. The algebra $\mathcal{A}$ becomes a Jordan algebra with the product $\circ$. We call this product *special*. A Jordan algebra is called *special* if it is isomorphic to, and hence identified with, a Jordan subalgebra of an associative algebra $\mathcal{A}$ with respect to the special Jordan product $\circ$. Otherwise, it is called *exceptional*.

It is often convenient to express the Jordan identity as an operator identity. Given an algebra $\mathcal{A}$ and $a \in \mathcal{A}$, we define a linear map $L_a : \mathcal{A} \longrightarrow \mathcal{A}$, called the *left multiplication by a*, as follows

$$L_a(x) = ax \qquad (x \in \mathcal{A}).$$

The Jordan identity can be expressed as

$$[L_a, L_{a^2}] = 0 \qquad (a \in \mathcal{A})$$

where $[\cdot, \cdot]$ is the usual commutator product of linear maps. Given $a, b \in \mathcal{A}$, we define the *quadratic operator* $Q_a : \mathcal{A} \longrightarrow \mathcal{A}$ and *box operator* $a \,\square\, b : \mathcal{A} \longrightarrow \mathcal{A}$ by

$$Q_a = 2L_a^2 - L_{a^2}, \qquad a \,\square\, b = L_{ab} + [L_a, L_b]. \qquad (2.1)$$

These operators are fundamental in Jordan theory, as well as the linearization of the quadratic operator:

$$Q_{a,b} = Q_{a+b} - Q_a - Q_b.$$

Let $\mathcal{A}$ be an algebra and let $a \in \mathcal{A}$. We define $a^0 = 1$ if $\mathcal{A}$ is unital,

$$a^1 = a, \quad a^{n+1} = aa^n \quad (n = 1, 2, \ldots).$$

The following power associative property depends on the assumption that the scalar field $\mathbb{F}$ for $\mathcal{A}$ is not of characteristic two.

Theorem 2.1.1. *A Jordan algebra $\mathcal{A}$ is power associative, that is,*

$$a^m a^n = a^{m+n} \quad (m, n = 1, 2, \ldots)$$

for each $a \in \mathcal{A}$. In fact, we have $[L_{a^m}, L_{a^n}] = 0$.

Proof. For any α, β in the underlying field $\mathbb{F}$, we have

$$[L_{a+\alpha b+\beta c}, L_{(a+\alpha b+\beta c)^2}] = 0$$

for all $a, b, c \in \mathcal{A}$. Expand the product, we find that the coefficient of the term $\alpha\beta$ is

$$2[L_a, L_{bc}] + 2[L_b, L_{ca}] + 2[L_c, L_{ab}]$$

which must be 0. Since $\mathbb{F}$ is not of characteristic 2, we have

$$[L_a, L_{bc}] + [L_b, L_{ca}] + [L_c, L_{ab}] = 0.$$

Applying the above operator identity to an element $x \in \mathcal{A}$ and using commutativity of the Jordan product yields

$$
\begin{aligned}
& (L_a L_{bc} + L_b L_{ca} + L_c L_{ab})(x) \\
={} & (L_{bc} L_a + L_{ca} L_b + L_{ab} L_c)(x) \\
={} & L_{bc} L_x(a) + L_{bx} L_c(a) + L_{xc} L_b(a) \\
={} & (L_{bc} L_x + L_{cx} L_b + L_{xb} L_c)(a) \\
={} & (L_x L_{bc} + L_b L_{cx} + L_c L_{xb})(a) \\
={} & (L_{((bc)a)} + L_b L_a L_c + L_c L_a L_b)(x).
\end{aligned}
$$

Put $b = a^n$ and $c = a$ in the above identity, we obtain a recursive formula

$$
L_{a^{n+2}} = 2 L_a L_{a^{n+1}} + L_{a^n} L_{a^2} - L_{a^n} L_a^2 - L_a^2 L_{a^n}
$$

which implies that each L_{a^n} is a polynomial in L_a and L_{a^2} which commute. It follows that L_{a^n} commutes with L_{a^m} for all $m, n \in \mathbb{N}$. In particular, we have

$$
L_{a^n} L_a(a^m) = L_a L_{a^n}(a^m)
$$

and power associativity follows from induction. $\square$

Corollary 2.1.2. *Let $\mathcal{A}$ be a Jordan algebra and let $a \in \mathcal{A}$. The subalgebra $\mathcal{A}(a)$ generated by a in $\mathcal{A}$ is associative.*

In fact, we have the following deeper result. We omit the proof which can be found, for instance, in the books [92, 128].

Shirshov-Cohn Theorem. *Let $\mathcal{A}$ be a Jordan algebra and let $a, b \in \mathcal{A}$. Then the Jordan subalgebra $\mathcal{B}$ generated by a, b (and $\mathbf{1}$, if $\mathcal{A}$ is unital) is special.*

One can use the Shirshov-Cohn theorem to establish various identities in Jordan algebras. For instance, in any Jordan algebra $\mathcal{A}$, we have the following identity:

$$2L_a^3 - 3L_{a^2}L_a + L_{a^3} = 0 \tag{2.2}$$

for each $a \in \mathcal{A}$. In other words, we have

$$2a(a(ab)) - 3a^2(ab) + a^3b = 0$$

for $a, b \in \mathcal{A}$. To see this, let $\mathcal{B}$ be the Jordan subalgebra of $\mathcal{A}$ generated by a and b. Then it is special and hence embeds in some associative algebra $(\mathcal{A}', \times)$ with

$$ab = \frac{1}{2}(a \times b + b \times a).$$

In $\mathcal{B}$, we have

$$
\begin{aligned}
2a(a(ab)) &= \frac{1}{4}(a^3 \times b + 3a^2 \times b \times a + 3a \times b \times a^2 + b \times a^3) \\
3a^2(ab) &= \frac{1}{4}(3a^3 \times b + 3a^2 \times b \times a + 3a \times b \times a^2 + 3b \times a^3)
\end{aligned}
$$

which, together with $2a^3b = a^3 \times b + b \times a^3$, verifies the identity.

Example 2.1.3. The Cayley algebra $\mathcal{O}$, known as the *Octonions*, is a complex non-associative algebra with a basis $\{e_0, e_1, \ldots, e_7\}$ and satisfies

$$a^2b = a(ab), \qquad ab^2 = (ab)b \qquad (a, b \in \mathcal{O}) \tag{2.3}$$

where e_0 is the identity of $\mathcal{O}$ and $e_j^2 = -e_0$ for $j \neq 0$.

We will denote by $\mathbb{O}$ the real Cayley algebra which is the real subalgebra of $\mathcal{O}$ with basis $\{e_0, \ldots, e_7\}$. Historically, octonions were discovered by a process of duplicating the real numbers $\mathbb{R}$. Indeed, the complex numbers arise from $\mathbb{R}$ as the product $\mathbb{R} \times \mathbb{R}$ with the multiplication

$$(a, b)(c, d) = (ac - db, bc + da) \qquad (a, b, c, d \in \mathbb{R}).$$

The real associative *quaternion* algebra $\mathbb{H}$ can be constructed by an analogous duplication process. One can define $\mathbb{H}$ as $\mathbb{C} \times \mathbb{C}$, with the multiplication

$$(a, b)(c, d) = (ac - \bar{d}b, b\bar{c} + da) \qquad (a, b, c, d \in \mathbb{C}),$$

which is isomorphic to the following real non-commutative algebra of 2×2 matrices:

$$\left\{ \begin{pmatrix} a & b \\ -\bar{b} & \bar{a} \end{pmatrix} : (a, b) \in \mathbb{C} \times \mathbb{C} \right\}. \tag{2.4}$$

In the identification with this algebra, $\mathbb{H}$ has a basis

$$\mathbf{1} = \begin{pmatrix} 1 & 0 \\ 0 & 1 \end{pmatrix}, \quad \mathbf{i} = \begin{pmatrix} i & 0 \\ 0 & -i \end{pmatrix}, \quad \mathbf{j} = \begin{pmatrix} 0 & 1 \\ -1 & 0 \end{pmatrix}, \quad \mathbf{k} = \begin{pmatrix} 0 & i \\ i & 0 \end{pmatrix}$$

satisfying

$$\mathbf{i}^2 = \mathbf{j}^2 = \mathbf{k}^2 = \mathbf{ijk} = -\mathbf{1}, \qquad \mathbf{ij} = -\mathbf{ji} = \mathbf{k}.$$

Likewise, $\mathbb{O}$ can be defined as the product $\mathbb{H} \times \mathbb{H}$ with the multiplication

$$(a, b)(c, d) = (ac - \bar{d}b, b\bar{c} + da) \qquad (a, b, c, d \in \mathbb{H})$$

where the *conjugate* $\bar{c}$ of a quaternion $c = \alpha\mathbf{1} + x\mathbf{i} + y\mathbf{j} + z\mathbf{k}$ is defined by

$$\bar{c} = \alpha\mathbf{1} - x\mathbf{i} - y\mathbf{j} - z\mathbf{k}$$

so that the *real part* of c is $\operatorname{Re} c = \frac{1}{2}(c + \bar{c}) = \alpha\mathbf{1}$. A *positive* quaternion is one of the form $\alpha\mathbf{1}$ for some $\alpha > 0$. The basis elements of $\mathbb{H} \times \mathbb{H}$ are

$$e_0 = (\mathbf{1}, 0), \ e_1 = (\mathbf{i}, 0), \ e_2 = (\mathbf{j}, 0), \ e_3 = (\mathbf{k}, 0),$$

$$e_4 = (0, \mathbf{1}), \ e_5 = (0, \mathbf{i}), \ e_6 = (0, \mathbf{j}), \ e_7 = (0, \mathbf{k}).$$

The algebras $\mathbb{C}$, $\mathbb{H}$ and $\mathbb{O}$ are *quadratic*, that is, each element x satisfies the equation $x^2 = \alpha x + \beta\mathbf{1}$ for some $\alpha, \beta \in \mathbb{R}$, where $\mathbf{1}$ denotes the identity of the algebra. If $x = (a_1, a_2) \in \mathbb{H} \times \mathbb{H}$ with

$$a_n = \alpha_n\mathbf{1} + x_n\mathbf{i} + y_n\mathbf{j} + z_n\mathbf{k} \qquad (n = 1, 2),$$

then we have

$$x^2 = 2\alpha_1 x - (x_1^2 + y_1^2 + z_1^2 + x_2^2 + y_2^2 + z_2^2)e_0.$$

Example 2.1.4. Let $\mathbb{F} = \mathbb{R}, \mathbb{C}$ or $\mathbb{H}$, with the usual conjugation $c \in \mathbb{F} \mapsto \bar{c} \in \mathbb{F}$ (which is just the identity map if $\mathbb{F} = \mathbb{R}$). An $n \times n$ matrix (a_{ij}) with $a_{ij} \in \mathbb{F}$ is called *hermitian* if

$$(a_{ij}) = (\bar{a}_{ji}).$$

Real *hermitian* matrices are usually called *symmetric*. For $n \geq 2$, the real vector space $H_n(\mathbb{F})$ of $n \times n$ hermitian matrices over $\mathbb{F}$ is a special Jordan algebra in the Jordan product

$$(a_{ij}) \circ (b_{ij}) = \frac{1}{2}((a_{ij})(b_{ij}) + (b_{ij})(a_{ij}))$$

where the product on the right-hand side is the usual matrix product.

Example 2.1.5. Besides the Jordan matrix algebras introduced in the previous example, another important class of special real Jordan algebras are the *real spin factors*. They are defined as a Hilbert space direct sum $H \oplus \mathbb{R}$, where H is a real Hilbert space with inner product $\langle \cdot, \cdot \rangle$, equipped with the Jordan product

$$(a \oplus \alpha)(b \oplus \beta) := (\beta a + \alpha b) \oplus (\langle a, b \rangle + \alpha\beta) \qquad (a, b \in H, \alpha, \beta \in \mathbb{R}).$$

We note that the Jordan matrix algebra $H_2(\mathbb{R})$ has a basis

$$\begin{pmatrix} 1 & 0 \\ 0 & -1 \end{pmatrix}, \quad \begin{pmatrix} 0 & 1 \\ 1 & 0 \end{pmatrix}, \quad \begin{pmatrix} 1 & 0 \\ 0 & 1 \end{pmatrix}$$

and the 3-dimensional spin factor $\mathbb{R}^2 \oplus \mathbb{R}$ is Jordan isomorphic to $H_2(\mathbb{R})$ via the isomorphism

$$(a, b) \oplus \alpha \mapsto a \begin{pmatrix} 1 & 0 \\ 0 & -1 \end{pmatrix} + b \begin{pmatrix} 0 & 1 \\ 1 & 0 \end{pmatrix} + \alpha \begin{pmatrix} 1 & 0 \\ 0 & 1 \end{pmatrix}.$$

Example 2.1.6. A well-known example in [8] of an exceptional Jordan algebra is the 27-dimensional real algebra

$$H_3(\mathbb{O}) = \{(a_{ij})_{1 \leq i,j \leq 3} : (a_{ij}) = (\tilde{a}_{ji}), a_{ij} \in \mathbb{O}\}$$

of 3×3 matrices over $\mathbb{O}$, hermitian with respect to the involution $\sim$ in $\mathbb{O}$ defined by

$$(\alpha_0 e_0 + \cdots + \alpha_7 e_7)^{\sim} = \alpha_0 e_0 - \cdots - \alpha_7 e_7.$$

The Jordan product is given by

$$A \circ B = \frac{1}{2}(AB + BA) \qquad (A, B \in H_3(\mathbb{O}))$$

where the multiplication on the right is the usual matrix multiplication. We refer to [92, 128] for a more detailed analysis of $H_3(\mathbb{O})$. The exceptionality of $H_3(\mathbb{O})$ involves the so-called *s-identities* which are valid in all *special* Jordan algebras, but not all Jordan algebras. One such identity was first found by Glennie [66]:

$$2Q_x(z)Q_{y,x}Q_z(y^2) - Q_x Q_z Q_{x,y} Q_y(z)$$
$$= 2Q_y(z)Q_{x,y}Q_z(x^2) - Q_y Q_z Q_{y,x} Q_x(z)$$

which does not hold in $H_3(\mathbb{O})$.

Definition 2.1.7. Two elements a and b in a Jordan algebra $\mathcal{A}$ are said to *operator commute* if the left multiplications L_a and L_b commute. The *centre* of $\mathcal{A}$ is the set $Z(\mathcal{A}) = \{z \in \mathcal{A} : L_z L_a = L_a L_z, \forall a \in \mathcal{A}\}$.

We observe that $L_a L_b = L_b L_a$ if, and only if, $(ax)b = a(xb)$ for all $x \in \mathcal{A}$. Evidently, the centre $Z(\mathcal{A}) = \{z \in \mathcal{A} : (za)b = z(ab), \forall a, b \in \mathcal{A}\}$ is an associative subalgebra of $\mathcal{A}$.

Definition 2.1.8. An element e in an algebra $\mathcal{A}$ is called an *idempotent* if $e^2 = e$. Two idempotents e and u are said to be *orthogonal* (to each other) if $eu = ue = 0$. An element $a \in \mathcal{A}$ is called *nilpotent* if $a^n = 0$ for some positive integer n.

Lemma 2.1.9. *Let $\mathcal{A}$ be a unital Jordan algebra with an idempotent e. Let $a \in \mathcal{A}$. The following conditions are equivalent.*

(i) *a and e operator commute.*

(ii) *$Q_e(a) = L_e a$.*

(iii) *a and e generate an associative subalgebra of $\mathcal{A}$.*

Proof. (i) $\Rightarrow$ (ii). We have

$$Q_e(a) = 2(L_e^2 - L_e)(a) = 2e(ea) - ea = 2e^2 a - ea = ea.$$

(ii) $\Rightarrow$ (iii). Let $\mathcal{B}$ be the subalgebra generated by a and e. By the Shirshov-Cohn theorem, $\mathcal{B}$ is isomorphic to a Jordan subalgebra $\mathcal{B}'$ of an associative algebra $(\mathcal{A}', \times)$ with respect to the special Jordan product. Identify a and e as elements in $\mathcal{B}'$. Then

$$L_e a = \frac{1}{2}(e \times a + a \times e) = Q_e(a) = e \times a \times e$$

since $e = e^2 = e \times e$. Multiply the above identity on the left by e, we get $e \times a = e \times a \times e$. Multiplying on the right by e gives $a \times e = e \times a \times e$. Hence $e \times a = a \times e$ and $ea = e \times a$. Hence $(\mathcal{B}', \times)$ is a commutative subalgebra of $(\mathcal{A}', \times)$ and the special Jordan product in $\mathcal{B}'$ is just the product $\times$ and is, in particular, associative.

(iii) $\Rightarrow$ (i). In the proof of Theorem 2.1.1, we have the operator identity

$$[L_e, L_{bc}] + [L_b, L_{ce}] + [L_c, L_{eb}] = 0$$

for all $b, c \in \mathcal{A}$. Put $c = e$, we have

$$[L_e, L_{be}] + [L_b, L_e] + [L_e, L_{eb}] = 0$$

which gives

$$2[L_e, L_{be}] = [L_e, L_b]. \tag{2.5}$$

Since $e^2 = e$, in the special Jordan algebra $\mathcal{A}(a, e, \mathbf{1})$ generated by a, e and $\mathbf{1}$, it can be verified easily that

$$a = Q_e(a) + Q_{1-e}(a)$$

and $Q_{1-e}(a)e = 0$ as well as $Q_e(a)e = Q_e(a)$. Substitute $Q_{1-e}(a)$ for b in (2.5), we get $[L_e, L_{Q_{1-e}(a)}] = 0$. Putting $b = Q_e(a)$ in (2.5) gives $[L_e, L_{Q_e(a)}] = 0$. It follows that

$$[L_e, L_a] = [L_e, L_{Q_e(a)}] + [L_e, L_{Q_{1-e}(a)}] = 0.$$

$\square$

Lemma 2.1.10. *Let $\mathcal{A}$ be a finite dimensional associative algebra containing an element a which is not nilpotent and not an identity. Then $\mathcal{A}$ contains a non-zero idempotent which is a polynomial in a, without constant term.*

Proof. We may assume $\mathcal{A}$ has an identity $\mathbf{1}$. Finite dimensionality implies that there is a non-zero polynomial p of the least degree and without constant term such that $p(a) = 0$. Write $p(x) = x^k q(x)$ where $k \geq 1$ and q is a polynomial such that $q(0) \neq 0$. The degree $\deg q$ of q is strictly positive since a is not nilpotent. There are then polynomials q_1 and q_2 with $\deg q_1 < \deg q$ and

$$x^k q_1(x) + q_2(x)q(x) = 1$$

where the non-zero polynomial $g(x) = x^k q_1(x)$ has no constant term and $\deg g < \deg p$. Hence $e = g(a) \neq 0$. We have $e^2 = e$ since $a^{2k} q_1(a) + a^k q_2(a)q(a) = a^k$ and

$$g(a)^2 - g(a) = a^{2k} q_1(a)^2 - a^k q_1(a) = a^k q_2(a)q(a)q_1(a) = 0.$$

$\square$

Lemma 2.1.11. *Let $\mathcal{A}$ be a Jordan algebra. Then an element $a \in \mathcal{A}$ is nilpotent if, and only if, the left multiplication $L_a : \mathcal{A} \longrightarrow \mathcal{A}$ is nilpotent.*

Proof. If L_a is nilpotent, then $a^{n+1} = L_a^n(a)$ implies that a is nilpotent. Conversely, for any $a \in \mathcal{A}$ with $a^n = 0$, we show that L_a is nilpotent by induction on the exponent n. The assertion is trivially true if $n = 1$. Given that the assertion is true for n, we consider $a^{n+1} = 0$. We have $(a^2)^n = 0 = (a^3)^n$ and therefore L_{a^2} and L_{a^3} are nilpotent by the inductive hypothesis. It follows from the identity

$$2L_a^3 = 3L_{a^2}L_a - L_{a^3}$$

that L_a is nilpotent.   □

Given an idempotent e in a Jordan algebra $\mathcal{A}$, the left multiplication $L_e : \mathcal{A} \longrightarrow \mathcal{A}$ satisfies the equation

$$2L_e^3 - 3L_e^2 + L_e = 0 \tag{2.6}$$

by the identity (2.2). Hence an eigenvalue α of L_e is a root of

$$2\alpha^3 - 3\alpha^2 + \alpha = 0$$

and is $0, \frac{1}{2}$ or 1. If $\mathcal{A}$ is associative, then $L_e^2 = L_e$ and $\frac{1}{2}$ is not an eigenvalue of L_e. Nevertheless, we denote the eigenspaces of $2L_e$ by

$$\mathcal{A}_k(e) = \{x \in \mathcal{A} : 2ex = kx\} \qquad (k = 0, 1, 2)$$

and call $\mathcal{A}_k(e)$ the *Peirce k-space* of e. The above remark implies $\mathcal{A}_1(e) = \{0\}$ if $\mathcal{A}$ is associative.

We define two linear operators $Q_e : \mathcal{A} \longrightarrow \mathcal{A}$ and $Q_e^\perp : \mathcal{A} \longrightarrow \mathcal{A}$ by

$$Q_e = 2L_e^2 - L_e, \quad Q_e^\perp = 4(L_e - L_e^2). \tag{2.7}$$

Evidently L_e commutes with both Q_e and $Q_e^\perp$. Using the equation (2.6), one can easily establish

$$L_e Q_e = Q_e = Q_e^2, \quad L_e Q_e^\perp = \frac{1}{2} Q_e^\perp = \frac{1}{2} (Q_e^\perp)^2, \quad L_e (I - Q_e - Q_e^\perp) = 0$$

where I is the identity operator on $\mathcal{A}$ and, Q_e and $Q_e^\perp$ are mutually orthogonal. It follows that

$$\mathcal{A}_2(e) = Q_e(\mathcal{A}), \quad \mathcal{A}_1(e) = Q_e^\perp(\mathcal{A}), \quad \mathcal{A}_0(e) = (I - Q_e - Q_e^\perp)(\mathcal{A}) \quad (2.8)$$

which gives rise to the following *Peirce decomposition* of $\mathcal{A}$:

$$\mathcal{A} = \mathcal{A}_0(e) \oplus \mathcal{A}_1(e) \oplus \mathcal{A}_2(e).$$

We will return to the Peirce decomposition with more details in the more general setting of Jordan triple systems. We note for the time being that the Peirce spaces $\mathcal{A}_0(e)$ and $\mathcal{A}_2(e)$ are Jordan subalgebras of $\mathcal{A}$ as shown below. We also note that $\mathcal{A}_2(e)$ never vanishes.

Lemma 2.1.12. *The Peirce spaces of an idempotent e in a Jordan algebra $\mathcal{A}$ satisfy*

$$\mathcal{A}_0(e)\mathcal{A}_0(e) \subset \mathcal{A}_0(e), \quad \mathcal{A}_1(e)\mathcal{A}_1(e) \subset \mathcal{A}_0(e) \oplus \mathcal{A}_2(e), \quad \mathcal{A}_2(e)\mathcal{A}_2(e) \subset \mathcal{A}_2(e).$$

Proof. We first prove the second inclusion. Let $x, y \in \mathcal{A}_1(e)$ and let $xy = a_0 + a_1 + a_2$ be the Peirce decomposition of xy. We have

$$
\begin{aligned}
0 &= [L_y, L_{ex}] + [L_e, L_{xy}] + [L_x, L_{ye}] \\
&= \frac{1}{2}[L_y, L_x] + [L_e, L_{xy}] + \frac{1}{2}[L_x, L_y] \\
&= [L_e, L_{xy}].
\end{aligned}
$$

In particular, $[L_e, L_{xy}](e) = 0$ gives

$$
\begin{aligned}
0 &= e(xy) - e(e(xy)) \\
&= a_2 + \frac{1}{2}a_1 - e\left(a_2 + \frac{1}{2}a_1\right) \\
&= a_2 + \frac{1}{2}a_1 - a_2 - \frac{1}{4}a_1 = \frac{1}{4}a_1.
\end{aligned}
$$

Hence $xy \in \mathcal{A}_0(e) \oplus \mathcal{A}_2(e)$.

Let $x, y \in \mathcal{A}_j(e)$ where $j = 0, 2$. Then we have $[L_e, L_{x^2}] = 0$ by the first equation in the proof above. Using this and expanding $((x + e)y)(x + e)^2 = (x + e)(y(x + e)^2)$, we obtain

$$2(xy)(xe) + (xy)e + 2(ey)(xe) = 2x(y(xe)) + x(ye) + 2(ey)(xe)$$

which gives $(xy)e = \frac{i}{2}xy$. This proves the first and the last inclusions.

$\square$

Definition 2.1.13. An idempotent e in a Jordan algebra $\mathcal{A}$ is called *maximal* if the Peirce 0-space $\mathcal{A}_0(e)$ is $\{0\}$. A non-zero idempotent e is called *primitive* if there are no non-zero orthogonal idempotents u and v satisfying $e = u + v$.

Lemma 2.1.14. *Let $\mathcal{A}$ be a finite dimensional Jordan algebra which contains no non-zero nilpotent element. Then $\mathcal{A}$ contains a maximal idempotent.*

Proof. Ignore the trivial case $\mathcal{A} = \{0\}$. Applying Lemma 2.1.10 to an associative subalgebra of $\mathcal{A}$ generated by a non-zero element, one finds a non-zero idempotent e. If $\mathcal{A}_0(e) \neq \{0\}$, then again one can pick a non-zero idempotent $u \in \mathcal{A}_0(e)$. Then $e' = e + u$ is an idempotent and $\mathcal{A}_0(e) \subset \mathcal{A}_0(e')$. Since $u \in \mathcal{A}_0(e') \backslash \mathcal{A}_0(e)$, we have $\dim \mathcal{A}_0(e) < \dim \mathcal{A}_0(e')$. By finite dimensionality of $\mathcal{A}$, this process of increasing dimension must stop, yielding a maximal idempotent. $\square$

Proposition 2.1.15. *Let $\mathcal{A}$ be a finite dimensional Jordan algebra which contains no non-zero nilpotent element. Then $\mathcal{A}$ has an identity.*

Proof. By Lemma 2.5.3, $\mathcal{A}$ contains a maximal idempotent e so that

$$\mathcal{A} = \mathcal{A}_1(e) \oplus \mathcal{A}_2(e).$$

We show $\mathcal{A}_1(e) = \{0\}$. Let $x \in \mathcal{A}_1(e)$. Then $x^2 \in \mathcal{A}_0(e) \oplus \mathcal{A}_2(e) = \mathcal{A}_2(e)$ and we have

$$\frac{1}{2}x^3 = (xe)x^2 = x(ex^2) = x^3$$

which implies $x = 0$. Hence $\mathcal{A} = \mathcal{A}_2(e)$ in which e is the identity. □

A real Jordan algebra $\mathcal{A}$ is called *formally real* if $a_1^2 + \cdots + a_k^2 = 0$ implies $a_1 = \cdots = a_k = 0$ for $a_1, \ldots, a_k \in \mathcal{A}$. By Proposition 2.1.15, a finite dimensional formally real Jordan algebra has an identity. The real spin factor $H \oplus \mathbb{R}$ and the Jordan matrix algebras $H_n(\mathbb{F})$ are formally real for $\mathbb{F} = \mathbb{R}, \mathbb{C}$ and $\mathbb{H}$.

Example 2.1.16. The exceptional Jordan algebra $H_3(\mathbb{O})$ is formally real since for each $a = (a_{ij}) \in H_3(\mathbb{O})$, we have

$$\text{Trace}(a^2) = \sum_{i=1}^{3} a_{ij}a_{ji} = \sum_{i=1}^{3} a_{ij}\widetilde{a}_{ij}$$

where $a_{ij}\widetilde{a}_{ij} = (\alpha_0^2 + \cdots + \alpha_7^2)e_0$ for $a_{ij} = \alpha_0 e_0 + \cdots + \alpha_7 e_7 \in \mathbb{O}$.

Theorem 2.1.17. *Let $\mathcal{A}$ be a finite dimensional real Jordan algebra. The following conditions are equivalent.*

(i) *$\mathcal{A}$ is formally real.*

(ii) *$a^2 + b^2 = 0 \Rightarrow a = b = 0$ for any $a, b \in \mathcal{A}$.*

(iii) *Each $x \in \mathcal{A}$ admits a decomposition $x = \alpha_1 e_1 + \cdots + \alpha_n e_n$ where $\alpha_1, \ldots, \alpha_n \in \mathbb{R}$ and $e_1, \ldots, e_n$ are mutually orthogonal idempotents in $\mathcal{A}$.*

(iv) *The bilinear form $(x, y) \in \mathcal{A}^2 \mapsto \text{Trace}(x \square y) \in \mathbb{R}$ is positive definite.*

Proof. (ii) $\Rightarrow$ (iii). Let $x \in \mathcal{A}$. Then the subalgebra $\mathcal{A}(x)$ generated by x is associative and has an identity e by Proposition 2.1.15. Finite dimensionality implies that

$$e = e_1 + \cdots + e_n$$

for some mutually orthogonal primitive idempotents in $\mathcal{A}(x)$. Hence we have, by associativity of $\mathcal{A}(x)$,

$$x = e_1 x e_1 + \cdots + e_n x e_n.$$

We show that the associative algebra $e_j \mathcal{A}(x) e_j$ reduces to $\mathbb{R} e_j$. Indeed, by primitivity, there is no non-zero idempotent in $e_j \mathcal{A}(x) e_j$ other than e_j, and it follows from Lemma 2.1.10 that each $a \in e_j \mathcal{A}(x) e_j \backslash \{0, e_j\}$ gives rise to a polynomial $g(a)$ without constant term satisfying $g(a) = e_j$ which implies that a is invertible in $e_j \mathcal{A}(x) e_j$. Hence $e_j \mathcal{A}(x) e_j$ is a field over $\mathbb{R}$ and must be either $\mathbb{R}$ or $\mathbb{C}$. Since $\mathbb{C}$ is not formally real, we conclude $e_j \mathcal{A}(x) e_j = \mathbb{R} e_j$ and therefore

$$x = \sum_j \alpha_j e_j$$

for some $\alpha_1, \ldots, \alpha_n \in \mathbb{R}$.

(iii) $\Rightarrow$ (iv). Given $x = \alpha_1 e_1 + \cdots + \alpha_n e_n$ for some mutually orthogonal idempotents $e_1, \ldots, e_n$, we have

$$\text{Trace}\,(x \,\square\, x) = \sum_j \alpha_j^2 \text{Trace}\,(e_j \,\square\, e_j) \geq 0.$$

If $\text{Trace}\,(x \,\square\, x) = 0$, then $\text{Trace}\,(e_j \,\square\, e_j) = 0$ and $e_j = 0$ since $e_j \,\square\, e_j = L_{e_j}$ has eigenvalues 0, $1/2$ or 1, for all j.

(iv) $\Rightarrow$ (i). Let $a_1^2 + \cdots + a_k^2 = 0$. Then $\sum_j \text{Trace}\,(a_j \,\square\, a_j) = \text{Trace}\,(a_1 \,\square\, a_1 + \cdots + a_k \,\square\, a_k) = 0$. Hence $\text{Trace}\,(a_j \,\square\, a_j) = 0$ implies $a_j = 0$ for all j. $\qquad\square$

An element a in a Jordan algebra $\mathcal{A}$ with identity e is called *invertible* if there exists an element $a^{-1} \in \mathcal{A}$ (which is necessarily unique) such that $aa^{-1} = e$ and $(a^2)a^{-1} = a$. This is equivalent to the invertibility of the quadratic operator Q_a, in which case $a^{-1} = Q_a^{-1}(a)$. If the left multiplication $L_a : \mathcal{A} \longrightarrow \mathcal{A}$ is invertible, then a is invertible with inverse $a^{-1} = L_a^{-1}(e)$.

A subspace J of a Jordan algebra $\mathcal{A}$ is called an *ideal* if $a \in \mathcal{A}$ and $x \in J$ imply $ax \in J$, in which case J is also an ideal in the unit extension $\mathcal{A}_1$ of $\mathcal{A}$ and, the quotient space $\mathcal{A}/J$ is a Jordan algebra with the natural product

$$(a + J)(b + J) = ab + J \qquad (a, b \in \mathcal{A}).$$

The kernel $\varphi^{-1}(0)$ of a Jordan homomorphism $\varphi : \mathcal{A} \longrightarrow \mathcal{B}$ is an ideal of $\mathcal{A}$. Given an ideal J of a Jordan algebra $\mathcal{A}$, the quotient map $q : \mathcal{A} \longrightarrow \mathcal{A}/J$ is a homomorphism with kernel $q^{-1}(0) = J$.

A Jordan algebra $\mathcal{A}$ with non-trivial multiplication, that is, $ab \neq 0$ for some $a, b \in \mathcal{A}$, is called *simple* if the only ideals of $\mathcal{A}$ are $\{0\}$ and $\mathcal{A}$ itself.

Finite dimensional formally real Jordan algebras have been classified in the seminal work of [94]. They are a finite direct sum of simple real Jordan algebras of the following types:

$$H_n(\mathbb{R}), \ H_n(\mathbb{C}), \ H_n(\mathbb{H}) \ (3 \leq n < \infty), \ H_3(\mathbb{O}), \ H \oplus \mathbb{R} \qquad (2.9)$$

where $\dim H < \infty$. Jordan algebras of any dimension have been classified by Zelmanov [183, 184].

An important connection of finite dimensional formally real Jordan algebras to geometry has been discovered by Koecher [109] and Vinberg [170]. They established the one-one correspondence between these algebras and a class of Riemannian symmetric spaces, namely, the linearly homogeneous self-dual cones (cf. Definition 2.4.11), which play a useful role in the study of automorphic functions on bounded homogeneous domains in complex spaces [144, 170]. Given a finite dimensional formally real Jordan algebra $\mathcal{A}$, the set

$$C = \{a^2 : a \in \mathcal{A}\}$$

forms a proper cone, its interior is the corresponding linearly homogeneous self-dual cone.

Here, by a *cone C* in a real vector space V, we mean a non-empty subset of V satisfying (i) $C + C \subset C$ and (ii) $\alpha C \subset C$ for all $\alpha > 0$. A cone C is called *proper* if $C \cap -C = \{0\}$. The partial ordering on V induced by a proper cone C will be denoted by $\leq_C$, or by $\leq$ if C is understood, so that $x \leq y$ whenever $y - x \in C$. Conversely, if V is equipped with a partial ordering $\leq$, we let $V_+ = \{v \in V : 0 \leq v\}$ denote the corresponding proper cone.

We will discuss the result of Koecher and Vinberg in more detail, as well as its infinite dimensional generalisation in Section 2.4.

Notes. The basic results of Jordan algebras presented in this section, as well as Jordan triple systems in the following section, are classical and overlap largely with those given in the author's book [37]. Further details can be found in the books by Braun and Koecher [23], Jacobson [92], Schafer [156] and McCrimmon [128].

2.2 Jordan triple systems

In this section, we introduce a generalisation of Jordan algebras, namely, the Jordan triple systems. These are real or complex vector spaces equipped with a Jordan triple structure. We discuss both real and complex cases although we will mainly be concerned with complex Jordan triple systems later.

Given a complex vector space V, we denote by $\overline{V}$ the *conjugate* of V, which is the complex vector space obtained from V itself, but with the scalar multiplication replaced by

$$(\lambda, x) \in \mathbb{C} \times \overline{V} \mapsto \lambda \cdot x := \bar{\lambda} x \in \overline{V}.$$

For a real vector space V, its conjugate $\overline{V}$ is defined to be itself.

A real or complex vector space V is called a (*real* or respectively, *complex*) *Jordan triple system* if it is equipped with a triple product

$$\{\cdot, \cdot, \cdot\} : V^3 \longrightarrow V$$

called a *Jordan triple product*, which is linear and symmetric in the outer variables, but conjugate linear in the middle variable, and satisfies the following identity:

$$\{a, b, \{x, y, z\}\} = \{\{a, b, x\}, y, z\} - \{x, \{b, a, y\}, z\} + \{x, y, \{a, b, z\}\} \tag{2.10}$$

where a conjugate linear map on a *real* vector space is just a linear map. By a *Jordan triple system*, we mean a real or complex Jordan triple system. We note that the Jordan triple product is trilinear in a real Jordan triple system. In the sequel, we will be concerned with mainly complex Jordan triple systems and henceforth, by a *Jordan triple*, we mean a *complex* Jordan triple system.

Remark 2.2.1. We should point out that a different terminology is used in [37], where the term *Hermitian Jordan triple* is used for a complex Jordan triple system defined here.

A vector subspace W of a Jordan triple system V is called a *subtriple* if $x, y, z \in W$ implies $\{x, y, z\} \in W$. Given subsets A, B and C of V, we define

$$\{A, B, C\} = \{\{a, b, c\} : a \in A, b \in B, c \in C\}.$$

A linear map $f : V \longrightarrow W$ between two Jordan triple systems V and W is called a (*Jordan*) *triple homomorphism* if it preserves the triple product:

$$f\{a, b, c\} = \{f(a), f(b), f(c)\} \qquad (a, b, c \in V).$$

A triple homomorphism is called a (*Jordan*) *triple monomorphism* if it is injective. A bijective triple homomorphism is called a (*Jordan*) *triple isomorphism.*

We call the identity (2.10) the *main triple identity* of a Jordan triple system. It is often written as

$$\{\{a, b, x\}, y, z\} - \{x, \{b, a, y\}, z\} = \{a, b, \{x, y, z\}\} - \{x, y, \{a, b, z\}\}.$$
$$(2.11)$$

Example 2.2.2. The complex space $\mathbb{C}^n$ can be given various Jordan triple structures, for instance, one can define a Jordan triple product by

$$\{x, y, z\} = x\bar{y}z \qquad (x, y, z \in \mathbb{C}^n)$$

where $\bar{y}$ is the complex conjugate of y and the multiplication on the right is the coordinatewise multiplication. We can also equip $\mathbb{C}^n$ with the Jordan triple product

$$\{x, y, z\} = \langle x, y \rangle z + \langle z, y \rangle x$$

where $\langle \cdot, \cdot \rangle$ denotes the usual inner product.

The concept of a Jordan triple system was originally derived from a generalisation of Jordan algebras relating to Lie algebras and differential geometry. We will show in Section 2.3 that they are in one-one correspondence with a class of Lie algebras via the Tits-Kantor-Koecher construction.

In a real Jordan algebra $\mathcal{A}$, we define a *canonical Jordan triple product* by

$$\{a, b, c\} = (ab)c + a(bc) - b(ac) \tag{2.12}$$

for $a, b, c \in \mathcal{A}$. If $\mathcal{A}$ is a complex Jordan algebra equipped with an algebra involution $*$, the *canonical Jordan triple product* is defined by

$$\{a, b, c\} = (ab^*)c + a(b^*c) - b^*(ac). \tag{2.13}$$

Equipped with one of the above triple products, the Jordan algebra $\mathcal{A}$ is a Jordan triple system.

We first prove three basic identities in a Jordan triple system.

Lemma 2.2.3. *Given* x, y, z *in a Jordan triple system, we have*

$$\{\{x, y, x\}, y, z\}\} = \{x, \{y, x, y\}, z\}. \tag{2.14}$$

Proof. This follows by putting $a = x$ and $b = y$ in the above triple identity. $\qquad\square$

Lemma 2.2.4. *Given* x, y, z *is a Jordan triple system, we have*

$$\{x, y, \{x, z, x\}\} = \{x, \{y, x, z\}, x\}. \tag{2.15}$$

Proof. Applying the triple identity repeatedly yields

$$
\begin{aligned}
\{x, y, \{x, z, x\}\} &= \{\{x, y, x\}, z, x\}\} - \{x, \{y, x, z\}, x\} + \{x, z, \{x, y, x\}\} \\
&= 2\{x, z, \{x, y, x\}\} - \{x, \{y, x, z\}, x\} \\
&= 2\{\{x, z, x\}, y, x\} - 2\{x, \{z, x, y\}, x\} \\
&\quad + 2\{x, y, \{x, z, x\}\} - \{x, \{y, x, z\}, x\} \\
&= 4\{x, y, \{x, z, x\}\} - 3\{x, \{y, x, z\}, x\} \\
&= \{x, \{y, x, z\}, x\}.
\end{aligned}
$$

$\qquad\square$

The above proof also yields the identity

$$\{x, y, \{x, z, x\}\} = \{\{x, y, x\}, z, x\}. \tag{2.16}$$

Lemma 2.2.5. *Given x, y, z is a Jordan triple system, we have*

$$\{\{x, y, x\}, z, \{x, y, x\}\} = \{x, \{y, \{x, z, x\}, y\}, x\}. \tag{2.17}$$

Proof. Adding the two triple identities

$$\{y, x, \{y, x, z\}\} = \quad \{\{y, x, y\}, x, z\} - \{y, \{x, y, x\}, z\} + \{y, x, \{y, x, z\}\}$$
$$\{z, x, \{y, x, y\}\} = \quad \{\{z, x, y\}, x, y\} - \{y, \{x, z, x\}, y\} + \{y, x, \{z, x, y\}\}$$

we obtain

$$\{y, \{x, y, x\}, z\} = 2\{y, x, \{y, x, z\}\} - \{y, \{x, z, x\}, y\}.$$

It follows that

$$\{\{x, y, x\}, z, \{x, y, x\}\}$$
$$= \quad 2\{\{\{x, y, x\}, z, x\}, y, x\} - \{x, \{z, \{x, y, x\}, y\}, x\}$$
$$= \quad 2\{\{\{x, y, x\}, z, x\}, y, x\} - 2\{x, \{y, x, \{y, x, z\}\}, x\}$$
$$\quad + \{x, \{y, \{x, z, x\}, y\}, x\}$$
$$= \quad \{x, \{y, \{x, z, x\}, y\}, x\}$$

where the last identity follows from repeated applications of (2.14):

$$\{\{\{x, y, x\}, z, x\}, y, x\} \quad = \quad \{\{x, \{y, x, z\}, x\}, y, x\}$$
$$= \quad \{x, \{\{y, x, z\}, x, y\}, x\}$$
$$= \quad \{x, \{y, \{x, z, x\}, y\}, x\}.$$

$\square$

In place of the triple identity (2.10), one can also use the basic identities (2.14), (2.15) and (2.17) as the defining identities for a Jordan triple system, as in [124], in which case the triple identity can be derived.

The arguments in the proof of Lemma 2.2.5 can be repeated to yield the identities below.

Lemma 2.2.6. *Given a, x, y, z in a Jordan triple system, we have*

$$2\{x, a, \{y, a, z\}\} = \{x, \{a, y, a\}, z\} + \{x, \{a, z, a\}, y\} \quad (2.18)$$
$$2\{a, x, \{a, y, z\}\} = \{\{a, x, a\}, y, z\} + \{a, \{x, z, y\}, a\} \quad (2.19)$$
$$2\{a, \{x, a, y\}, z\} = \{\{a, x, a\}, y, z\} + \{\{a, y, a\}, x, z\}. \quad (2.20)$$

Proof. The first identity follows from adding the two triple identities

$$\{y, a, \{x, a, z\}\} = \{\{y, a, x\}, a, z\} - \{x, \{a, y, a\}, z\} + \{x, a, \{y, a, z\}\}$$

$$\{z, a, \{x, a, y\}\} = \{\{z, a, x\}, a, y\} - \{x, \{a, z, a\}, y\} + \{x, a, \{z, a, y\}\}.$$

We obtain the third identity by adding the two triple identities

$$\{a, x, \{a, y, z\}\} = \{\{a, x, a\}, y, z\} - \{a, \{x, a, y\}, z\} + \{a, y, \{a, x, z\}\}$$
$$\{a, y, \{a, x, z\}\} = \{\{a, y, a\}, x, z\} - \{a, \{y, a, x\}, z\} + \{a, x, \{a, y, z\}\}.$$

The second identity follows from the triple identity

$$\{\{a, x, a\}, y, z\} = \{a, x, \{a, y, z\}\} - \{a, \{x, z, y\}, a\} + \{\{a, y, z\}, x, a\}.$$

$\square$

In a Jordan triple system V, we define the odd powers of an element x by induction as follows:

$$x^1 = x, \quad x^3 = \{x, x, x\}, \quad x^{2n+1} = \{x, x^{2n-1}, x\} \quad (n = 2, 3, \ldots).$$

By (2.15) and induction, we have

$$x^{2n+1} = \{x^{2n-1}, x, x\}. \qquad (2.21)$$

One often makes use of the following *polarization* in a Jordan triple system V:

$$2\{x, y, z\} = \{x + z, y, x + z\} - \{x, y, x\} - \{z, y, z\}. \qquad (2.22)$$

It follows that the triple product in a Jordan triple system is completely determined by the *symmetrized* product $\{x, y, x\}$.

If V is a complex Jordan triple system, we also have the polarization identity

$$4\{z, y, z\} = (y + z)^3 + (y - z)^3 - (y + iz)^3 - (y - iz)^3 \qquad (2.23)$$

and the triple product in V is determined by the cubes x^3.

Given a real Jordan triple system $(V, \{\cdot, \cdot, \cdot\})$, it can be complexified to a complex Jordan triple system. Let $V_c = V \oplus iV$ be the complexification of the vector space V. Then the following triple product turns V_c into a complex Jordan triple system:

$$\{x \oplus iu, y \oplus iv, x \oplus iu\}_c = (\{x, y, x\} - \{u, y, u\} + 2\{x, v, u\})$$
$$\oplus i\left(-\{x, v, x\} + \{u, v, u\} + 2\{x, y, u\}\right). \quad (2.24)$$

We shall call $(V_c, \{\cdot, \cdot, \cdot\}_c)$ the *complexification* of the Jordan triple system V.

Jordan triple systems are equivalent to Jordan pairs with involution. The concept of a Jordan pair was introduced by Loos [124] in connection with a pair of mutually dual Riemannian symmetric spaces. A pair (V_{-1}, V_1) of real or complex vector spaces is called a *Jordan pair* if there are two trilinear maps

$$\{\cdot, \cdot, \cdot\}_\alpha : V_\alpha \times V_{-\alpha} \times V_\alpha \longrightarrow V_\alpha \qquad (\alpha = \pm 1)$$

which are symmetric in the outer variables and satisfy

$$\{a, b, \{x, y, z\}_\alpha\}_\alpha$$
$$= \{\{a, b, x\}_\alpha, y, z\}_\alpha - \{x, \{b, a, y\}_{-\alpha}, z\}_\alpha + \{x, y, \{a, b, z\}_\alpha\}_\alpha \quad (2.25)$$

for $a, x, z \in V_\alpha$ and $b, y \in V_{-\alpha}$.

An *involution* θ of a Jordan pair (V_{-1}, V_1) is a conjugate linear isomorphism $\theta : V_{-1} \to V_1$ satisfying

$$\theta\{x, \theta y, x\}_{-1} = \{\theta x, y, \theta x\}_1 \qquad (x, y \in V_{-1}).$$

Of course, θ is linear for a real Jordan pair. Evidently, if (V_{-1}, V_1) is a Jordan pair with involution θ, then V_{-1} is a Jordan triple system with triple product $\{x, \theta y, z\}_{-1}$ for $x, y, z \in V_{-1}$.

Conversely, given a real Jordan triple system $(V, \{\cdot, \cdot, \cdot\})$, let $V_{-1} = V_1 = V$. Then (V_{-1}, V_1) is a Jordan pair with trilinear maps $\{x, y, z\}_\alpha = \{x, y, z\}$ for $\alpha = \pm 1$ and the identity map $\theta : V_{-1} \to V_1$ is an involution.

For a complex Jordan triple system $(V, \{\cdot, \cdot, \cdot\})$, we let $V_{-1} = V$ and $V_1 = \overline{V}$, the latter being the conjugate of V defined at the beginning of this section. Then the identity map $\theta : V_{-1} \longrightarrow V_1$ is a conjugate linear involution of the Jordan pair (V_{-1}, V_1) with the trilinear map $\{x, y, z\}_\alpha = \{x, \theta^{\pm\alpha} y, z\}$.

Example 2.2.7. Let $M_{m,n}(\mathbb{C})$ be the complex vector space of $m \times n$ complex matrices. Then $(M_{m,n}(\mathbb{C}), M_{n,m}(\mathbb{C}))$ is a Jordan pair with trilinear maps

$$\{A, B, C\}_\alpha = \frac{1}{2}(ABC + CBA) \qquad (A, C \in V_{-\alpha}, B \in V_\alpha)$$

where $V_{-1} = M_{m,n}(\mathbb{C})$ and $V_1 = M_{n,m}(\mathbb{C})$. The pair has an involution

$$\theta : M_{m,n}(\mathbb{C}) \longrightarrow M_{n,m}(\mathbb{C})$$

given by $\theta(B) = B^*$, where $B^* = (\overline{b_{ji}})$ is the adjoint of the matrix $B = (b_{ij})$.

The complex vector space $M_{m,n}(\mathbb{C})$ is a Jordan triple with triple product

$$\{A, B, C\} = \{A, \theta(B), C\}_{-1} = \frac{1}{2}(AB^*C + CB^*A)$$

and $M_{mm}(\mathbb{C})$ is a Jordan algebra with involution * and Jordan product $A \circ B = (AB + BA)/2$. We shall write $M_m(\mathbb{C})$ for $M_{mm}(\mathbb{C})$ in the sequel.

One can consider an infinite dimensional extension of the above example. Let H and K be complex Hilbert spaces and let $L(H, K)$ be the Banach space of bounded linear operators between H and K. Then it is a complex Jordan triple in the triple product

$$\{R, S, T\} = \frac{1}{2}(RS^*T + TS^*R) \qquad (R, S, T \in L(H, K))$$

where $S^* : K \longrightarrow H$ is the adjoint of S. Henceforth we will write $L(H)$ for $L(H, H)$.

In fact, for any real associative algebra $\mathcal{A}$ with special Jordan product $a \circ b = (ab + ba)/2$, the canonical Jordan triple product of the Jordan algebra $(\mathcal{A}, \circ)$ is given by

$$\{a, b, c\} = \frac{1}{2}(abc + cba).$$

If $\mathcal{A}$ is over $\mathbb{C}$ and equipped with an involution *, then the canonical Jordan triple product is given by

$$\{a, b, c\} = \frac{1}{2}(ab^*c + cb^*a).$$

Example 2.2.8. Let $M_{1,2}(\mathcal{O}) = \{(z_1, z_2) : z_1, z_2 \in \mathcal{O}\}$ be the complex vector space of 1×2 matrices over $\mathcal{O}$. Given $y = (y_1, y_2)$ with $y_1 =$

$\sum_{0}^{7} \alpha_{1,k} e_k$ and $y_2 = \sum_{0}^{7} \alpha_{2,k} e_k$, we define $y_j^* = \overline{\alpha}_{j,0} e_0 - \sum_{1}^{7} \overline{\alpha}_{j,k} e_k$ for $j = 1, 2$, and

$$y^* = \begin{pmatrix} y_1^* \\ y_2^* \end{pmatrix}.$$

The space $M_{1,2}(\mathcal{O})$ is a 16-dimensional Jordan triple in the following triple product:

$$\{x, y, z\} = \frac{1}{2}(x(y^*z) + z(y^*x)).$$

Example 2.2.9. Let $H_3(\mathcal{O})$ be the complex vector space of 3×3 matrices over the Cayley algebra $\mathcal{O}$, hermitian with respect to the standard involution in $\mathcal{O}$, that is, (a_{ij}) belongs to $H_3(\mathcal{O})$ if, and only if,

$$(a_{ij}) = (\tilde{a}_{ji})$$

where the usual linear involution $\sim$ on $\mathcal{O}$ is defined by

$$\left(\sum_{k=0}^{7} \alpha_k e_k \right)^{\sim} = \alpha_0 e_0 - \sum_{k=1}^{7} \alpha_k e_k$$

for $\sum_{k=0}^{7} \alpha_k e_k \in \mathcal{O}$. We always equip $H_3(\mathcal{O})$ with the Jordan product

$$A \circ B = \frac{1}{2}(AB + BA) \qquad (A, B \in H_3(\mathcal{O}))$$

where the multiplication on the right is the usual matrix multiplication. The product $\circ$ makes $H_3(\mathcal{O})$ into a complex exceptional Jordan algebra. There is a natural conjugate linear involution $\natural$ on $\mathcal{O}$ defined by

$$\left(\sum_{k=0}^{7} \alpha_k e_k \right)^{\natural} = \sum_{k=0}^{7} \overline{\alpha}_k e_k$$

which induces a conjugate linear involution $*$ on $H_3(\mathcal{O})$:

$$(a_{ij})^* := (a_{ij}^{\natural}).$$

We always equip $H_3(\mathcal{O})$ with the triple product

$$\{A, B, C\} = (A \circ B^*) \circ C + A \circ (B^* \circ C) - B^* \circ (A \circ C).$$

With this triple product, $H_3(\mathcal{O})$ becomes a Jordan triple.

The exceptional real Jordan algebra $H_3(\mathbb{O})$ in Example 2.1.6 is the *real form* of $H_3(\mathcal{O})$ with respect to the involution $* : H_3(\mathcal{O}) \longrightarrow H_3(\mathcal{O})$, that is,

$$H_3(\mathbb{O}) = \{(a_{ij}) \in H_3(\mathcal{O}) : (a_{ij})^* = (a_{ij})\}$$

and $H_3(\mathcal{O}) = H_3(\mathbb{O}) + iH_3(\mathbb{O})$ is the complexification of $H_3(\mathbb{O})$.

We can embed $M_{1,2}(\mathcal{O})$ as a subtriple of $H_3(\mathcal{O})$ via the triple monomorphism

$$(z_1, z_2) \in M_{1,2}(\mathcal{O}) \mapsto \begin{pmatrix} 0 & z_1 & z_2 \\ \tilde{z}_1 & 0 & 0 \\ \tilde{z}_2 & 0 & 0 \end{pmatrix} \in H_3(\mathcal{O}).$$

For each element a in a Jordan triple system V, we define a binary product $\circ_a$ in V by

$$x \circ_a y = \{x, a, y\} \qquad (x, y \in V).$$

The above product is clearly commutative. It also satisfies the Jordan identity which follows from the identities (2.10), (2.18) and (2.15):

$$\begin{aligned}
x^2 \circ_a (x \circ_a y) &= \{\{x, a, x\}, a, \{x, a, y\}\} \\
&= \{\{\{x, a, x\}, a, x\}, a, y\} - \{x, \{a, \{x, a, x\}, a\}, y\} \\
&\quad + \{x, a, \{\{x, a, x\}, a, y\}\} \\
&= x \circ_a (x^2 \circ_a y)
\end{aligned}$$

where, by identities (2.18) and (2.15), we have

$$\{x, \{a, \{x, a, x\}, a\}, y\}$$
$$= 2\{\{x, a, \{x, a, x\}\}, a, y\} - \{\{x, a, x\}, \{a, x, a\}, y\}$$
$$= 2\{\{x, a, \{x, a, x\}\}, a, y\} - 2\{x, \{\{a, x, a\}, x, a\}, y\}$$
$$\quad + \{\{x, \{a, x, a\}, x\}, a, y\}$$
$$= 3\{\{x, a, \{x, a, x\}\}, a, y\} - 2\{x, \{a, \{x, a, x\}, a\}, y\}$$
$$= \{\{x, a, \{x, a, x\}\}, a, y\}.$$

Definition 2.2.10. Let V be a Jordan triple system and $a \in V$. The a-*homotope* $V^{(a)}$ of V is defined to be the Jordan algebra $(V, \circ_a)$.

Using identities (2.14) and (2.18), we see that the canonical triple product $\{\cdot, \cdot, \cdot\}_a$ in the a-homotope $V^{(a)} = (V, \circ_a)$ is given by

$$\{x, y, x\}_a = 2(x \circ_a y) \circ_a x - (x \circ_a x) \circ_a y$$
$$= 2\{\{x, a, y\}, a, x\} - \{\{x, a, x\}, a, y\}$$
$$= \{x, \{a, y, a\}, x\}.$$

In particular, if $\{a, y, a\} = y$ for all $y \in V$, then we have $\{x, y, x\}_a = \{x, y, x\}$ for all $x, y \in V$ and the Jordan triple system $(V^{(a)}, \{\cdot, \cdot, \cdot\}_a)$ is just V itself. Therefore Jordan triple systems generalise Jordan algebras in that unital real Jordan algebras and complex Jordan algebras with involution are Jordan triples containing a *unit element*, that is, an element e satisfying $\{e, y, e\} = y$ for all elements y.

One can deduce power associativity in V via homotopes.

Lemma 2.2.11. *Let V be a Jordan triple system and let $a \in V$. For odd natural numbers m, n and p, we have*

$$\{a^m, a^n, a^p\} = a^{m+n+p}.$$

Proof. Let $V^{(a)}$ be the a-homotope and $x^{(n)}$ the n-th power of x in the Jordan algebra $(V^{(a)}, \circ_a)$. By (2.21) and induction, we have

$$a^{2n-1} = a^{(n)}.$$

By (2.14) and power associativity in Jordan algebras, we have

$$\{a^m, \{a, a, a\}, a^p\} = \{\{a^m, a, a\}, a, a^p\} = a^{(m+3/2)} \circ_a a^{(p+1/2)}$$
$$= a^{(m+3+p+1/2)} = a^{m+3+p}.$$

One concludes the proof by induction, using the identity

$$\{a^m, \{a, a^{2k-1}, a\}, a^p\} = \{\{a, a, a^m\}, a^{2k-1}, a^p\}$$
$$+ \{a^m, a^{2k-1}, \{a, a, a^p\}\} - \{a, a, \{a^m, a^{2k-1}, a^p\}\}.$$

$\square$

Definition 2.2.12. Let V be a Jordan triple system and $x, y \in V$. Extending the definition of a box operator on a Jordan algebra, we define the box operator $x \,\square\, y : V \longrightarrow V$ by

$$(x \,\square\, y)(v) = \{x, y, v\} \qquad (v \in V).$$

Given two subsets A and B of V, we shall write

$$A \,\square\, B = \{a \,\square\, b : a \in A, b \in B\}.$$

The triple identity (2.11) can be written in terms of Lie brackets of box operators:

$$[a \,\square\, b, x \,\square\, y] = \{a, b, x\} \,\square\, y - x \,\square\, \{y, a, b\}. \qquad (2.26)$$

Definition 2.2.13. A Jordan triple system V is called *abelian* or *commutative* if

$$[a \,\square\, b, x \,\square\, y] = 0$$

for all $a, b, x, y \in V$, in other words, if the operators $a \square b$ and $x \square y$ commute for all $a, b, x, y \in V$. A subspace F of V is called *flat* if $x \square y = y \square x$ for all $x, y \in F$.

Lemma 2.2.14. *Let V be a Jordan triple system. Then V is abelian if and only if*

$$\{a, b, \{x, y, z\}\} = \{a, \{b, x, y\}, z\} = \{\{a, b, x\}, y, z\}$$

for all $a, b, x, y, z \in V$.

Proof. By the triple identity (2.26), the above identities are equivalent to

$$[x \square y, z \square b](a) = 0 = [x \square b, a \square y](z)$$

for all $a, b, x, y, z \in V$. $\qquad\qquad\square$

Example 2.2.15. Let V be a Jordan triple system and let $a \in V$. It is plain from power associativity in Lemma 2.2.11 that the linear span $V(a)$ of odd powers of a is the subtriple of V generated by a, that is, the smallest subtriple of V containing a. Moreover, power associativity implies that $V(a)$ is abelian. If V is real, then $V(a)$ is flat, a consequence of power associativity again.

Example 2.2.16. Flat Jordan triple systems must be abelian. However, an abelian Jordan triple system need not be flat. For instance, $\mathbb{C}$ with the triple product $\{x, y\, z\} = x\overline{y}z$ is abelian but not flat.

Apart from the box operator, there are two important operators on Jordan triple systems, namely, the *quadratic operator* and the *Bergman operator*. Let V be a Jordan triple system and let $a, b \in V$. The quadratic operator $Q_a : V \longrightarrow V$, induced by a, is defined by

$$Q_a(x) = \{a, x, a\} \qquad (x \in V) \qquad\qquad (2.27)$$

which is linear if V is real, but conjugate linear if V is a complex Jordan triple system.

The Bergman operator $B(a, b) : V \longrightarrow V$, induced by (a, b), is defined by

$$B(a, b)(x) = x - 2(a \,\square\, b)(x) + Q_a Q_b(x) \qquad (x \in V). \qquad (2.28)$$

In terms of quadratic operators, the identity (2.15) can be formulated as

$$Q_a(b \,\square\, a) = (a \,\square\, b)Q_a \qquad (a, b \in V). \qquad (2.29)$$

One also deduces from (2.15) that

$$B(a, b)Q_a = Q_{a - Q_a(b)} \qquad (a, b \in V).$$

The identity (2.17) can be formulated as

$$Q_{Q_a(b)} = Q_a Q_b Q_a \qquad (a, b \in V). \qquad (2.30)$$

We need to derive a few more identities for later applications. Given $a, b \in V$, let us define the map $Q(a, b) : V \longrightarrow V$ by

$$Q(a, b)(x) = \{a, x, b\} \qquad (x \in V).$$

We have of course $Q_a = Q(a, a)$ and it is easy to verify that

$$Q(a + tx, a + tx) = Q_a + 2tQ(a, x) + t^2 Q_x$$

for all scalars t. Hence we have, from the identity (2.30),

$$Q(Q_{a+tx}(b), Q_{a+tx}(b)) = Q_{a+tx} Q_b Q_{a+tx}$$
$$= (Q_a + 2tQ(a, x) + t^2 Q_x)(Q_b Q_a + 2tQ_b Q(a, x) + t^2 Q_b Q_x).$$

Comparing the coefficients of t on both sides above, we obtain

$$2Q(Q_a(b), \{a, b, x\}) = Q(a, x)Q_b Q_a + Q_a Q_b Q(a, x). \qquad (2.31)$$

Also, comparing the coefficients of t^2 on both sides of the previous identity, we obtain

$$2Q(Q_a(b), Q_x(b)) + 4Q(\{a, b, x\}, \{a, b, x\})$$
$$= Q_a Q_b Q_x + Q_x Q_b Q_a + 4Q(a, x) Q_b Q(a, x). \qquad (2.32)$$

Replacing b by $b + c$ in (2.31) and expand, we get

$$Q(Q_a(b), \{a, c, x\}) + Q(Q_a(c), \{a, b, x\})$$
$$= Q_a Q(b, c) Q(a, x) + Q(a, x) Q(b, c) Q_a. \qquad (2.33)$$

Lemma 2.2.17. *For a, c and x in a Jordan triple system V, we have*

$$2Q(Q_a Q_c(x), \{a, c, x\}) = Q_a Q_c Q_x(c \,\square\, a) + (a \,\square\, c) Q_x Q_c Q_a.$$

Proof. Substitute $Q_c(x)$ for b in (2.33), we have

$$2Q(Q_a Q_c(x), \{a, c, x\}) = 2Q_a Q(Q_c(x), c) Q(a, x)$$
$$+ 2Q(a, x) Q(Q_c(x), c) Q_a - 2Q(Q_a(c), \{a, Q_c(x), x\})$$
$$= 2Q_a Q_c(x \,\square\, c) Q(a, x) + 2Q(a, x)(c \,\square\, x) Q_c Q_a$$
$$- 2Q(Q_a(c), \{a, c, Q_x(c)\})$$

where the second identity follows from (2.14). Applying (2.19) and (2.20) to the first two terms, the last formula becomes

$$Q_a Q_c(Q(Q_c(x), a) + Q_x(c \,\square\, a)) + (Q(Q_c(x), a) + (a \,\square\, c) Q_x) Q_c Q_a$$
$$- 2Q(Q_a(c), \{a, c, Q_x(c)\})$$
$$= Q_a Q_c Q_x(c \,\square\, a) + (a \,\square\, c) Q_x Q_c Q_a + Q_a Q_c Q(Q_c(x), a)$$
$$+ Q(Q_c(x), a) Q_c Q_a - 2Q(Q_a(c), \{a, c, Q_x(c)\})$$
$$= Q_a Q_c Q_x(c \,\square\, a) + (a \,\square\, c) Q_x Q_c Q_a$$

where the last three terms in the second formula sum to 0 by (2.31). $\square$

Lemma 2.2.18. *For a, c and x in a Jordan triple system V, we have*

$$2Q(x, Q_a Q_b(x)) + 4Q(\{a, b, x\}, \{a, b, x\})$$
$$= Q_a Q_b Q_x + Q_x Q_b Q_a + 4(a \,\square\, b) Q_x(b \,\square\, a).$$

Proof. Consider the last term above and apply (2.20), we have

$$2(a \,\square\, b) Q_x(b \,\square\, a) = 4Q(x, a)(b \,\square\, x)(b \,\square\, a) - 2Q(Q_x(b), a)(b \,\square\, a)$$

which, by (2.19), equals

$$2Q(x, a)(Q_b(x) \,\square\, a) + 2Q(x, a)Q_b Q(x, a) - 2Q(Q_x(b), a)(b \,\square\, a)$$

and by (2.20) again, the above is equal to

$$\begin{aligned} Q(x, Q_a Q_b(x)) \quad &+ \quad (x \,\square\, Q_b(x))Q_a + 2Q(x, a)Q_b Q(x, a) \\ &- \quad Q(Q_a(b), Q_x(b)) - (Q_x(b) \,\square\, b)Q_a \\ = Q(x, Q_a Q_b(x)) \quad &+ \quad 2Q(x, a)Q_b Q(x, a) - Q(Q_a(b), Q_x(b)) \end{aligned}$$

where the last identity follows from (2.14). It follows from (2.32) that

$$\begin{aligned} &4(a \,\square\, b) Q_x(b \,\square\, a) \\ = \quad &2Q(x, Q_a Q_b(x)) + 4Q(x, a)Q_b Q(x, a) - 2Q(Q_a(b), Q_x(b)) \\ = \quad &2Q(x, Q_a Q_b(x)) + 4Q(\{a, b, x\}, \{a, b, x\}) - Q_a Q_b Q_x - Q_x Q_b Q_a \end{aligned}$$

which completes the proof. $\qquad\square$

We are now ready to prove an important identity for the Bergman operator.

Theorem 2.2.19. *Let V be a Jordan triple system and let $x, y, z \in V$. The Bergman operator $B(x, y)$ satisfies*

$$Q(B(x, y)z, B(x, y)z) = B(x, y)Q_z B(y, x). \tag{2.34}$$

Proof. The identity can be proved by comparing the expansions of both sides. Indeed, the left-hand side is equal to

$$Q_z \; - \; 4Q(z, \{x, y, z\}) + 2Q(z, Q_x Q_y(z)) + 4Q(\{x, y, z\}, \{x, y, z\})$$
$$- \; 4Q(Q_x Q_y(z), \{x, y, z\}) + Q(Q_x Q_y(z), Q_x Q_y(z))$$

whereas the right-hand side equals

$$Q_z - 2(x \square y)Q_z - 2Q_z(y \square x) + Q_x Q_y Q_z + Q_z Q_y Q_x + 4(x \square y)Q_z(y \square x)$$
$$- \; 2Q_x Q_y Q_z(y \square x) - 2(x \square y)Q_z Q_y Q_x + Q_x Q_y Q_z Q_y Q_x$$

which is identical to the left-hand side by the triple identity (2.10), Lemma 2.2.5, Lemma 2.2.17 and Lemma 2.2.18. □

Definition 2.2.20. A Jordan triple system V is called *non-degenerate* if

$$Q_a = 0 \implies a = 0$$

for each $a \in V$.

Lemma 2.2.21. *Let V be a non-degenerate Jordan triple system and let $a \in V$. If $x \square a = 0$ for all $x \in V$, then $a = 0$.*

Proof. We have

$$
\begin{aligned}
0 &= \{x, a, \{y, a, y\}\} \\
&= \{\{x, a, y\}, a, y\} - \{y, \{a, x, a\}, y\} + \{y, a, \{x, a, y\}\} \\
&= -\{y, \{a, x, a\}, y\}
\end{aligned}
$$

for all $y \in V$ which implies $Q_{\{a,y,a\}} = 0$ for all $y \in V$. Hence $\{a, y, a\} = 0$ for all $y \in V$ and $a = 0$. □

Lemma 2.2.22. *Let V be a non-degenerate Jordan triple system and let $a \in V$. If $\{x, a, x\} = 0$ for all $x \in V$, then $a = 0$.*

Proof. We have $\{x + v, a, x + v\} = 0$ for all $x, v \in V$ which implies $x \,\square\, a = 0$ for all $x \in V$. Hence $a = 0$. $\qquad\square$

Let V be a finite dimensional Jordan triple system over $\mathbb{F} = \mathbb{R}, \mathbb{C}$. We define a bilinear form $\langle \cdot, \cdot \rangle : V \times V \longrightarrow \mathbb{F}$ by

$$\langle x, y \rangle = \text{Trace}\,(x \,\square\, y).$$

If V is a finite dimensional complex Jordan triple system, then

$$\langle x, y \rangle = \text{Trace}\,(x \,\square\, y)$$

is a complex sesquilinear form. We call $\langle \cdot, \cdot \rangle$ the *trace form* of V.

If the trace form of a Jordan triple system V is non-degenerate, that is, $\langle x, y \rangle = 0$ for all $y \in V$ implies $x = 0$, then every linear map $T : V \longrightarrow V$ has an adjoint $T^* : V \longrightarrow V$ with respect to the trace form:

$$\langle Tx, y \rangle = \langle x, T^*y \rangle \qquad (x, y \in V).$$

In this case, the quadratic form $q(x) = \text{Trace}\,(x \,\square\, x)$ is also called the trace form as q determines $\langle \cdot, \cdot \rangle$ completely. We call q, or $\langle \cdot, \cdot \rangle$, *positive definite* if $q(x) > 0$ for all $x \neq 0$.

Lemma 2.2.23. *Let V be Jordan triple system which admits a non-degenerate trace form. Then we have, for every $a, b \in V$,*

$$(a \,\square\, b)^* = b \,\square\, a$$

and hence the trace form $\langle \cdot, \cdot \rangle$ is symmetric if V is real. If V is complex, the trace form $\langle \cdot, \cdot \rangle$ is Hermitian.

Proof. By triple identity (2.26), we have

$$\langle (a \,\square\, b)x, y \rangle - \langle x, (b \,\square\, a)y \rangle = \text{Trace}\,((a \,\square\, b)x \,\square\, y) - \text{Trace}\,(x \,\square\, (b \,\square\, a)y)$$
$$= \text{Trace}\,[a \,\square\, b, x \,\square\, y] = 0.$$

Hence $\langle a, b \rangle = \text{Trace}\,(a \square\, b) = \text{Trace}\,(a \square\, b)^* = \text{Trace}\,(b \square\, a) = \langle b, a \rangle$ if V is real, whereas $\text{Trace}\,(a \square\, b) = \overline{\text{Trace}\,(a \square\, b)^*} = \overline{\text{Trace}\,(b \square\, a)}$ if V is complex. $\qquad\qquad\qquad\qquad\qquad\qquad\qquad\qquad\qquad\qquad\qquad\qquad\square$

Lemma 2.2.24. *Let V be a real Jordan triple system. The following conditions are equivalent.*

(i) *The trace form is non-degenerate.*

(ii) *The bilinear form $(x, y) \in V^2 \mapsto \text{Trace}\,(x \square\, y + y \square\, x)$ is non-degenerate.*

Proof. (i) $\Rightarrow$ (ii). This follows from Lemma 2.2.23.

(ii) $\Rightarrow$ (i). The bilinear form $\ll x, y \gg = \text{Trace}\,(x \square\, y + y \square\, x)$ on V is non-degenerate and symmetric. Using $\text{Trace}\,[x \square\, y, u \square\, v] = 0$ and the triple identity (2.26) as before, we have

$$
\begin{aligned}
\ll (x \square\, y)u, v \gg &= \text{Trace}\,((x \square\, y)u \square\, v + v \square\, (x \square\, y)u) \\
&= \text{Trace}\,(u \square\, (y \square\, x)v + (y \square\, x)v \square\, u) \\
&= \ll u, (y \square\, x)v \gg .
\end{aligned}
$$

Hence the box operator $y \square\, x$ is the adjoint of $x \square\, y$ with respect to the bilinear form $\ll \cdot, \cdot \gg$. Therefore $\text{Trace}\,(y \square\, x) = \text{Trace}\,(x \square\, y)$ and (i) follows. $\qquad\qquad\qquad\qquad\qquad\qquad\qquad\qquad\qquad\qquad\qquad\qquad\square$

Definition 2.2.25. A Jordan triple system V is called *semisimple* if for each $a \in V$, we have

$$a \square\, x \text{ is nilpotent for all } x \in V \Longrightarrow a = 0.$$

Definition 2.2.26. A Jordan triple system V is called *anisotropic* if $\{x, x, x\} = 0$ implies $x = 0$ for each $x \in V$.

Lemma 2.2.27. *Let V be a Jordan triple system which admits a positive definite trace form. Then V is anisotropic.*

Proof. Let $x \in V$ and $\{x, x, x\} = 0$. Using the identities (2.18) and (2.16), we deduce that the box operator $x \,\square\, x$ is nilpotent, in fact, $(x \,\square\, x)^3 = 0$. Therefore $\mathrm{Trace}\,(x \,\square\, x) = 0$ which implies $x = 0$ since the trace form is positive definite. $\square$

For finite dimensional Jordan triple systems, semisimplicity is equivalent to non-degeneracy of the trace form.

Lemma 2.2.28. *Let V be a finite-dimensional Jordan triple system. The following conditions are equivalent.*

(i) *V is semisimple.*

(ii) *The trace form of V is non-degenerate.*

The above conditions imply that V is non-degenerate.

Proof. We show that

$$\{a \in V : a \,\square\, x \text{ is nilpotent } \forall x \in V\}$$
$$= \{a \in V : \mathrm{Trace}\,(a \,\square\, x) = 0 \,\forall x \in V\}$$

from which the equivalence of (i) and (ii) follows immediately.

Indeed, if $a \,\square\, x$ is nilpotent, then $\mathrm{Trace}\,(a \,\square\, x) = 0$. If the operator $a \,\square\, x$ is not nilpotent for some $x \in V$, then the left multiplication $L_a : V^{(x)} \longrightarrow V^{(x)}$ on the x-homotope $V^{(x)}$ is not nilpotent as the two operators are identical. Therefore a is not nilpotent in the Jordan algebra $V^{(x)}$ by Lemma 2.1.11. From Lemma 2.1.10, there is an idempotent e in the subalgebra $\mathcal{A}(a)$ of $V^{(x)}$ generated by a, and e is of the form $e = \sum_k \alpha_k a^{(n_k)}$. It follows that $\mathrm{Trace}\,(e \,\square\, x) = \mathrm{Trace}\,L_e > 0$ and the triple identity gives

$$
\begin{aligned}
\mathrm{Trace}\,(e \,\square\, x) &= \left\langle \sum_k \alpha_k a^{(n_k)}, x \right\rangle \\
&= \sum_k \alpha_k \langle (a^{(n_k-1)} \,\square\, x)(a), x \rangle \\
&= \left\langle a, \sum_k \alpha_k (x \,\square\, a^{(n_k-1)})(x) \right\rangle \\
&= \mathrm{Trace}\left(a \,\square\, \sum_k \alpha_k (x \,\square\, a^{(n_k-1)})(x) \right)
\end{aligned}
$$

and hence a does not belong to the above set on the right. This proves the equality of the two sets.

Assume condition (i) and let $Q_a = 0$ for some $a \in V$. For every $x \in V$, we have

$$
2(a \,\square\, x)^2 = Q_a(x) \,\square\, x + Q_a Q_x = 0
$$

by the identity (2.19). Hence $a \,\square\, x$ is nilpotent for all $x \in V$ and $a = 0$ by semisimplicity. This proves non-degeneracy of V. $\square$

Definition 2.2.29. A finite dimensional Jordan triple system V is called *positive* if the box operator $x \,\square\, x : V \longrightarrow V$ has a non-negative spectrum for each $x \in V$.

Example 2.2.30. A positive Jordan triple system need not be non-degenerate. Let

$$
V = \left\{ \begin{pmatrix} 0 & \alpha \\ 0 & \beta \end{pmatrix} : \alpha, \beta \in \mathbb{R} \right\}
$$

be equipped with the triple product

$$
\{a, b, c\} = abc + cba \qquad (a, b, c \in V)
$$

where the product on the right is the matrix product. Then V is a positive real Jordan triple system. Indeed, for $x = \begin{pmatrix} 0 & \alpha \\ 0 & \beta \end{pmatrix}$, the eigenvalues of the box operator $x \,\square\, x$ are β^2 and $2\beta^2$. However, $Q_a = 0$ for $a = \begin{pmatrix} 0 & 1 \\ 0 & 0 \end{pmatrix}$.

The role of idempotents in a Jordan algebra is played by tripotents in a Jordan triple system.

Definition 2.2.31. An element e in a Jordan triple system V is called a *tripotent* if $e = \{e, e, e\}$. A tripotent e is said to be *orthogonal* to a tripotent f if $e \,\square\, f = 0$.

For tripotents e and f in a Jordan triple system V, it will be shown that the condition $e \,\square\, f = 0$ is equivalent to $f \,\square\, e = 0$. Two elements $a, b \in V$ are defined to be *orthogonal to each other* if $a \,\square\, b = b \,\square\, a = 0$.

Lemma 2.2.32. *Let V be an anisotropic Jordan triple system and let $a, b \in V$. Then a and b are orthogonal to each other if, and only if, $a \,\square\, b = 0$.*

Proof. Let $a \,\square\, b = 0$. By the triple identity (2.26), we have

$$x \,\square\, \{b, a, y\} = \{a, b, x\} \,\square\, y - [a \,\square\, b, x \,\square\, y] = 0 \qquad (x, y \in V)$$

which gives

$$\{b \,\square\, a(y), b \,\square\, a(y), b \,\square\, a(y)\} = 0 \qquad (y \in V).$$

Hence $b \,\square\, a = 0$ by anisotropicity.   □

Lemma 2.2.33. *Let e be a tripotent in a Jordan triple system V. Then the box operator $e \,\square\, e : V \longrightarrow V$ has eigenvalues in $\{0, 1/2, 1\}$.*

Proof. Let $V^{(e)}$ be the e-homotope of V. Then e is an idempotent in the Jordan algebra $V^{(e)}$ with the Jordan product

$$a \circ_e b = \{a, e, b\}.$$

The box operator $e \square e : V \longrightarrow V$ is the left multiplication operator $L_e : V^{(e)} \longrightarrow V^{(e)}$ which has eigenvalues 0, 1/2 or 1 by (2.6). □

We now show the existence of tripotents.

Theorem 2.2.34. *Let V be a finite dimensional Jordan triple system. The following conditions are equivalent.*

(i) *V is semisimple and positive.*

(ii) *The trace form $q(x) = \text{Trace}\,(x \square x)$ is positive definite.*

(iii) *Each non-zero $x \in V$ admits a unique decomposition $x = \alpha_1 e_1 + \cdots + \alpha_n e_n$ where $0 < \alpha_1 < \cdots < \alpha_n$ and $e_1, \ldots, e_n$ are mutually orthogonal tripotents in V.*

Proof. (i) $\Rightarrow$ (ii). We have $q(x) \geq 0$ by positivity of V. If $\text{Trace}\,(x \square x) = 0$, then all eigenvalues of $x \square x$ are zero and we must have $x \square x = 0$ since $x \square x$ is self-adjoint with respect to the trace form on V, by Lemma 2.2.23. It follows that x is nilpotent in the a-homotope $V^{(a)}$ for each $a \in V$ since in the Jordan algebra $V^{(a)}$, we have

$$
\begin{aligned}
x^{(4)} &= \{\{x, a, \{x, a, x\}\}, a, x\} \\
&= \{(\{x, a\{x, x, a\}\} + \{x, \{a, x, x\}, a\} - \{x, x, \{x, a, a\}\}), a, x\} \\
&= 0.
\end{aligned}
$$

Hence the left multiplication $L_x : V^{(a)} \longrightarrow V^{(a)}$ is nilpotent by Lemma 2.1.11. It follows that $\text{Trace}\,(x \square a) = \text{Trace}\,L_x = 0$ and $x = 0$ by semisimplicity of V.

(ii) $\Rightarrow$ (iii). Let q be positive definite and let $x \in V \backslash \{0\}$. Let $V(x)$ be the subtriple of V generated by x. Then $V(x)$ is a finite dimensional inner product space with the trace form q. Let $\mathbb{L}$ be the real linear span of the box operators $\{a \square b|_{V(x)} : a, b \in V(x)\}$. Then $\mathbb{L}$ is a commutative algebra with respect to composition as $V(x)$ is abelian, as noted in Example 2.2.15. By Lemma 2.2.23, $\mathbb{L}$ consists of self-adjoint operators on $V(x)$ and can therefore be simultaneously diagonalised. In other words, there is a basis $\{v_1, \ldots, v_n\}$ in $V(x)$ such that $Lv_k \in \mathbb{R}v_k$ for all $L \in \mathbb{L}$ and $k = 1, \ldots, n$. In particular, we have

$$(v_k \square v_k)(v_k) = \lambda_k v_k \qquad (k = 1, \ldots, n)$$

where $\lambda_k \neq 0$ since V is anisotropic by Lemma 2.2.27. Moreover v_k/λ_k is an idempotent in the v_k-homotope $V^{(v_k)}$ on which the left multiplication L_{v_k/λ_k} has eigenvalues 0, $1/2$ or 1. The latter says the same of the box operator $(v_k \square v_k)/\lambda_k$. It follows that $\text{Trace}\,(v_k \square v_k)$ is a positive multiple of λ_k and hence $\lambda_k > 0$.

Let $e_k = v_k/\sqrt{\lambda_k}$ for $k = 1, \ldots, n$. Then $e_1, \ldots, e_n$ are mutually orthogonal tripotents in V since

$$(e_i \square e_j)(e_k) = (e_k \square e_j)(e_i) \in \mathbb{R}e_k \cap \mathbb{R}e_i = \{0\}$$

for $i \neq k$.

After permutation and sign change, we can write

$$x = \alpha_1 e_1 + \cdots + \alpha_n e_n \qquad (0 \leq \alpha_1 \leq \cdots \leq \alpha_n).$$

By orthogonality, we have

$$x^p = \alpha_1^p e_1 + \cdots + \alpha_n^p e_n$$

for odd powers x^p of x. Since $V(x)$ is spanned by these odd powers, we must have

$$0 < \alpha_1 < \cdots < \alpha_n.$$

To see uniqueness of the decomposition, let

$$x = \mu_1 u_1 + \cdots + \mu_m u_m$$

where $0 < \mu_1 < \cdots < \mu_m$ and $u_1, \ldots, u_m$ are mutually orthogonal tripotents. Then we have

$$\mu_1^{2k+1} u_1 + \cdots + \mu_m^{2k+1} u_m = x^{(2k+1)} \in V(x)$$

for $k = 1, 2, \ldots$. Hence each u_j is in $V(x)$ and $u_1, \ldots, u_m$ form a basis of $V(x)$ with $m = n$. We have $u_j = \sum_k \beta_k e_k$ and

$$u_j = \{u_j, u_j, u_j\} = \sum_k \beta_k^3 e_k$$

which imply $\beta_k = 0$ or ± 1. It follows that u_j equals $\pm e_k$ for some k; but the inequalities on the coefficients imply $u_j = e_j$ and $\mu_j = \alpha_j$ for each j .

(iii) $\Rightarrow$ (i). Given $x = \alpha_1 e_1 + \cdots + \alpha_n e_n$ with orthogonal tripotents $e_1, \ldots, e_n$, we have

$$x \square x = \sum_{k=1}^{n} \alpha_k^2 (e_k \square e_k)$$

where each $e_k \square e_k$ has spectrum in $\{0, 1/2, 1\}$. Hence $x \square x$ has non-negative spectrum. If $\text{Trace}(x \square x) = 0$, then $\alpha_k^2 = 0$ for all k and $x = 0$. This proves semisimplicity of V. $\qquad \square$

Definition 2.2.35. Let V be a semisimple positive finite dimensional Jordan triple system and let $x \in V$. The decomposition

$$x = \alpha_1 e_1 + \cdots + \alpha_n e_n$$

in Theorem 2.2.34 (iii) is called the *spectral decomposition* of x. We define the *triple spectrum* of x to be the set

$$s(x) = \{\alpha_1, \ldots, \alpha_n\}.$$

We now discuss the Peirce decomposition of a Jordan triple system V induced by a tripotent $u \in V$. Let $V^{(u)}$ be the u-homotope of V. As noted before, u is an idempotent in the Jordan algebra $V^{(u)}$ with Jordan product $a \circ_u b = \{a, u, b\}$ and the box operator $u \square u : V \longrightarrow V$ is the left multiplication $L_u : V^{(u)} \longrightarrow V^{(u)}$, with eigenvalues in $\{0, 1/2, 1\}$.

Definition 2.2.36. Let u be a tripotent in a Jordan triple system V. The eigenspaces

$$V_k(u) = \left\{ z \in V : (u \square u)(z) = \frac{k}{2}z \right\} \qquad (k = 0, 1, 2)$$

are called the Peirce k-spaces of u, and the eigenspace decomposition

$$V = V_0(u) \oplus V_1(u) \oplus V_2(u)$$

is called the *Peirce decomposition* of V.

By (2.7) and (2.8), the Peirce k-space is the range of the Peirce k-projection

$$P_k(u) : V \longrightarrow V$$

given by, using (2.18),

$$
\begin{aligned}
P_2(u)(z) &= 2L_u^2(z) - L_u(z) = 2\{u, u, \{u, u, z\}\} - \{u, u, z\} = Q_u^2(z) \\
P_1(u) &= 4(L_u - L_u^2) = 2(L_u - (2L_u^2 - L_u)) = 2(u \square u - Q_u^2) \\
P_0(u) &= I - 2(u \square u - Q_u^2) - Q_u^2 = I - 2u \square u + Q_u^2 = B(u, u)
\end{aligned}
$$

where $I : V \longrightarrow V$ is the identity operator.

Trivially, if $u = 0$, then $P_0(u) = I$ and $V = V_0(u)$. We usually consider non-zero tripotents in Peirce decompositions.

By Lemma 2.1.12, the Peirce 2-space $V_2(u)$ of a tripotent $u \in V$ is a Jordan subalgebra of the u-homotope $V^{(u)}$, containing the identity u, with Jordan product

$$a \circ_u b = \{a, u, b\} \qquad (a, b \in V).$$

Lemma 2.2.37. *Given a tripotent u in an abelian Jordan triple system V, we have*

$$V = V_0(u) \oplus V_2(u)$$

where $(V_2(u), \circ_u)$ is an (abelian) associative algebra.

Proof. Since V is abelian, the box operator $u \square u : V \longrightarrow V$ is a projection, that is $(u \square u)^2 = u \square u$ and it follows that $2\{u, u, v\} = v$ implies $v = 0$. Hence $V_1(u) = \{0\}$. The associativity of $(V_2(u), \circ_u)$ follows directly from the abelian condition on the triple product. $\square$

Definition 2.2.38. A non-zero tripotent u in a Jordan triple system V is called *maximal* or *complete* if $V_0(u) = \{0\}$. It is called *minimal* if $V_2(u) = \mathbb{C}u$, and called *unitary* if $V_2(u) = V$, given a complex Jordan triple V.

Both maximal and unitary tripotents cannot be 0 unless $V = \{0\}$. We see from Lemma 2.2.37 that an abelian Jordan triple system becomes an abelian associative algebra if it admits a maximal tripotent.

Example 2.2.39. Consider the Jordan triple $M_2(\mathbb{C})$ of 2×2 complex matrices. For the tripotent

$$u = \begin{pmatrix} 1 & 0 \\ 0 & 0 \end{pmatrix}$$

the Peirce k-projections are given by

$$P_0(u) \begin{pmatrix} a & b \\ c & d \end{pmatrix} = \begin{pmatrix} 0 & 0 \\ 0 & d \end{pmatrix}$$

$$P_1(u) \begin{pmatrix} a & b \\ c & d \end{pmatrix} = \begin{pmatrix} 0 & b \\ c & 0 \end{pmatrix}$$

$$P_2(u) \begin{pmatrix} a & b \\ c & d \end{pmatrix} = \begin{pmatrix} a & 0 \\ 0 & 0 \end{pmatrix}.$$

Example 2.2.40. In the Jordan triple $H_3(\mathcal{O})$, introduced in Example 2.2.9, the Peirce 1-space of the tripotent

$$u = \begin{pmatrix} 1 & 0 & 0 \\ 0 & 0 & 0 \\ 0 & 0 & 0 \end{pmatrix}$$

is none other than

$$M_{1,2}(\mathcal{O}) = P_1(u)(H_3(\mathcal{O})).$$

More generally, consider the Jordan triple $L(H, K)$ in Example 2.2.7, an operator $u \in L(H, K)$ is a tripotent if, and only if, $u = uu^*u$, that is, u is a partial isometry. The operators $\ell = uu^*$ and $r = u^*u$ are projections on the Hilbert spaces K and H respectively. They can be represented, with suitable orthonormal bases, by square block matrices

$$\ell = \begin{pmatrix} 1_K & O \\ O & O \end{pmatrix}, \quad r = \begin{pmatrix} 1_H & O \\ O & O \end{pmatrix}$$

where 1_H and 1_K are identities. In this representation, each operator in the Peirce 2-space

$$L(H, K)_2(u) = P_2(u)L(H, K) = \{\ell Tr : T \in L(H, K)\}$$

has a rectangular matrix representation

$$P_2(u)T = \ell Tr = \begin{pmatrix} [\ell Tr] & O \\ O & O \end{pmatrix}.$$

The other two Peirce projections of u are given by

$$P_0(u)T = (1_K - \ell)T(1_H - r) = \begin{pmatrix} O & O \\ O & [(1_K - \ell)T(1_H - r)] \end{pmatrix},$$

$$P_1(u)T = \ell T(1_H - r) + (1_K - \ell)Tr = \begin{pmatrix} O & [\ell T(1_H - r)] \\ [(1_K - \ell)Tr] & O \end{pmatrix}.$$

The matrix form of the Peirce decomposition of $T \in L(H, K)$ is given by

$$T = \begin{pmatrix} [\ell Tr] & [\ell T(\mathbf{1}_H - r)] \\ [(\mathbf{1}_K - \ell)Tr] & [(\mathbf{1}_K - \ell)T(\mathbf{1}_H - r)] \end{pmatrix}.$$

Let u be a tripotent in a Jordan triple system V. For any real scalar $t \neq 0$, the Bergman operator

$$B(u, (1 - t)u) : V \longrightarrow V$$

is invertible. Indeed, a simple calculation gives

$$B(u, (1 - t)u) = P_0(u) + tP_1(u) + t^2 P_2(u) \tag{2.35}$$

and therefore $B(u, (1 - t)u)$ has inverse

$$B(u, (1 - t^{-1})u) = P_0(u) + \frac{1}{t}P_1(u) + \frac{1}{t^2}P_2(u)$$

by mutual orthogonality of the Peirce projections.

Since

$$B(u, (1 - t)u)v = \sum_{k=0}^{2} t^k P_k(u)v \tag{2.36}$$

for each $v \in V$, it follows that, for $t \neq 1$, we have $B(u, (1 - t)u)v = t^k v$ if, and only if, $v \in V_k(u)$ for $k = 0, 1, 2$. However, for $k \notin \{0, 1, 2\}$, we have $B(u, (1 - t)u)v = t^k v$ if, and only if, $v = 0$.

We now derive the basic Peirce multiplication rules.

Theorem 2.2.41. *Let u be a tripotent of a Jordan triple system V. Then the Peirce k-spaces $V_k(u)$ satisfy*

$$\{V_0(u), V_2(u), V\} = \{V_2(u), V_0(u), V\} = \{0\}$$

$$\{V_i(u), V_j(u), V_k(u)\} \subset V_{i-j+k}(u) \tag{2.37}$$

where $V_\alpha(u) = \{0\}$ for $\alpha \notin \{0, 1, 2\}$.

Proof. Let $x \in V_2(u)$ and $y \in V_0(u)$. We first observe that

$$\{u, y, u\} = Q_u^3 y = Q_u P_2(u) y = Q_u P_2(u) P_0(u) y = 0.$$

Hence by (2.18), we have

$$\{z, u, y\} = \{z, \{u, u, u\}, y\} = 2\{z, u, \{u, u, y\}\} - \{z, \{u, y, u\}, u\} = 0$$

for any $z \in V$. It follows from (2.20) that

$$\{x, y, z\} = \{\{u, \{u, x, u\}, u\}, y, z\}$$
$$= 2\{u, \{y, u, \{u, x, u\}\}, z\} - \{\{u, y, u\}, \{u, x, u\}, z\} = 0$$

for all $z \in V$. Likewise $\{V_0(u), V_2(u), V\} = \{0\}$.

To show (2.37), we make use of invertibility of $B(u, (1-t)u)$ for real scalars $t \neq 0$, with inverse $B(u, (1 - t^{-1})u)$, and deduce from Theorem 2.61 that

$$B(u, (1-t)u)\{z, x, z\} = \{B(u, (1-t)u)z, B(u, (1-t^{-1})u)x, B(u, (1-t)u)z\}$$

for $x, z \in V$. By polarization in (2.22), we have

$$B(u, (1 - t)u)\{x, y, z\}$$
$$= \{B(u, (1 - t)u)x, B(u, (1 - t^{-1})u)y, B(u, (1 - t)u)z\} \quad (2.38)$$

for $x, y, z \in V$. In particular, for $v_\alpha \in V_\alpha(u)$, the above remarks imply

$$B(u, (1 - t)u)\{v_i, v_j, v_k\}$$
$$= \{B(u, (1 - t)u)v_i, B(u, (1 - t^{-1})u)v_j, B(u, (1 - t)u)v_k\}$$
$$= \{t^i v_i, t^{-j} v_j, t^k v_k\} = t^{i-j+k}\{v_i, v_j, v_k\}$$

and $\{v_i, v_j, v_k\} \in V_{i-j+k}(u)$. $\square$

The Peirce multiplication rules reveal immediately that the Peirce k-spaces $V_k(u)$ of a tripotent u in a Jordan triple system V are subtriples of V. These rules also entail the following useful results.

Corollary 2.2.42. *Let u, e be tripotents in a Jordan triple system V such that $u \in V_2(e)$. Then we have $V_2(u) \subset V_2(e)$ and $V_0(e) \subset V_0(u)$.*

Proof. We have $V_2(u) = P_2(u)(V)$. Each $x \in V$ has a Peirce decomposition $x = x_0 + x_1 + x_2 \in V_0(e) \oplus V_1(e) \oplus V_2(e)$ with respect to e and the Peirce rules imply $P_2(u)(x) = P_2(u)(x_2) \in V_2(e)$.

The Peirce rules also imply $(u \square u)(V_0(e)) = \{0\}$. $\qquad \square$

Corollary 2.2.43. *Let u, v be tripotents in a Jordan triple system V. The following conditions are equivalent.*

(i) *u and v are orthogonal to each other.*

(ii) *$v \square u = 0$.*

(iii) *$\{u, u, v\} = 0$.*

(iv) *$\{v, v, u\} = 0$.*

Proof. This follows easily from Theorem 2.2.41. Indeed, orthogonality implies (iv) which in turn implies that u is in the Peirce 0-space $V_0(v)$ of v and therefore $v \square u = 0$ by Theorem 2.2.41. By the same token, (ii) is equivalent to (iii). $\qquad \square$

Given two mutually orthogonal tripotents e_1 and e_2 in a Jordan triple system V, it is evident that $e_1 + e_2$ is also a tripotent. One can form a *joint* Peirce decomposition of V with respect to e_1 and e_2. By orthogonality, we have

$$B(e_j, te_j)e_k = e_k \qquad (j, k \in \{1, 2\}, j \neq k \text{ and } t \in \mathbb{R})$$

and a direct computation using (3.57) yields

$$B(e_1, (1-t)e_1)B(e_2, (1-s)e_2) = B(e_2, (1-s)e_2)B(e_1, (1-t)e_1). \quad (2.39)$$

Since $B(e_1, (1-t)e_1) = P_0(e_1) + tP_1(e_1) + t^2 P_2(e_1)$ and

$$B(e_2, (1-s)e_2) = P_0(e_2) + sP_1(e_2) + s^2 P_2(e_2),$$

comparing coefficients in the equation (2.39), one finds that the Peirce projections $P_j(e_1)$ and $P_k(e_2)$ commute. Therefore we have the decomposition

$$V = \bigoplus_{0 \le j \le k \le 2} V_{j,k} = V_{0,0} \oplus V_{0,1} \oplus V_{0,2} \oplus V_{1,1} \oplus V_{1,2} \oplus V_{2,2}$$

where $V_{k,k} = V_2(e_k)$ for $k \ne 0$ and $V_{0,0} = V_0(e_1) \cap V_0(e_2)$ is the range of the projection $P_0(e_1)P_0(e_2)$, and

$$V_{0,1} = V_0(e_2) \cap V_1(e_1), \quad V_{0,2} = V_0(e_1) \cap V_1(e_2), \quad V_{1,2} = V_1(e_1) \cap V_1(e_2)$$

are ranges of mutually orthogonal projections $P_j(e_{j'})P_k(e_{k'})$ for suitably chosen indices j, j', k and k'.

More generally, given a family $\{e_1, \ldots, e_n\}$ of mutually orthogonal tripotents in a Jordan triple system V, one can form the *joint* Peirce decomposition of V as follows. For $i, j \in \{0, 1, \ldots, n\}$, the *joint* Peirce space V_{ij} is defined by

$$V_{ij} := V_{ij}(e_1, \ldots, e_n)$$
$$= \{z \in V \ : \ 2\{e_k, e_k, z\} = (\delta_{ik} + \delta_{jk})z \text{ for } k = 1, \ldots, n\},$$

where δ_{ij} is the Kronecker delta and $V_{ij} = V_{ji}$. The decomposition

$$V = \bigoplus_{0 \le i \le j \le n} V_{ij}$$

is called a *joint* Peirce decomposition. More verbosely,

$$V_{00} = V_0(e_1) \cap \cdots \cap V_0(e_n),$$

$$V_{ii} = V_2(e_i) \qquad\qquad\qquad (i = 1, \ldots, n),$$

$$V_{ij} = V_{ji} = V_1(e_i) \cap V_1(e_j) \qquad (1 \le i < j \le n),$$

$$V_{i0} = V_{0i} = V_1(e_i) \cap \bigcap_{j \ne i} V_0(e_j) \qquad (i = 1, \ldots, n).$$

The Peirce multiplication rules

$$\{V_{ij}, V_{jk}, V_{k\ell}\} \subset V_{i\ell} \quad \text{and} \quad V_{ij} \,\square\, V_{pq} = \{0\} \quad \text{for} \quad i, j \notin \{p, q\}$$

hold. The contractive projection $P_{ij}(e_1, \ldots, e_n)$ from V onto $V_{ij} = V_{ij}(e_1, \ldots, e_n)$ is called a *joint Peirce projection* which satisfies

$$P_{ij}(e_1, \ldots, e_n)(e_k) = \begin{cases} 0 & (i \neq j) \\ \delta_{ik} e_k & (i = j). \end{cases} \tag{2.40}$$

We shall simplify the notation $P_{ij}(e, \ldots, e_n)$ to P_{ij} if the tripotents $e_1, \ldots, e_n$ are understood. For a single tripotent $e \in V$, we have $P_{11}(e) = P_2(e)$, $P_{10}(e) = P_1(e)$ and $P_{00}(e) = P_0(e)$.

Let $M = \{0, 1, \ldots, n\}$ and $N \subset \{1, \ldots, n\}$. The Peirce k-spaces of the tripotent $e_N = \sum_{i \in N} e_i$ are given by

$$V_2(e_N) = \bigoplus_{i,j \in N} V_{ij}, \tag{2.41}$$

$$V_1(e_N) = \bigoplus_{\substack{i \in N \\ j \in M \setminus N}} V_{ij}, \tag{2.42}$$

$$V_0(e_N) = \bigoplus_{i,j \in M \setminus N} V_{ij}. \tag{2.43}$$

More details of the construction of the preceding joint Peirce decomposition can be found in [124, 5.14].

A subspace J of a Jordan triple system V is called a *triple ideal*, or simply, an *ideal*, if it satisfies the condition

$$\{J, V, V\} + \{V, J, V\} \subset J.$$

If a subspace $J \subset V$ satisfies only

$$\{J, V, J\} \subset J,$$

then it is called an *inner ideal*. The concept of an inner ideal is important in modern Jordan structure theory. Inner ideals are substitutes for *one-sided ideals*, the latter are absent in Jordan triple systems. Actually, every left or right ideal, or their intersection, in an associative algebra $\mathcal{A}$ is an inner ideal in the special Jordan algebra $(\mathcal{A}, \circ)$ with the Jordan product $a \circ b = (ab + ba)/2$. So is any subspace of the form $a\mathcal{A}b$. In a Jordan triple system V, the subspace $\{v, V, v\}$ is an inner ideal, called the *principal inner ideal* determined by v.

Given an ideal J of a Jordan triple system V, the quotient space V/J is naturally a Jordan triple system with the triple product

$$\{x + J, y + J, z + J\} = \{x, y, z\} + J.$$

The kernel $\varphi^{-1}(0)$ of a triple homomorphism $\varphi : V \longrightarrow W$ is an ideal of V. On the other hand, an ideal J of a Jordan triple system V is the kernel of the quotient map $q : V \longrightarrow V/J$ which is a triple homomorphism.

Let u be a tripotent in a Jordan triple system V. Applying the Peirce multiplication rules in Theorem 2.2.41 to the Peirce decomposition $V = V_0(u) \oplus V_1(u) \oplus V_2(u)$, one deduces the following fact readily.

Proposition 2.2.44. *Given a tripotent u of a Jordan triple system V, the Peirce spaces $V_0(u)$ and $V_2(u)$ are inner ideals of V.*

2.3   Lie algebras and Tits-Kantor-Koecher construction

We now show an important connection between Jordan triple systems and Lie algebras via the Tits-Kantor-Koecher construction. Lie algebras play an important role in geometry and this connection provides us with a useful link to apply Jordan theory to geometry. We will only be concerned with real or complex Lie algebras which, however, can be infinite dimensional.

In what follows, a *Lie algebra* is a real or complex vector space $\mathfrak{g}$ of any dimension, with a bilinear multiplication, called the *Lie brackets*,

$$(x, y) \in \mathfrak{g} \times \mathfrak{g} \mapsto [x, y] \in \mathfrak{g}$$

satisfying $[x, x] = 0$ and the Jacobi identity

$$[[x, y], z] + [[y, z], x] + [[z, x], y] = 0$$

for all $x, y, z \in \mathfrak{g}$. We note that the multiplication is not associative but is anticommutative:

$$[x, y] = -[y, x].$$

On any associative algebra $\mathfrak{a}$, one can define the Lie brackets by *commutation*:

$$[x, y] = xy - yx$$

where the product on the right-hand side is the original product in $\mathfrak{a}$. Then $(\mathfrak{a}, [\cdot, \cdot])$ is a Lie algebra. Unlike Jordan algebras, it follows from a theorem of Poincaré, Birkhoff and Witt [21, I.2.7] that *any* Lie algebra can be obtained in this way from an associative algebra.

Given subspaces h and k of a Lie algebra $\mathfrak{g}$, we define

$$[h, k] = \{[x_1, y_1] + \cdots + [x_n, y_n] : x_1, \ldots, x_n \in h; y_1, \ldots, y_n \in k\}$$

which is a subspace of $\mathfrak{g}$. We note that $[h, k] = [k, h]$. A Lie algebra $\mathfrak{g}$ is called *abelian* if $[\mathfrak{g}, \mathfrak{g}] = 0$. An *ideal* of $\mathfrak{g}$ is a subspace $\mathfrak{h}$ of $\mathfrak{g}$ satisfying $[\mathfrak{g}, \mathfrak{h}] \subset \mathfrak{h}$. For instance, $[\mathfrak{g}, \mathfrak{g}]$ is an ideal of $\mathfrak{g}$. Given an ideal $\mathfrak{h}$ of a Lie algebra $\mathfrak{g}$, the quotient space $\mathfrak{g}/\mathfrak{h}$ is a Lie algebra in the product

$$[x + \mathfrak{h}, y + \mathfrak{h}] = [x, y] + \mathfrak{h} \qquad (x + \mathfrak{h}, y + \mathfrak{h} \in \mathfrak{g}/\mathfrak{h}).$$

A (Lie) *homomorphism* between two Lie algebras $\mathfrak{g}$ and $\mathfrak{h}$ is a linear map $\theta : \mathfrak{g} \longrightarrow \mathfrak{h}$ satisfying $\theta[x, y] = [\theta x, \theta y]$ for all $x, y \in \mathfrak{g}$.

Given an ideal $\mathfrak{h}$ of $\mathfrak{g}$, the quotient map $\theta : x \in \mathfrak{g} \mapsto x + \mathfrak{h} \in \mathfrak{g}/\mathfrak{h}$ is a homomorphism. A bijective homomorphism between Lie algebras is called an *isomorphism* or a Lie isomorphism. An isomorphism from $\mathfrak{g}$ onto itself is called an *automorphism*, or a Lie automorphism, of $\mathfrak{g}$.

Let $\overline{\mathfrak{g}}$ be the conjugate of the vector space $\mathfrak{g}$. Then $\overline{\mathfrak{g}}$ is a Lie algebra with the multiplication of $\mathfrak{g}$. An automorphism $\theta : \mathfrak{g} \longrightarrow \overline{\mathfrak{g}}$ is called *involutive* if θ^2 is the identity map. In other words, for a complex Lie algebra $\mathfrak{g}$, an involutive automorphism is a conjugate linear bijection $\theta : \mathfrak{g} \longrightarrow \mathfrak{g}$ which preserves the Lie brackets and θ^2 is the identity map. An involutive automorphism of a Lie algebra $\mathfrak{g}$ is also called an *involution*. We will adopt this terminology although an algebra involution defined at the beginning of Section 2.1 is meant to be an antiautomorphism. However, confusion should be unlikely from the context.

Definition 2.3.1. Let $\mathfrak{g}$ be a complex Lie algebra and let $\mathfrak{g}_{\mathbb{R}}$ be its real restriction, that is, $\mathfrak{g}$ itself considered as a Lie algebra over the real field $\mathbb{R}$. A *real form* of $\mathfrak{g}$ is a subalgebra $\mathfrak{g}_r$ of $\mathfrak{g}_{\mathbb{R}}$ such that

$$\mathfrak{g} = \mathfrak{g}_r + i\mathfrak{g}_r.$$

If $\mathfrak{g}_r$ is the real form of $\mathfrak{g}$, the map $\sigma : X + iY \in \mathfrak{g}_r + i\mathfrak{g}_r \mapsto X - iY \in \mathfrak{g}_r + i\mathfrak{g}_r$ is called the *conjugation* of $\mathfrak{g}$ with respect to $\mathfrak{g}_r$. It is readily seen that σ is conjugate linear on $\mathfrak{g}$ and a Lie automorphism of the real restriction $\mathfrak{g}_{\mathbb{R}}$.

Definition 2.3.2. Let $\mathfrak{g}$ be a Lie algebra. The set Aut $\mathfrak{g}$ of all automorphisms $\theta : \mathfrak{g} \longrightarrow \mathfrak{g}$ forms a group with composition as group product, called the *automorphism group* of $\mathfrak{g}$.

A *derivation* of a Lie algebra $\mathfrak{g}$ is a linear map $\delta : \mathfrak{g} \longrightarrow \mathfrak{g}$ satisfying

$$\delta[x, y] = [\delta x, y] + [x, \delta y] \qquad (x, y \in \mathfrak{g}).$$

The vector space aut $\mathfrak{g}$ of all derivations of $\mathfrak{g}$ is a Lie algebra in the Lie brackets

$$[\delta, \gamma] = \delta\gamma - \gamma\delta.$$

For each element $x \in \mathfrak{g}$, the map $\mathrm{ad}(x) : \mathfrak{g} \longrightarrow \mathfrak{g}$ defined by

$$\mathrm{ad}(x)(y) = [x, y] \qquad (y \in \mathfrak{g})$$

is a derivation of $\mathfrak{g}$ and the Jacobi identity implies that the map

$$\mathrm{ad} : \mathfrak{g} \longrightarrow \mathrm{aut}\,\mathfrak{g}$$

is a homomorphism, called the *adjoint representation* of $\mathfrak{g}$. The kernel of ad is the *centre* of $\mathfrak{g}$:

$$\mathfrak{z}(\mathfrak{g}) = \{x \in \mathfrak{g} : [x, y] = 0 \; \forall y \in \mathfrak{g}\}.$$

The range of ad, denoted by $\mathrm{ad}\,\mathfrak{g}$, is an ideal of aut $\mathfrak{g}$ since

$$[\delta, \mathrm{ad}(x)] = \mathrm{ad}(\delta x) \qquad (\delta \in \mathrm{aut}\,\mathfrak{g}, x \in \mathfrak{g}).$$

The elements of $\mathrm{ad}\,\mathfrak{g}$ are called the *inner derivations* of $\mathfrak{g}$.

A Lie algebra $\mathfrak{g}$ is called *solvable* if its *derived series*

$$\mathfrak{g} \supset \mathfrak{g}^{(1)} = [\mathfrak{g}, \mathfrak{g}] \supset \mathfrak{g}^{(2)} = [\mathfrak{g}^{(1)}, \mathfrak{g}^{(1)}] \supset \cdots \supset \mathfrak{g}^{(n+1)} = [\mathfrak{g}^{(n)}, \mathfrak{g}^{(n)}] \supset \cdots$$

eventually terminates, that is, $\mathfrak{g}^{(n)} = \{0\}$ for some n.

Given a finite-dimensional Lie algebra $\mathfrak{g}$, the symmetric bilinear form

$$\beta : \mathfrak{g} \times \mathfrak{g} \longrightarrow \mathbb{F} \qquad (\mathbb{F} = \mathbb{R} \text{ or } \mathbb{C})$$

defined by

$$\beta(x, y) = \mathrm{Trace}\,(\mathrm{ad}(x)\mathrm{ad}(y))$$

is called the *Killing form* of $\mathfrak{g}$. The Killing form β is *invariant*, that is,

$$\beta([x, y], z) = \beta(x, [y, z]) \qquad (x, y, z \in \mathfrak{g})$$

which is equivalent to

$$\beta([x,y],z) + \beta(y,[x,z]) = 0 \qquad (x,y,z \in \mathfrak{g}).$$

The latter condition says that $\mathrm{ad}(x)$ is skew-symmetric with respect to β.

A Lie algebra $\mathfrak{g}$ is called *semisimple* if $\mathfrak{g}$ contains no non-zero abelian ideal which is equivalent to the condition that $\mathfrak{g}$ contains no non-zero solvable ideal. We prove below the Cartan-Killing criterion for semisimplicity.

Theorem 2.3.3. *A finite dimensional Lie algebra $\mathfrak{g}$ is semisimple if and only if its Killing form β is non-degenerate.*

Proof. Let the Killing form

$$\beta(x,y) = \mathrm{Trace}\,(\mathrm{ad}(x)\mathrm{ad}(y))$$

be non-degenerate. Let $\mathfrak{h}$ be an abelian ideal of $\mathfrak{g}$. We show that $\mathfrak{h} = \{0\}$. Let $x \in \mathfrak{h}$ and let $y \in \mathfrak{g}$. Since $\mathfrak{h}$ is an ideal, we have $\mathrm{ad}(x)\mathrm{ad}(y)(\mathfrak{g}) \subset \mathfrak{h}$ and hence

$$(\mathrm{ad}(x)\mathrm{ad}(y))^2(\mathfrak{g}) \subset \mathrm{ad}(x)\mathrm{ad}(y)(\mathfrak{h}) \subset [x,\mathfrak{h}] = \{0\}$$

since $\mathfrak{h}$ is abelian. This shows that the linear map $\mathrm{ad}(x)\mathrm{ad}(y) : \mathfrak{g} \longrightarrow \mathfrak{g}$ is nilpotent and therefore $\mathrm{Trace}\,(\mathrm{ad}(x)\mathrm{ad}(y)) = 0$. Non-degeneracy of β gives $x = 0$.

Conversely, let $\mathfrak{g}$ be semisimple. Let

$$\mathfrak{k} = \{x \in \mathfrak{g} : \beta(x,\mathfrak{g}) = \{0\}\}.$$

We need to show $\mathfrak{k} = \{0\}$. Since β is invariant, $\mathfrak{k}$ is an ideal of $\mathfrak{g}$. We have $\mathrm{Trace}\,(AB) = 0$ for $A, B \in \mathrm{ad}(\mathfrak{k})$. It follows from Cartan's solvability criterion that $\mathrm{ad}(\mathfrak{k})$ is a solvable ideal in $\mathrm{aut}\,\mathfrak{g}$. Semisimplicity of $\mathfrak{g}$ implies that the homomorphism ad has zero kernel and therefore $\mathfrak{k}$ is a solvable ideal in $\mathfrak{g}$. Hence $\mathfrak{k} = \{0\}$ by semisimplicity again. $\qquad\square$

The Tits-Kantor-Koecher construction originally relates Jordan algebras to finitely graded Lie algebras. This construction has been extended to Jordan triple systems by Meyberg [130]. We will describe the construction for Jordan triple systems. Let $\mathbb{Z}$ be the ring of integers. By a $\mathbb{Z}$-grading of a Lie algebra $\mathfrak{g}$, we mean a decomposition of $\mathfrak{g}$ into a direct sum of vector subspaces:

$$\mathfrak{g} = \bigoplus_{n \in \mathbb{Z}} g_n$$

such that $[g_n, g_m] \subset g_{n+m}$. The grading is said to be *finite* if the set $\{n : g_n \neq 0\}$ is finite. It is said to be *non-trivial* if $\oplus_{n \neq 0} g_n \neq 0$. Lie algebras with a non-trivial finite $\mathbb{Z}$-grading have been classified by Zelmanov [185] in which the Tits-Kantor-Koecher construction plays an important part. If $A \subset \mathbb{Z}$, a Lie algebra

$$\mathfrak{g} = \bigoplus_{\alpha \in A} \mathfrak{g}_\alpha$$

is said to be graded if $[\mathfrak{g}_\alpha, \mathfrak{g}_\beta] \subset \mathfrak{g}_{\alpha+\beta}$ where $\mathfrak{g}_{\alpha+\beta} = \{0\}$ if $\alpha + \beta \notin A$. A Lie homomorphism $\psi : \bigoplus_{\alpha \in A} \mathfrak{g}_\alpha \longrightarrow \bigoplus_{\alpha \in A} \mathfrak{h}_\alpha$ is called *graded* if it respects the grading, that is $\psi(\mathfrak{g}_\alpha) \subset \mathfrak{h}_\alpha$.

There is a one-to-one correspondence between 3-graded Lie algebras $\mathfrak{g}_{-1} \oplus \mathfrak{g}_0 \oplus \mathfrak{g}_1$ and Jordan pairs [128]. The 3-graded Lie algebras with an involution correspond to Jordan triple systems.

By an *involutive Lie algebra* $(\mathfrak{g}, \theta)$, we mean a Lie algebra $\mathfrak{g}$ equipped with an involutive automorphism θ. We will always denote by $\mathfrak{k}$ the 1-eigenspace of θ, and by $\mathfrak{p}$ the (-1)-eigenspace of θ so that $\mathfrak{g}$ has the decomposition $\mathfrak{g} = \mathfrak{k} \oplus \mathfrak{p}$ where

$$[\mathfrak{k}, \mathfrak{k}] \subset \mathfrak{k}, \quad [\mathfrak{p}, \mathfrak{p}] \subset \mathfrak{k} \quad \text{and} \quad [\mathfrak{k}, \mathfrak{p}] \subset \mathfrak{p}.$$

If $(\mathfrak{g}, \theta)$ is finite dimensional, the involution θ is called a *Cartan invo-*

lution if the symmetric bilinear form β_θ defined by

$$\beta_\theta(x, y) = -\beta(x, \theta y) \qquad (x, y \in \mathfrak{g})$$

is positive definite.

Example 2.3.4. Let V be a normed vector space and $\mathfrak{gl}(V)$ the normed algebra of continuous linear self-maps on V. Then $\mathfrak{gl}(V)$ is a Lie algebra in the usual Lie brackets

$$[X, Y] = XY - YX \qquad (X, Y \in \mathfrak{gl}(V)).$$

If $n = \dim V < \infty$, we often denote $\mathfrak{gl}(V)$ as $\mathfrak{gl}(n, \mathbb{F})$ where $\mathbb{F} = \mathbb{R}$ or $\mathbb{C}$.

If V is a Hilbert space, then the subspace $\mathfrak{gl}_{hs}(V)$ of $\mathfrak{gl}(V)$, consisting of all Hilbert-Schmidt operators, is an ideal and is equipped with a natural complete inner product

$$\langle X, Y \rangle_2 = \mathrm{Trace}\,(XY^*) \qquad (X, Y \in \mathfrak{gl}_{hs}(V)).$$

Of course, $\mathfrak{gl}_{hs}(V) = \mathfrak{gl}(V)$ if $\dim V < \infty$.

We can define an involution $\theta : \mathfrak{gl}(V) \longrightarrow \mathfrak{gl}(V)$ by

$$\theta(X) = -X^* \qquad (X \in \mathfrak{gl}(V))$$

where $X^* : V \longrightarrow V$ denotes the adjoint operator of X. If $\dim V < \infty$, then θ is a Cartan involution since

$$-\beta(X, \theta X) = -\mathrm{Trace}\,(\mathrm{ad}(X)\mathrm{ad}(\theta X))$$

$$= \mathrm{Trace}\,(\mathrm{ad}(X)\mathrm{ad}(X^*)) = \mathrm{Trace}\,(\mathrm{ad}(X)\mathrm{ad}(X)^*) \geq 0$$

where $\mathrm{ad}\,(X)^* : \mathfrak{gl}(V) \longrightarrow \mathfrak{gl}(V)$ is the adjoint operator of $\mathrm{ad}\,(X)$ with respect to the inner product $\langle \cdot, \cdot \rangle_2$.

In fact, the Killing form of $\mathfrak{gl}(V)$ can be computed explicitly, it is given by

$$\beta(X, Y) = 2n\,\mathrm{Trace}\,(XY) - 2\mathrm{Trace}\,(X)\mathrm{Trace}\,(Y) \qquad (X, Y \in \mathfrak{gl}(V)).$$

Definition 2.3.5. A graded Lie algebra $\mathfrak{g} = \mathfrak{g}_{-1} \oplus \mathfrak{g}_0 \oplus \mathfrak{g}_1$ is called a *Tits-Kantor-Koecher Lie algebra* or *TKK Lie algebra* if $\mathfrak{g}$ admits an involution θ, which is *negatively graded*, that is

$$\theta(\mathfrak{g}_\alpha) = \mathfrak{g}_{-\alpha}.$$

We call θ the TKK-involution of $\mathfrak{g}$ and note that the map $x \in \overline{\mathfrak{g}_{-1}} \mapsto \theta(x) \in \mathfrak{g}_1$ is a linear isomorphism. We call $\mathfrak{g}$ *canonical* if $[\mathfrak{g}_{-1}, \mathfrak{g}_1] = \mathfrak{g}_0$.

We define the *canonical* part of $\mathfrak{g}$ to be the Lie subalgebra

$$\mathfrak{g}^c = \mathfrak{g}_{-1} \oplus [\mathfrak{g}_{-1}, \mathfrak{g}_1] \oplus \mathfrak{g}_1$$

which is also a TKK Lie algebra with the restriction of θ as the TKK-involution.

There is a one-one correspondence between Jordan triple systems and TKK Lie algebras. To show this, we prove a lemma first.

Lemma 2.3.6. *Let V be a non-degenerate Jordan triple system and let $\sum_j a_j \,\square\, b_j = \sum_k u_k \,\square\, v_k$. Then we have $\sum_j b_j \,\square\, a_j = \sum_k v_k \,\square\, u_k$.*

Proof. We have

$$\left[\sum_j a_j \,\square\, b_j \,,\, x \,\square\, y \right] = \left(\sum_j (a_j \,\square\, b_j) x \right) \square\, y - x \,\square\, \left(\sum_j b_j \,\square\, a_j \right) y$$

$$= \left[\sum_k u_k \,\square\, v_k \,,\, x \,\square\, y \right] = \left(\sum_k (u_k \,\square\, v_k) x \right) \square\, y - x \,\square\, \left(\sum_k v_k \,\square\, u_k \right) y$$

which gives $x \,\square\, \left(\sum_j b_j \,\square\, a_j \right) y = x \,\square\, \left(\sum_k v_k \,\square\, u_k \right) y$ for all $x, y \in V$. By Lemma 2.2.21, we conclude $\sum_j b_j \,\square\, a_j = \sum_k v_k \,\square\, u_k$. $\square$

We now show the Tits-Kantor-Koecher construction of a Lie algebra from a non-degenerate Jordan triple system.

Theorem 2.3.7. *Let V be a non-degenerate Jordan triple system. Then there is a canonical Tits-Kantor-Koecher Lie algebra $\mathfrak{L}(V)$ with grading*

$$\mathfrak{L}(V) = \mathfrak{L}(V)_{-1} \oplus \mathfrak{L}(V)_0 \oplus \mathfrak{L}(V)_1$$

and an involution θ such that $\mathfrak{L}(V)_{-1} = V = \mathfrak{L}(V)_1$ and

$$\{x, y, z\} = [[x, \theta y], z]$$

for $x, y, z \in \mathfrak{L}(V)_{-1}$.

Proof. Form the algebraic direct sum

$$\mathfrak{L}(V) = V_{-1} \oplus V_0 \oplus V_1$$

where $V_{-1} = V$ and $V_1 = \overline{V}$, which is the conjugate of V and is just V if it is real, and V_0 is the linear span of $V \,\square\, V$ in the space $L(V)$ of linear self-maps on V. The Jordan triple identity (2.26) implies that V_0 is a Lie algebra in the bracket product

$$[h, k] = hk - kh.$$

By Lemma 2.3.6, the mapping

$$x \,\square\, y \in V \,\square\, V \mapsto y \,\square\, x \in V \,\square\, V$$

is well-defined and extends to a conjugate linear map $^\natural : V_0 \longrightarrow V_0$ satisfying

$$[x \,\square\, y, u \,\square\, v]^\natural = -[y \,\square\, x, v \,\square\, u].$$

This enables us to define an involutive conjugate linear map $\theta : \mathfrak{L}(V) \longrightarrow \mathfrak{L}(V)$ by

$$\theta(x \oplus h \oplus y) = y \oplus -h^\natural \oplus x \qquad (x \oplus h \oplus y \in V_{-1} \oplus V_0 \oplus V_1)$$

where we also write (x, h, y) for $x \oplus h \oplus y$.

Evidently, θ^2 is the identity map and also, θ is conjugate linear since

$$\theta(\alpha(x, a \square b, y)) = \theta(\alpha x, \alpha(a \square b), \alpha \cdot y)$$
$$= (\overline{\alpha} y, -\overline{\alpha}(b \square a), \overline{\alpha} \cdot x)$$
$$= \overline{\alpha}(y, -(b \square a), x).$$

By identifying V_α naturally as subspaces of $\mathfrak{L}(V)$, we see immediately that $\theta(V_\alpha) = V_{-\alpha}$ for $\alpha = 0, \pm 1$.

Equip $\mathfrak{L}(V)$ with the multiplication

$$[x \oplus h \oplus y, u \oplus k \oplus v]$$
$$= (h(u) - k(x), [h, k] + x \square v - u \square y, k^\natural(y) - h^\natural(v)). \quad (2.44)$$

With this multiplication, one can show that $\mathfrak{L}(V)$ becomes a Lie algebra. The proof of this has been given in [37, Theorem 1.3.8]. We suppress the details. It is routine to verify that θ is an automorphism.

Given $x \in V_{-1}$ and $y \in V_1$, we have $[x, y] = [(x, 0, 0), (0, 0, y)] = (0, x \square y, 0)$ and hence $[V_{-1}, V_1] = V_0$, that is, $\mathfrak{L}(V)$ is canonical. We also have $[V_{-1}, V_{-1}] = [V_1, V_1] = 0$. $\square$

To facilitate the computation of the Lie product (2.44) in the Tits-Kantor-Koecher construction, one can make use of matrix notation (cf. [185]). The TKK Lie algebra

$$\mathfrak{L}(V) = V_{-1} \oplus V_0 \oplus V_1$$

can be written in the following matrix form:

$$\mathfrak{L}(V) = \left\{ \begin{pmatrix} \sum_j a_j \square b_j & x \\ y & -\sum_j b_j \square a_j \end{pmatrix} : a_j, b_j, x, y \in V \right\}$$

with the Lie multiplication

$$\left[\begin{pmatrix} 0 & x \\ y & 0 \end{pmatrix}, \begin{pmatrix} 0 & u \\ v & 0 \end{pmatrix}\right] := \begin{pmatrix} x \square v - u \square y & 0 \\ 0 & -v \square x + y \square u \end{pmatrix} \quad (2.45)$$

$$\left[\begin{pmatrix} 0 & x \\ y & 0 \end{pmatrix}, \begin{pmatrix} a \square b & 0 \\ 0 & -b \square a \end{pmatrix}\right] := \begin{pmatrix} 0 & -a \square b(x) \\ b \square a(y) & 0 \end{pmatrix}. \quad (2.46)$$

The construction in Theorem 2.3.7 translates the non-degeneracy of a Jordan triple V into the following property of its TKK Lie algebra $\mathfrak{L}(V)$:

$$[[a, \theta y], a] = 0 \quad \text{for all } a, y \in \mathfrak{L}(V)_{-1} \Longrightarrow a = 0$$

which is equivalent to the condition

$$(\operatorname{ad} a)^2 = 0 \Longrightarrow a = 0 \qquad (a \in \mathfrak{L}(V)_{-1}) \qquad (2.47)$$

since $(\operatorname{ad} a)^2 (x \oplus h \oplus y) = -Q_a(y)$ for $a \in \mathfrak{L}(V)_{-1}$ and $x \oplus h \oplus y \in \mathfrak{L}(V)$.

Definition 2.3.8. A TKK Lie algebra $\mathfrak{g} = \mathfrak{g}_{-1} \oplus \mathfrak{g}_0 \oplus \mathfrak{g}_1$ is called *non-degenerate* if $(\operatorname{ad} a)^2 = 0 \Longrightarrow a = 0$ for $a \in \mathfrak{g}_{-1}$. Two TKK Lie algebras $(\mathfrak{g}, \theta)$ and $(\mathfrak{g}', \theta')$ are said to be *isomorphic* if there is a graded isomorphism $\psi : \mathfrak{g} \longrightarrow \mathfrak{g}'$ which commutes with involutions:

$$\psi\theta = \theta'\psi.$$

Given a TKK Lie algebra $\mathfrak{g} = \mathfrak{g}_{-1} \oplus \mathfrak{g}_0 \oplus \mathfrak{g}_1$ with TKK involution θ, there is a natural way to construct a Jordan triple system V such that $\mathfrak{L}(V) = \mathfrak{g}_{-1} \oplus [\mathfrak{g}_{-1}, \mathfrak{g}_1] \oplus \mathfrak{g}_1$, which is the canonical part $\mathfrak{g}^c$ of $\mathfrak{g}$. Indeed, it suffices to take $V = \mathfrak{g}_{-1}$ and define a triple product on V by

$$\{a, b, c\} := [[a, \theta(b)], c] \qquad (a, b, c \in V).$$

Then the Jacobi identity in $\mathfrak{g}$ and the condition $[\mathfrak{g}_{-1}, \mathfrak{g}_1] = 0$ implies that V, together with the above triple product, is a Jordan triple system. Further, $\mathfrak{g}$ is non-degenerate if and only if $\mathfrak{L}(V)$ is so.

Hence the preceding construction $V \mapsto \mathfrak{L}(V)$ establishes the correspondence between non-degenerate Jordan triple systems and TKK-Lie algebras. This correspondence is one-to-one in the sense that two Jordan triple systems V and V' are triple isomorphic if and only if the corresponding TKK-Lie algebras $(\mathfrak{L}(V), \theta)$ and $(\mathfrak{L}(V'), \theta')$ are isomorphic, that is, there is a graded Lie isomorphism

$$\widetilde{\varphi} : \mathfrak{L}(V) \longrightarrow \mathfrak{L}(V')$$

satisfying

$$\widetilde{\varphi}\theta = \theta'\widetilde{\varphi}.$$

Indeed, given a triple isomorphism $\varphi : V \longrightarrow V'$, one can define the corresponding graded isomorphism $\widetilde{\varphi} : \mathfrak{L}(V) \longrightarrow \mathfrak{L}(V')$ by

$$\widetilde{\varphi}(a \oplus h \oplus b) = \varphi a \oplus \varphi h \varphi^{-1} \oplus \varphi b \qquad (a \oplus h \oplus b \in \mathfrak{L}(V)).$$

Conversely, given a graded isomorphism $\psi : \mathfrak{L}(V) \longrightarrow \mathfrak{L}(V')$ satisfying $\psi\theta = \theta'\psi$, the restriction $\psi|_V : V \longrightarrow V'$ defines a triple isomorphism.

Notes. The construction of Lie algebras from Jordan algebras was discovered independently by Tits [165], Kantor [115, 97] and Koecher [110, 111]. Meyberg introduced the concept of a Jordan triple system in [130] and extended the construction of Koecher to the wider class of Jordan triple systems. The TKK construction in this section is essentially the same as the one given in [37, Theorem 1.3.8]. However, the construction in Theorem 2.3.7 includes the case of complex Jordan triples, which are not considered in [37, Theorem 1.3.8], where the triple product in a *'Jordan triple'* as defined there is linear in the middle variable.

2.4   Jordan and Lie structures in Banach spaces

To apply Jordan and Lie theory to bounded symmetric domains of all dimensions, we need to consider infinite dimensional Jordan triples and Lie algebras which carry the structure of a Banach space.

Unless stated otherwise, all Banach spaces in the sequel are over the complex field $\mathbb{C}$.

Definition 2.4.1. A complex Jordan triple system V is called a *JB*-triple* if it is a complex Banach space in which the Jordan triple product $\{\cdot, \cdot, \cdot\}$ is continuous and for each $a \in V$, the continuous box operator $a \square a : V \longrightarrow V$ satisfies the following conditions:

(i) $a \square a$ is hermitian, that is, $\| \exp it(a \square a) \| = 1$ for all $t \in \mathbb{R}$;

(ii) $a \square a$ has a non-negative spectrum;

(iii) $\| a \square a \| = \| a \|^2$

where the linear exponential operator $\exp it(a \square a) : V \longrightarrow V$ is defined by

$$\exp it(a \square a)(x) = x + \sum_{n=1}^{\infty} \frac{1}{n!}(it)^n (a \square a)^n (x) \qquad (x \in V)$$

and $(a \square a)^n$ is the n-fold product of $a \square a$ in the Banach algebra $L(V)$ of bounded linear operators on V. By [20, p. 46], $a \square a$ is a hermitian element in $L(V)$ if and only if it has *real* numerical range. The spectrum of $a \square a$ is the set

$$\sigma(a \square a) := \{\lambda \in \mathbb{C} : \lambda \mathbf{1} - a \square a \text{ is not invertible in } L(V)\}$$

and condition (ii) states that $\sigma(a \square a) \subset [0, \infty)$.

Condition (iii) in Definition 2.4.1 can be replaced by

$$\| \{a, a, a\} \| = \| a \|^3 \tag{2.48}$$

(cf. [37, Lemma 3.1.3]). It follows that JB*-triples are non-degenerate and anisotropic. The fundamental role of JB*-triples in bounded symmetric domains will unfold in ensuing sections. Here are some examples.

Example 2.4.2. The complex Euclidean space $\mathbb{C}^n$ is a JB*-triple in the Euclidean norm and coordinatewise triple product

$$\{\mathbf{x}, \mathbf{y}, \mathbf{z}\} := (x_1 \overline{y}_1 z_1, \ldots, x_n \overline{y}_n z_n)$$

for $\mathbf{x} = (x_1, \ldots, x_n), \mathbf{y} = (y_1, \ldots, y_n), \mathbf{z} = (z_1, \ldots, z_n) \in \mathbb{C}^n$, where '$-$' denotes the complex conjugate as usual.

Example 2.4.3. Let $\{V_\alpha\}_{\alpha \in \Lambda}$ be a family of normed vector spaces. We define their ℓ_∞-sum $\overset{\ell_\infty}{\underset{\alpha \in \Lambda}{\bigoplus}} V_\alpha$ to be the direct sum

$$\overset{\ell_\infty}{\underset{\alpha \in \Lambda}{\bigoplus}} V_\alpha = \{(v_\alpha) \in \times_\alpha V_\alpha : \sup_\alpha \|v_\alpha\| < \infty\},$$

equipped with the ℓ_∞-norm

$$\|(v_\alpha)\|_\infty := \sup_\alpha \|v_\alpha\|.$$

If $\Lambda = \{1, 2, \ldots\}$ and $V_\alpha = F$ for all $\alpha \in \Lambda_\alpha$, where $F = \mathbb{C}$ or $\mathbb{R}$, then $\overset{\ell_\infty}{\underset{\alpha \in \Lambda}{\bigoplus}} V_\alpha$ is just the Banach space ℓ_∞ of bounded sequences in F.

Let V_α be a JB*-triple for each $\alpha \in \Lambda$. One can verify that the ℓ_∞-sum $\overset{\ell_\infty}{\underset{\alpha}{\bigoplus}} V_\alpha$, with the ℓ_∞-norm, is a JB*-triple with the coordinatewise triple product.

Infinite dimensional JB*-triples include C*-algebras.

Example 2.4.4. A C*-algebra is a norm closed subalgebra $\mathcal{A}$ of the Banach *-algebra $L(H)$ of bounded linear operators on a Hilbert space

H such that $a \in \mathcal{A}$ implies $a^* \in \mathcal{A}$, where a^* is the adjoint of the operator a. We refer to [154] for a comprehensive exposition of C*-algebras.

In a C*-algebra $\mathcal{A}$, an operator $a \in \mathcal{A}$ is hermitian, i.e. $\|\exp ita\| = 1$ for all $t \in \mathbb{R}$, if and only if it is self-adjoint, i.e. $a = a^*$ (cf. [20, p. 47]). Since $\mathcal{A}$ is an associative algebra with involution, it is a complex Jordan algebra with involution in the special Jordan product

$$a \circ b = \frac{1}{2}(ab + ba) \qquad (a \in \mathcal{A})$$

and also a complex Jordan triple system in the canonical Jordan triple product

$$\{a, b, c\} = \frac{1}{2}(ab^*c + cb^*a). \tag{2.49}$$

Given $a \in \mathcal{A}$, we can write the box operator $a \square a : \mathcal{A} \longrightarrow \mathcal{A}$ as the sum

$$2a \square a = L_{aa^*} + R_{a^*a}$$

of the left multiplication $L_{aa^*} : x \in \mathcal{A} \mapsto aa^*x \in \mathcal{A}$ and the right multiplication $R_{a^*a} : x \in \mathcal{A} \mapsto xa^*a \in \mathcal{A}$. We have

$$\|\exp itL_{aa^*}\| = \sup\{\|\exp it(aa^*x)\| : \|x\| \leq 1\} \leq \|\exp it(aa^*)\| = 1$$

for all $t \in \mathbb{R}$. Hence L_{aa^*} is a hermitian operator on $\mathcal{A}$ and likewise, R_{a^*a} is hermitian. It follows that $2a \square a$ is hermitian since its numerical range is contained in the sum of those of L_{aa^*} and R_{a^*a} (cf. [20, p. 15]). Further, the spectrum $\sigma(L_{a^*a})$ of $L_{a^*a} : \mathcal{A} \longrightarrow \mathcal{A}$ coincides with the spectrum of a^*a in the C*-algebra $\mathcal{A}$, which is contained in $[0, \infty)$. Likewise $\sigma(R_{aa^*}) \subset [0, \infty)$. By [20, p. 53], both L_{aa^*} and R_{a^*a} have positive numerical range, which implies that $a \square a$ has positive numerical range. This implies that $\sigma(a \square a) \subset [0, \infty)$ since the spectrum of an operator is contained in its numerical range (cf. [20, p. 19]). It follows that, in the canonical triple product (2.49), a C*-algebra $\mathcal{A}$ is JB*-triple since the identity (2.48) is just the C*-identity

$$\|\{a, a, a\}\|^2 = \|aa^*a\|^2 = \|(aa^*a)(a^*aa^*)\| = \|aa^*\|^3 = \|a\|^6.$$

Tripotents in a C*-algebra $\mathcal{A}$ are precisely the partial isometries, which are elements $e \in \mathcal{A}$ such that $e = ee^*e$, equivalently, ee^* (and hence e^*e) is a projection. The Peirce k-spaces of the tripotent e are given by

$$\mathcal{A}_0(e) = (1 - ee^*)\mathcal{A}(1 - e^*e),$$
$$\mathcal{A}_1(e) = (1 - ee^*)\mathcal{A}e^*e + ee^*\mathcal{A}(1 - e^*e),$$
$$\mathcal{A}_2(e) = ee^*\mathcal{A}e^*e.$$

Example 2.4.5. We will denote by $C_0(\Omega)$ the Banach algebra of complex continuous functions vanishing at infinity on a locally compact Hausdorff space Ω, equipped with the involution defined by pointwise complex conjugation:

$$f^*(\omega) := \overline{f(\omega)} \qquad (f \in C_0(\Omega), \omega \in \Omega).$$

It is well-known that $C_0(\Omega)$ is isometrically *-isomorphic to a commutative C*-algebra in some $L(H)$. Hence $C_0(\Omega)$ is a complex Jordan triple system in the canonical Jordan triple product

$$\{f, g, h\}(\omega) = f(\omega)\overline{g(\omega)}h(\omega) \qquad (f, g, h \in C_0(\Omega), \omega \in \Omega).$$

The celebrated Gelfand-Naimark theorem asserts that, via an isometric *-isomorphism, commutative C*-algebras are of the form $C_0(\Omega)$ and also, a Banach algebra $\mathcal{A}$ with an involution * satisfying $\|x^*x\| = \|x\|^2$ for all $x \in \mathcal{A}$ is (isometrically *-isomorphic to) a C*-algebra.

Given a normed vector space E, we will denote its dual space by E^* as usual.

Definition 2.4.6. A JB*-triple V is called a *JBW*-triple* if it is a *dual Banach space*, that is, $V = E^*$ for some complex Banach space E, in which case E is called a *predual* of V. In fact, the predual E is *unique* in the sense that, if $V = E^* = F^*$, then $E = F$ when they are canonically embedded into V^* (cf. [87, (3.21)]). Henceforth, we will denote by V_* unambiguously *the* predual of a JBW*-triple V and refer to the weak topology $w(V, V_*)$ as *the weak* topology* of V.

The triple product of a JBW*-triple V is separately weak* continuous (cf. [37, Theorem 3.3.9]) and a Jordan triple isomorphism between two JBW*-triples is necessarily weak* continuous (cf. [87, (3.22)]). The second dual V^{**} of a JB*-triple V is a JBW*-triple and V identifies as a subtriple of V^{**} via the canonical embedding $V \hookrightarrow V^{**}$ [37, Corollary 3.3.5].

Example 2.4.7. A C*-algebra $\mathcal{A}$ is called a *von Neumann algebra* if it has a predual $\mathcal{A}_*$, in which case $\mathcal{A}_*$ is unique [154, Corollary 1.13.3]. In the canonical triple product (2.49), von Neumann algebras are JBW*-triples. In particular, $L(H)$ is a JBW*-triple. Its predual is the Banach space $T(H)$ of trace-class operators on H. A von Neumann algebra $\mathcal{A}$ always contains an identity and an abundance of *projections*, which are elements $p \in \mathcal{A}$ satisfying $p = p^* = p^2$. In fact, $\mathcal{A}$ is the norm closed linear span of its projections.

Lemma 2.4.8. *A closed subtriple of a JB*-triple is itself a JB*-triple in the inherited triple product.*

Proof. Let W be a closed subtriple of a JB*-triple V and let $a \in W$. The restriction $a \,\square\, a|_W$ to W of the box operator $a \,\square\, a : V \longrightarrow V$ is the box operator $x \in W \mapsto \{a, a, x\} \in W$ on W in the inherited triple product and is hermitian since

$$\| \exp it(a \,\square\, a|_W) \| \leq \| \exp it(a \,\square\, a) \| = 1$$

for all $t \in \mathbb{R}$. Let

$$\mathcal{A} := \{T \in L(V) : T(W) \subset W\}.$$

Then $\mathcal{A}$ is a closed subalgebra of the Banach algebra $L(V)$ and $a \square a \in \mathcal{A}$. As an element in $L(V)$, the spectrum $\sigma(a \square a)$ is contained in $[0, \infty)$. In particular, the set $\mathbb{C}\backslash\sigma(a \square a)$ is connected and by [153, p. 239], the spectrum $\sigma_{\mathcal{A}}(a \square a)$ of $a \square a$ in the algebra $\mathcal{A}$ coincides with $\sigma(a \square a)$. Since the map

$$T \in \mathcal{A} \mapsto T|_W \in L(W)$$

is an algebra homomorphism, we have

$$\sigma(a \square a|_W) \subset \sigma_{\mathcal{A}}(a \square a) \subset [0, \infty).$$

Finally, the identity $\|\{a, a, a\}\| = \|a\|^3$ clearly holds in the subtriple W. This concludes the proof. $\square$

Example 2.4.9. The Jordan triple $L(H, K)$ of bounded linear operator between Hilbert spaces H and K is a JBW*-triple. Indeed, $L(H, K)$ is the dual space of the Banach space of trace-class operators between H and K. Moreover, $L(H, K)$ can be identified as a closed subtriple of the JBW*-triple $L(H \oplus K)$ of bounded linear operators on the direct sum $H \oplus K$ of H and K, via the embedding

$$a \in L(H, K) \mapsto \begin{pmatrix} 0 & 0 \\ a & 0 \end{pmatrix} \in L(H \oplus K)$$

which is an isometric triple homomorphism.

Given a Hilbert space $(H, \langle \cdot, \cdot \rangle)$, the linear isometry $x \in L(\mathbb{C}, H) \mapsto x(1) \in H$ identifies the two spaces and induces a JB*-triple structure on H. The adjoint x^* of $x \in L(\mathbb{C}, H)$ is given by $x^*(h) = \langle h, x(1) \rangle$ for $h \in H$ and hence the triple product in H can be expressed as

$$\{x, y, z\} = \frac{1}{2}(\langle x, y \rangle z + \langle z, y \rangle x).$$

For any $a, b \in H$, the adjoint of the box operator $a \,\square\, b : H \longrightarrow H$ is $b \,\square\, a$, by the *associativity* of the inner product $\langle \cdot, \cdot \rangle$, that is,

$$\langle \{a, b, x\}, y \rangle = \langle x, \{b, a, y\} \rangle \qquad (x, y \in H). \tag{2.50}$$

Besides C*-algebras, there is another class of Jordan algebras which carry the structure of a JB*-triple. We first introduce the *real forms* of these algebras. A real Jordan algebra $\mathcal{B}$ is called a *JB-algebra* if it is also a Banach space and the norm satisfies

$$\|ab\| \leq \|a\|\|b\|, \quad \|a^2\| = \|a\|^2, \quad \|a^2\| \leq \|a^2 + b^2\|$$

for all $a, b \in \mathcal{B}$. Idempotents in JB-algebras are also called *projections*.

A JB-algebra $\mathcal{A}$ is called a *JBW-algebra* if it is a dual Banach space in which case the predual of $\mathcal{A}$ is unique [77, Theorem 4.4.16], the weak* topology on $\mathcal{A}$ is unambiguous and $\mathcal{A}$ must have an identity [77, Lemma 4.1.7]. The second dual $\mathcal{A}^{**}$ of a JB-algebra $\mathcal{A}$ is a JBW-algebra in which A identifies as a closed Jordan subalgebra [77, Theorem 4.4.3].

JB-algebras are formally real Jordan algebras [77, Corollary 3.3.8]. In particular, finite dimensional JB-algebras are Hilbert spaces in the trace form

$$\langle a, b \rangle = \text{Trace}(a \,\square\, b)$$

by Theorem 2.1.17.

A real Jordan algebra $\mathcal{H}$ is called a *JH-algebra* if it is also a real Hilbert space in which the inner product $\langle \cdot, \cdot \rangle$ is *associative*, that is,

$$\langle ab, c \rangle = \langle b, ac \rangle \qquad (a, b, c \in \mathcal{H}).$$

Real spin factors $H \oplus \mathbb{R}$ are JH-algebras. Finite dimensional JH-algebras are exactly the *Euclidean Jordan algebras* introduced in [60].

Lemma 2.4.10. *Let $\mathcal{A}$ be a finite dimensional real Jordan algebra. The following conditions are equivalent.*

(i) $\mathcal{A}$ *is formally real.*

(ii) $\mathcal{A}$ *is a JH-algebra with an identity* **1**.

(iii) $\mathcal{A}$ *is a JB-algebra.*

Proof. (i) $\Rightarrow$ (ii). By Lemma 2.2.23, a finite dimensional formally real Jordan algebra, equipped with the trace form as an inner product, is a JH-algebra.

(ii) $\Rightarrow$ (i). Given $a^2 + b^2 = 0$ in a JH-algebra with identity **1**, we have

$$\langle a, a \rangle + \langle b, b \rangle = \langle a^2, \mathbf{1} \rangle + \langle b^2, \mathbf{1} \rangle = \langle a^2 + b^2, \mathbf{1} \rangle = 0$$

which gives $a = b = 0$.

(i) $\Rightarrow$ (iii). In a finite dimensional formally real Jordan algebra $\mathcal{A}$, the set $\{a^2 : a \in \mathcal{A}\}$ forms a proper cone, which induces a partial ordering $\leq$ in $\mathcal{A}$ and the following norm

$$\|a\| := \inf\{\lambda > 0 : -\lambda\mathbf{1} \leq a \leq \lambda\mathbf{1}\}$$

makes $\mathcal{A}$ into a JB-algebra [77, Corollary 3.1.7].   □

The celebrated result of Koecher and Vinberg, stated at the end of Section 2.1, establishes the one-one correspondence between *finite dimensional* formally real Jordan algebras and linearly homogeneous self-dual cones. In view of Lemma 2.4.10, an appropriate infinite dimensional generalisation of formally real Jordan algebras is the notion of a *unital JH-algebra*. Indeed, it has been shown recently in [40] that unital JH-algebras are in one-one correspondence with linearly homogeneous self-dual cones of any dimension (see Theorem 2.4.14 below), generalising the Koecher-Vinberg result to infinite dimension.

Definition 2.4.11. Let V be a real Hilbert space with inner product $\langle \cdot, \cdot \rangle$. An open cone $\Omega \subset V$ is called *self-dual homogeneous* if it satisfies the following conditions:

(i) (self-duality) $\Omega = \Omega^*$, where

$$\Omega^* := \{v \in V : \langle v, x \rangle > 0 \,\, \forall x \in \overline{\Omega} \backslash \{0\}\}$$

is called the *dual cone* of Ω;

(ii) (homogeneity) Ω is *linearly homogeneous*, that is, given $x, y \in \Omega$, there is a continuous linear isomorphism $h : V \longrightarrow V$ such that $h(\Omega) = \Omega$ and $h(x) = y$.

Evidently, the preceding concept of homogeneity can be defined for cones in real Banach spaces, which will be discussed in Section 3.6.

Remark 2.4.12. A self-dual homogeneous cone is also called a *symmetric cone* in literature (e.g. [60]). We use the former terminology in this book lest the latter be confused with the notion of a symmetric domain in complex Banach spaces.

Remark 2.4.13. To avoid confusion with the notion of homogeneity introduced in Definition 1.3.21 for bounded domains, we shall henceforth adopt the terminology *'linearly homogeneous self-dual cone'* instead of *'self-dual homogeneous cone'*.

We refer to [40] for a proof of the following extension of the aforementioned result of Koecher and Vinberg.

Theorem 2.4.14. *Let Ω be an open cone in a real Hilbert space. Then Ω is a linearly homogeneous self-dual cone if, and only if, it is of the form*

$$\Omega = \text{int} \, \{a^2 : a \in \mathcal{H}\} \tag{2.51}$$

for a unique unital JH-algebra $\mathcal{H}$.

Remark. The finite dimensional result of Koecher and Vinberg has also been shown in [60, 155]. The derivation of the Jordan identity in the proof of Theorem 3.1 in [40, p. 369] can be replaced by verbatim arguments in [60, p. 50], using the associativity of the inner product.

Example 2.4.15. Let Ω be the interior of the closed cone

$$\{x^2 : x \in H \oplus \mathbb{R}\} = \{a \oplus \alpha : \alpha \geq \|a\|\}$$

in a real spin factor $H \oplus \mathbb{R}$. Then it is self-dual and linearly homogeneous. For $H = \mathbb{R}^n$ $(n \geq 2)$, the cone Ω is known as the *Lorentz cone*, it can be written as

$$\Omega = \{(x_1, \ldots, x_{n+1}) \in \mathbb{R}^{n+1} : x_{n+1} > 0 \text{ and } x_{n+1}^2 > x_1^2 + \cdots + x_n^2\}.$$

It is also known as the *second-order cone* in optimization theory. Interestingly, for $1 < p < \infty$ and $p \neq 2$, the so-called *p-order cone*

$$C_p := \{(x_1, \ldots, x_{n+1}) \in \mathbb{R}^{n+1} : x_{n+1} > 0 \text{ and } x_{n+1}^p > x_1^p + \cdots + x_n^p\}$$

is not linearly homogeneous in $\mathbb{R}^n \oplus \mathbb{R}$, nor self-dual [91].

Let Ω be a linearly homogeneous self-dual cone in a real Hilbert space, identified with int $\{a^2 : a \in \mathcal{H}\}$ in a unital JH-algebra $\mathcal{H}$ with inner product $\langle \cdot, \cdot \rangle$. Then it follows from Corollary 2.4.20 and Lemma 3.6.12, shown later, that each element in Ω is invertible and, as shown in [37, Theorem 2.3.19], one can define a Riemannian metric on Ω by

$$g_\omega(u, v) = \langle \{\omega^{-1}, u, \omega^{-1}\}, v \rangle \qquad (\omega \in \Omega, u, v \in \mathcal{H})$$

which turns Ω into a Riemannian symmetric space. For a finite dimensional *irreducible* linearly homogeneous self-dual cone Ω, where irreducibility means that Ω is not a direct product of two non-trivial linearly homogeneous self-dual cones, this metric is proportional to the one defined by the so-called *characteristic function* of the cone Ω, as follows (see, for example, [60]).

Let λ be the Euclidean measure on a finite dimensional Hilbert space V with inner product $\langle \cdot, \cdot \rangle$, and Ω an open cone in V. The *characteristic function* $\varphi : \Omega \longrightarrow [0, \infty)$ is defined by

$$\varphi(x) = \int_{\Omega^*} \exp - \langle x, y \rangle d\lambda(y) \qquad (x \in \Omega).$$

One can define a Riemannian metric on Ω by

$$g_x(u, v) = D_u D_v \log \varphi(x) \qquad (x \in \Omega, u, v \in V) \qquad (2.52)$$

where D_u and D_v are directional derivatives. The metric in (2.52) is called the *canonical Riemannian metric* on Ω.

Positive cones will feature in the discussion of Siegel domains in the next chapter, where a classification of unital JH-algebras is given.

We now introduce the complexification of JB-algebras, which are complex Jordan algebras with involution and form a class of JB*-triples in the canonical Jordan triple product.

Definition 2.4.16. A complex Jordan algebra $\mathcal{B}$ with an involution $*$ is called a *JB*-algebra* or *Jordan C*-algebra* if it is also a Banach space in which the norm satisfies

$$\|ab\| \leq \|a\|\|b\|, \qquad \|a^*\| = \|a\|, \qquad \|\{a, a, a\}\| = \|a\|^3$$

for all $a, b \in \mathcal{B}$, where $\{\cdot, \cdot, \cdot\}$ denotes the canonical Jordan triple product defined in (2.13). A JB*-algebra is called a *JBW*-algebra* if it has a predual, which is necessarily unique.

Evidently, a C*-algebra is a JB*-algebra in the special Jordan product. Given a JB*-algebra $\mathcal{B}$, its *self-adjoint part*

$$\mathcal{B}_{sa} = \{a \in \mathcal{B} : a^* = a\}$$

forms a JB-algebra in the inherent Jordan product [77, 3.8.2] and we have

$$\mathcal{B} = \mathcal{B}_{sa} + i\mathcal{B}_{sa}.$$

Conversely, it has been shown in [179] that a unital JB-algebra A can be complexified to a JB*-algebra $\mathcal{A} = A + iA$ so that A identifies with the self-adjoint part $\mathcal{A}_{sa}$ of $\mathcal{A}$. The norm on $\mathcal{A}$ is given by the Minkowski functional of the convex hull of the set

$$\{\exp ia : a \in A\} \subset \mathcal{A}$$

and it coincides with the original norm on A.

Example 2.4.17. The exceptional real Jordan algebra $H_3(\mathbb{O})$ in Example 2.1.16 is formally real and by Lemma 2.4.10, it is a unital JB-algebra in the order-unit norm

$$\|a\| = \inf\{\lambda > 0 : -\lambda\mathbf{1} \le a \le \lambda\mathbf{1}\}.$$

Its complexification is the exceptional Jordan algebra $H_3(\mathcal{O}) = H_3(\mathbb{O}) + iH_3(\mathbb{O})$ which is a JB*-algebra.

Example 2.4.18. For any JB*-triple V which admits a non-zero tripotent u, the u-homotope of the Peirce 2-space $P_2(u)V$ is a JB*-algebra in the inherited norm, with involution defined by

$$a^* := \{u, a, u\}$$

and u becomes the identity of the algebra.

Unital JB*-algebras are in one-one correspondence with the class of bounded symmetric domains realisable as tube domains. We will discuss the detail in Section 3.6.

Lemma 2.4.19. *A JB*-algebra $\mathcal{B}$, with the canonical Jordan triple product*

$$\{a, b, c\} = (ab^*)c + a(b^*c) - b^*(ac),$$

is a JB*-triple. In particular, a JBW*-algebra is a JBW*-triple.

Proof. For each $a \in \mathcal{B}$, it has been shown in [181, Theorem 6] that $a = a^*$ if, and only if, the left multiplication $L_a : \mathcal{B} \longrightarrow \mathcal{B}$ is hermitian.

Given $x \in \mathcal{B}$, we can write $x = a + ib$ with $a = a^*$ and $b = b^*$. Since

$$x \,\square\, x = a \,\square\, a + b \,\square\, b + 2i(L_b L_a - L_a L_b)$$

where the left multiplications L_a and L_b are hermitian, $x \,\square\, x$ is hermitian (cf. [20, p. 47]).

The closed subalgebra $\mathcal{A}$ generated by a in $\mathcal{B}$ is associative and is therefore an abelian C*-algebra since each $z \in \mathcal{A}$ satisfies

$$\|z\|^3 = \|(zz^*)z\| \leq \|zz^*\|\|z\| \leq \|z\|^2\|z^*\| = \|z\|^3.$$

Identifying $\mathcal{A}$ as the algebra $C_0(\Omega)$ of complex continuous functions vanishing at infinity on a locally compact Hausdorff space Ω (cf. Example 2.4.5), it is readily seen that the operator

$$a \,\square\, a|_{\mathcal{A}} : \mathcal{A} \longrightarrow \mathcal{A}$$

has non-negative spectrum since it is just the left multiplication by $|a|^2$. The closed subtriple $V(a)$ generated by a in $\mathcal{A}$ is a JB*-triple by Lemma 2.4.8 and hence $\sigma(a \,\square\, a|_{V(a)}) \subset [0, \infty)$.

The JB*-algebra $\mathcal{B}$ is a complex Jordan triple system with continuous triple product. Hence by Lemma 3.2.10 (cf. Section 3.2), we have

$$\sigma(a \,\square\, a) \subset \frac{1}{2}(S + S) \subset [0, \infty)$$

where $S = \sigma(a \,\square\, a|_{V(a)}) \cup \{0\}$.

Finally, observe that

$$x^* \square\, x^* = *\circ (x\square\, x)\circ *.$$

Hence $\sigma(x^*\square\, x^*) = \sigma(x\square\, x)$. Now

$$x\square\, x + x^*\square\, x^* = 2a\square\, a + 2b\square\, b$$

implies $\sigma(x\square\, x) \subset [0,\infty)$. $\qquad\qquad\qquad\square$

By the preceding lemma, the exceptional Jordan algebra $H_3(\mathcal{O})$ is a JBW*-triple. The finite dimensional Jordan triple $M_{1,2}(\mathcal{O})$ identifies as a closed subtriple of $H_3(\mathcal{O})$ and is therefore, by Lemma 2.4.8, a JBW*-triple.

Given a JB-algebra A, the set $\{x^2 : x \in A\}$ forms a proper closed cone [77, Lemma 3.3.7] which induces a partial ordering $\leq$ in A. If A has an identity e, then the norm of A satisfies

$$\|a\| = \inf\{\lambda > 0 : -\lambda e \leq a \leq \lambda e\} \qquad (a \in A)$$

(cf. [77, Proposition 3.3.10]). The identity e lies in the interior

$$\mathrm{int}\,\{x^2 : x \in A\}$$

of the cone since $\|a - e\| < 1/2$ implies $-\frac{1}{2}e \leq a - e \leq \frac{1}{2}e$ and in particular, $0 \leq \frac{1}{2}e \leq a$.

Corollary 2.4.20. *Let A be a unital JB-algebra and*

$$C = \mathrm{int}\,\{x^2 : x \in A\}.$$

Then each element $z \in A + iC$ is invertible in the JB*-algebra $\mathcal{A} = A + iA$.

Proof. Let $z = v + ix^2 \in A + iC$. By a remark following Theorem 2.1.17 in Section 2.1, it suffices to show that the left multiplication $L_z : A \longrightarrow A$ is invertible in the unital Banach algebra $L(A)$ of bounded linear operators on A, that is, the spectrum $\sigma(L_z)$ does not contain 0. Let $\mathbf{1}$ be the identity of $L(A)$ and let $B \subset L(A)$ be a maximal abelian subalgebra containing L_z. Then the spectrum $\sigma_B(L_z)$ of L_z in B coincides with $\sigma(L_z)$. Hence for each $\lambda \in \sigma(L_z)$, there is a unital algebra homomorphism $\varphi : B \longrightarrow \mathbb{C}$ such that $\varphi(L_z) = \lambda$. We have $\|\varphi\| = 1$ and by the Hahn-Banach theorem, φ extends to a continuous linear functional $\widetilde{\varphi} \in L(A)^*$ with $\|\widetilde{\varphi}\| = 1$.

Observe that for each element $c \in C$, the left multiplication $L_c : A \longrightarrow A$ is invertible. Indeed, $c - C$ is an open neighbourhood of $0 \in A$ and hence the identity $e \in A$ is in $\alpha(c - C)$ for some $\alpha > 0$, which gives $c = \alpha^{-1} + a^2$ for some $a \in A$ and $L_c = \alpha^{-1}\mathbf{1} + L_{a^2}$. We have $\sigma(L_{a^2}) = \sigma(a \square a) \subset [0, \infty)$ by Lemma 2.4.19 and therefore $\sigma(L_c) \subset \alpha^{-1} + [0, \infty)$.

We have $L_z = L_v + iL_{x^2}$. As noted in the proof of Lemma 2.4.19, L_v is a hermitian operator on A as well as L_{x^2}. Hence the numerical range of L_{x^2} is the convex hull of its spectrum $\sigma(L_{x^2})$ [20, p. 58]. It follows that for each $\psi \in L(A)^*$ with $\psi(\mathbf{1}) = 1 = \|\psi\|$, we have $\psi(L_v) \in \mathbb{R}$ [20, p. 46] and $\psi(L_z) = \psi(L_v) + i\psi(L_{x^2}) \in \mathbb{R} + i(0, \infty) \neq \{0\}$, which gives $0 \notin \sigma(L_z)$. $\qquad\square$

In the proof of Lemma 2.4.19, one can actually say more about the closed subtriple $V(a)$ in A. We refer to [37, Theorem 3.1.12] for a proof of the following fundamental result.

Theorem 2.4.21. *Let V be a JB*-triple and $a \in V\backslash\{0\}$. Then there is a locally compact Hausdorff space $S_a \subset (0, \infty)$, the triple spectrum of a, such that the closed subtriple $V(a)$ generated by a is isometrically triple isomorphic to the JB*-triple $C_0(S_a)$ of continuous functions on*

S_a vanishing at infinity.

We now discuss Lie algebras which carry Banach space structures. A real or complex Lie algebra $\mathfrak{g}$ is called a *normed Lie algebra* if $\mathfrak{g}$ is a normed vector space and the Lie product is continuous:

$$\|[X, Y]\| \leq C\|X\|\|Y\| \qquad (X, Y \in \mathfrak{g})$$

for some $C > 0$. A *Banach Lie algebra* is a normed Lie algebra $\mathfrak{g}$, which is also a Banach space in the given norm.

There are many natural examples of Banach Lie algebras. For instance, a C*-algebra $\mathcal{A}$ is a Banach Lie algebra in the usual Lie brackets

$$[a, b] = ab - ba \qquad (a, b \in \mathcal{A}).$$

In what follows, we will only be concerned with those Banach Lie algebras induced by bounded symmetric domains. They are related to the TKK Lie algebras of JB*-triples.

Let V be a JB*-triple and let

$$\mathfrak{L}(V) = V_{-1} \oplus V_0 \oplus V_1$$

be its TKK Lie algebra, as constructed in Section 2.3. We will equip $\mathfrak{L}(V)$ with the norm

$$\|(x, h, y)\| := \|x\| + \|h\| + \|y\|, \qquad (x, h, y) \in V_{-1} \oplus V_0 \oplus V_1$$

which makes $\mathfrak{L}(V)$ into a normed Lie algebra. We recall that $V_{-1} = V$, $V_1 = \overline{V}$ and

$$V_0 = \left\{ \sum_{j=1}^{k} a_j \,\square\, b_j : a_1, \ldots, a_k, b_1, \ldots, b_k \in V \right\}.$$

If V is finite dimensional, then $\mathfrak{L}(V)$ is a finite dimensional Banach Lie algebra.

We conclude this section by showing that the Lie algebra of a Banach Lie group is a Banach Lie algebra. We show that, in local coordinates, the Lie product can be expressed by continuous bilinear maps and is therefore continuous.

Theorem 2.4.22. *The Lie algebra $\mathfrak{g}$ of a Banach Lie group G is a Banach Lie algebra.*

Proof. We already know that $\mathfrak{g} = T_e G$ is a Lie algebra and also, it identifies with the Banach space V in a local chart $(\mathcal{U}, \varphi, V)$ at e. We show that the Lie product $(X_e, Y_e) \in T_e G \times T_e G \mapsto [X_e, Y_e] \in T_e G = V$ is continuous.

Let $X_e = [\alpha] \in T_e G$ and $Y_e = [\beta] \in T_e G$. By Lemma 1.2.22, We have

$$
\begin{aligned}
[X_e, Y_e] &= ad([\alpha])(Y_e) \\
&= \lim_{t \to 0} \frac{1}{t}\left(Ad(\alpha(t))Y_e - Ad(\alpha(0))Y_e\right) \\
&= \frac{d}{dt}\Big|_{t=0} \left(Ad(\alpha(t))Y_e\right) \\
&= \frac{d}{dt}\Big|_{t=0} \left(d(r_{\alpha(t)^{-1}}\ell_{\alpha(t)})_e Y_e\right) \\
&= \frac{d}{dt}\Big|_{t=0} \left(\frac{d}{ds}\Big|_{s=0} (\varphi(\alpha(t)\beta(s)\alpha(t)^{-1}))\right)
\end{aligned}
$$

where a tangent vector $[\gamma] \in T_e G$ identifies with $(\varphi \circ \gamma)'(0) \in V$.

Considering $\varphi(\alpha(t)\beta(s)\alpha(t)^{-1}) = F(t, s)$ as a function from some open neighbourhood of $(0,0) \in \mathbb{R}^2$ to V, we have the Taylor expansion

$$
\begin{aligned}
F(t,s) &= F(0,0) + F'(0,0)(t,s) + \frac{1}{2!}F''(0,0)((t,s),(t,s)) + \cdots \\
&= D_1 F(0,0)t + D_2 F(0,0)s + \frac{1}{2}\left(D_1 D_1 F(0,0)(t,t) \right. \\
&\quad + 2D_1 D_2 F(0,0)(t,s) + D_2 D_2 F(0,0)(s,s)) + \cdots \\
&= (\varphi \circ \beta)'(0)s + \frac{1}{2}\left(D_1 D_1 F(0,0)(t,t) + 2D_1 D_2 F(0,0)(t,s)\right. \\
&\quad + D_2 D_2 F(0,0)(s,s)) + \cdots.
\end{aligned}
$$

It follows that

$$
\left.\frac{d}{dt}\right|_{t=0}\left(\left.\frac{d}{ds}\right|_{s=0} F(t,s)\right) = D_1 D_2 F(0,0)(1,1) = f((\varphi \circ \alpha)'(0), (\varphi \circ \beta)'(0))
$$

where $f : V \times V \longrightarrow V$ is a continuous bilinear map. It follows that $[X_e, Y_e] = f(X_e, Y_e)$ and the Lie product is continuous. $\qquad\square$

Remark 2.4.23. In contrast to the finite dimensional case, a Banach Lie algebra need not arise as the Lie algebra of a Banach Lie group [173]. However, if the centre

$$
\mathfrak{z}(\mathfrak{g}) = \{X \in \mathfrak{g} : [X, \mathfrak{g}] = 0\}
$$

of a Banach Lie algebra $(\mathfrak{g}, [\cdot, \cdot])$ is trivial, then $\mathfrak{g}$ is the Lie algebra of a connected Banach Lie group. We refer to [149] for a proof.

Example 2.4.24. A Banach algebra $\mathfrak{A}$ with identity $\mathbf{1}$ is a Banach Lie algebra in the commutator product

$$
[a, b] = ab - ba \qquad (a, b \in \mathfrak{A}).
$$

The set G of invertible elements is open in $\mathfrak{A}$ and forms a group in the associative multiplication of $\mathfrak{A}$, and G is a Banach Lie group with Lie algebra $T_1 G = \mathfrak{A}$.

Example 2.4.25. The infinite dimensional analogue of the general linear groups is the group $GL(H)$ of invertible elements in the Banach algebra $L(H)$ of bounded linear operators on a Hilbert space H. The group $GL(H)$ is norm open in $L(H)$ and is a Banach Lie group modelled on $L(H)$ which is a Banach Lie algebra in the commutator product. The exponential map $\exp : L(H) \longrightarrow GL(H)$ is the usual one:

$$\exp A = 1 + A + \frac{A^2}{2!} + \cdots \qquad (A \in L(H))$$

where 1 is the identity operator in $L(H)$. The two-sided ideal $K(H)$ of compact operators in $L(H)$ is a Lie ideal and is the Lie algebra of the Banach Lie group

$$GL_c(H) = \{A \in GL(H) : 1 - A \in K(H)\}.$$

In the operator norm topology, the group

$$GL_2(H) = \{A \in GL(H) : 1 - A \text{ is Hilbert-Schmidt}\}$$

is a Banach Lie group and its Lie algebra is the Hilbert space $L_2(H)$ of Hilbert-Schmidt operators, equipped with the commutator product and the Hilbert-Schmidt norm

$$\|A\|_2 = \left(\sum_\alpha \|A(e_\alpha)\|^2 \right)^{1/2} \qquad (A \in L_2(H))$$

where $\{e_\alpha\}$ is an orthonormal basis of H. If H is complex, the unitary group

$$U(H) = \{u \in GL(H) : uu^* = u^*u = 1\}$$

is a real Banach Lie group in the norm topology and its Lie algebra is the real Banach space

$$\mathfrak{u}(H) = \{A \in L(H) : A + A^* = 0\}.$$

of skew-hermitian operators, equipped with the commutator product. For a real Hilbert space, the orthogonal group

$$O(H) = \{t \in GL(H) : tt^* = t^*t = 1\}$$

is a Banach Lie group with Lie algebra

$$\mathfrak{o}(H) = \{A \in L(H) : A + A^* = 0\},$$

consisting of skew-symmetric operators.

Notes. We refer the readers to the recent book [30] for a comprehensive list of references for JB*-algebras and JB*-triples. One can find more details of JB*-triples in the books [37] and [64], the latter contains a chapter devoted to JBW*-triples and their classification.

2.5 Cartan factors

We introduce in this section a fundamental class of JB*-triples, called the *Cartan factors*, which are the building blocks of JB*-triples. We will see in Sections 3.3 and 3.8 that finite dimensional Cartan factors are the classifying spaces of finite dimensional bounded symmetric domains. There are six types of Cartan factors, listed below.

Type I $L(H, K)$, $(\dim H \leq \dim K)$

Type II $\{z \in L(H) : z^t = -z\}$,

Type III $\{z \in L(H) : z^t = z\}$,

Type IV   spin factor,

Type V $M_{1,2}(\mathcal{O}) = \{1 \times 2 \text{ matrices over the Cayley algebra } \mathcal{O}\}$,

Type VI $H_3(\mathcal{O}) = \{3 \times 3 \text{ hermitian matrices over } \mathcal{O}\}$, where $L(H, K)$ is the JBW*-triple of bounded linear operators between Hilbert spaces H and K, and z^t denotes the transpose of z in the JBW*-triple $L(H) := L(H, H)$ of bounded linear operators on H. The Jordan triple product in the first three types is given by

$$\{x, y, z\} = \frac{1}{2}(xy^*z + zy^*x)$$

where y^* denotes the adjoint of y.

The transpose on $L(H)$ is defined by a *conjugation* $j : H \longrightarrow H$, which is a conjugate linear isometry such that j^2 is the identity map. Given $z \in L(H)$, the transpose z^t is defined by $z^t = jz^*j$.

It is evident that the Type II and Type III Cartan factors are weak* closed subtriple of $L(H)$ and hence they are JBW*-triple by Lemma 2.4.8. In fact, $L(H)$ and Type III Cartan factors are JBW*-algebra in the special Jordan product $x \circ y = (xy + yx)/2$.

A *spin factor* is a JB*-triple V equipped with a complete inner product $\langle \cdot, \cdot \rangle$ and a conjugation $* : V \to V$ satisfying

$$\langle x^*, y^* \rangle = \langle y, x \rangle \text{ and } \{x, y, z\} = \frac{1}{2}\left(\langle x, y \rangle z + \langle z, y \rangle x - \langle x, z^* \rangle y^*\right). \quad (2.53)$$

We have already seen that $H_3(\mathcal{O})$ is a JB*-algebra with Jordan product

$$x \cdot y = \frac{1}{2}(xy + yx)$$

where the product on the right-hand side is the usual matrix product. The Cartan factor $M_{1,2}(\mathcal{O})$ can be identified as a subtriple of $H_3(\mathcal{O})$.

The open unit balls of the finite dimensional Cartan factors are exactly the six types of irreducible bounded symmetric domains in É. Cartan's classification. This explains the etymology of *Cartan factor*. The open unit ball of a spin factor is known as a *Lie ball*. For J = I, II, III, IV, V and VI, a bounded symmetric domain is called a *Type J domain* if it is biholomorphic to the open unit ball of a Type J Cartan factor.

We will discuss the classification in more detail in Section 3.8. For now, we study the structures of spin factors, which play an important role in physics (cf. [64]).

Let V be a spin factor and let $\| \cdot \|_h$ be the inner product norm of V. The spin factor norm $\| \cdot \|$ satisfies

$$\|a\|^3 = \|\{a, a, a\}\| = \|\langle a, a\rangle a - \langle a, a^*\rangle a^*/2\| \qquad (2.54)$$

where $|\langle a, a^*\rangle| \leq \|a\|_h \|a^*\|_h = \|a\|_h^2$. Actually, the two norms $\| \cdot \|$ and $\| \cdot \|_h$ are equivalent. Indeed, we have

$$\frac{1}{2}\|a\|_h^2\|a\| \leq \|a\|_h^2\|a\| - \frac{1}{2}|\langle a, a^*\rangle|\|a^*\| \leq \|a\|^3 \leq \frac{3}{2}\|a\|_h^2\|a\|$$

and

$$\|a\|_h \leq \sqrt{2}\|a\| \leq \sqrt{3}\|a\|_h \qquad (a \in V). \qquad (2.55)$$

In particular, $(V, \| \cdot \|)$ is a reflexive Banach space and V is a JBW*-triple. We denote by V_h the Hilbert space $(V, \langle \cdot, \cdot \rangle)$ equipped with the triple product

$$\{a, b, c\}_h = \frac{1}{2}(\langle a, b\rangle c + \langle c, b\rangle a) \qquad (a, b, c \in V).$$

The two Banach spaces V and V_h have very different geometry. Let D and D_h be the open unit balls of V and V_h respectively. By (2.55) and (2.58) below, we have $D_h \subset D \subset \sqrt{2}D_h$.

Lemma 2.5.1. *Let V be a spin factor and V_h the underlying Hilbert space. Then*

   (i) *the tripotents of V are either minimal or maximal;*

   (ii) *the minimal tripotents of V, with $\dim V > 1$, are exactly the extreme points v of the closed unit ball $\overline{D}_h$ of V_h, satisfying $\langle v, v^*\rangle = 0$.*

Also, we have $v \square v^* = 0$ *if* v *is a minimal tripotent.*

Proof. (i) Let $a \in V$ be a tripotent. Then we have

$$a = \{a, a, a\} = \langle a, a \rangle a - \frac{1}{2}\langle a, a^* \rangle a^*.$$

If $\langle a, a^* \rangle = 0$, then $\|a\|_h^2 = \langle a, a \rangle = 1$ and $\{a, V, a\} = \mathbb{C}V$. Hence a is a minimal tripotent of V and an extreme point of $\overline{D}_h$.

If $\langle a, a^* \rangle \neq 0$, then $a^* = \lambda a$ where

$$|\lambda| = \left| \frac{2(\langle a, a \rangle - 1)}{\langle a, a^* \rangle} \right| = 1.$$

Let $e = \sqrt{\lambda}a$. Then we have $a = \sqrt{\overline{\lambda}}e$ and

$$e^* = \sqrt{\overline{\lambda}}a^* = \sqrt{\overline{\lambda}}\lambda a = \sqrt{\lambda}a = e.$$

If $v \in Z_0(a)$, then $\{a, a, v\} = 0$ and

$$0 = 2\{e, e, v\} = \langle e, e \rangle v + \langle v, e \rangle e - \langle e, v^* \rangle e^* = \langle e, e \rangle v$$

implies $v = 0$. Hence a is a maximal tripotent of V.

(ii) Given an extreme point $v \in \overline{D}_h$ with $\langle v, v^* \rangle = 0$, we have $\langle v, v \rangle = 1$ and therefore

$$\{v, v, v\} = \langle v, v \rangle v - \frac{1}{2}\langle v, v^* \rangle v^* = v.$$

Hence v is a tripotent of V and is minimal since $\{v, V, v\} = \mathbb{C}v$.

Conversely, given a minimal tripotent $v \in V$, we first show $\langle v, v^* \rangle = 0$. Since $\dim V > 1$, we can pick $v_\perp \in V \backslash \{0\}$ such that $\langle v, v_\perp \rangle = 0$. By minimality, there exists $\alpha \in \mathbb{C}$ such that

$$\alpha v = \{v, v_\perp, v\} = \langle v, v_\perp \rangle v - \frac{1}{2}\langle v, v^* \rangle v_\perp^* = -\frac{1}{2}\langle v, v^* \rangle v_\perp^*$$

and hence

$$\alpha \langle v, v^* \rangle = -\frac{1}{2}\langle v, v^* \rangle \langle v_\perp^*, v^* \rangle = 0.$$

It follows that $\langle v, v^* \rangle = 0$. Further, the proof in (i) reveals that $\|v\|_h^2 = \langle v, v \rangle = 1$, that is, v is an extreme point of $\overline{D}_h$.

Finally, given a minimal tripotent v and any $x \in V$, we have

$$\{v, v^*, x\} = \frac{1}{2}(\langle v, v^* \rangle x + \langle x, v^* \rangle v - \langle v, x^* \rangle v) = 0.$$

$\square$

Triple orthogonality in V should not be confused with orthogonality with respect to the inner product $\langle \cdot, \cdot \rangle$. However, two triple orthogonal tripotents e and u must be orthogonal in the Hilbert space V_h since $\langle e, u \rangle = \langle \{e, e, e\}, u \rangle = \langle e, \{e, e, u\} \rangle = 0$.

Given two triple orthogonal tripotents e and u in V, we note that u is a scalar multiple of e^*. Indeed, $\{e, e, u\} = 0$ implies

$$\langle e, e \rangle u - \langle e, u^* \rangle e^* = 0$$

and $u = \langle e, u^* \rangle e^*$. It follows that in a spin factor V, there are at most two mutually triple orthogonal tripotents.

Let $a \in V \backslash \{0\}$. By Theorem 2.4.21, the closed subtriple $V(a)$ generated by a identifies with the JB*-triple $C_0(S_a)$ of continuous functions vanishing on the triple spectrum S_a. Since $V(a)$ is reflexive, S_a must be a finite set in which case, each indicator function χ_t of $\{t\} \subset S_a$, where $\chi_t(t) = 1$ and $\chi_t(x) = 0$ for $x \neq t$, is a tripotent in $C_0(S_a)$ and these indicator functions are mutually triple orthogonal. It follows from the above remark that S_a reduces to a set of at most two points. We therefore infer that there are two mutually orthogonal minimal tripotents $e, u \in V$ such that

$$a = \alpha e + \beta u \qquad (\alpha \geq \beta \geq 0) \tag{2.56}$$

and $\|a\| = \alpha$. This decomposition is called a *spectral representation* of a. As u is a scalar multiple of e^*, it can be written as

$$a = \alpha_1 e + \alpha_2 e^* \qquad (\alpha_2 \in \mathbb{C}, \|a\| = \alpha_1 \geq |\alpha_2|). \tag{2.57}$$

This representation is unique if $\alpha_1 > |\alpha_2|$. We note that α_2 can be 0 and

$$\|a\|_h^2 = \alpha_1^2 + |\alpha_2|^2 \geq \alpha_1^2 = \|a\|^2. \tag{2.58}$$

Also,

$$\{a, a, a\} = \alpha_1|\alpha_1|^2 e + \alpha_2|\alpha_2|^2 e^*$$

implies that a is a tripotent if, and only if, $\alpha_1 = \alpha_1^3$ and $\alpha_2 = \alpha_2|\alpha_2|^2$.

We can express the spin factor norm in terms of the inner product. Given $v = \alpha_1 e + \alpha_2 e^*$ in (2.57), we have $\langle v, v \rangle = \alpha_1^2 + |\alpha_2|^2$ and $\langle v, v^* \rangle = 2\alpha_1\alpha_2$. This gives

$$\langle v, v \rangle + \sqrt{\langle v, v \rangle^2 - |\langle v, v^* \rangle|^2} = 2\alpha_1^2 = 2\|v\|^2.$$

The Lie ball D can therefore be written as

$$D = \{v \in V : \frac{1}{2}(\langle v, v \rangle + \sqrt{\langle v, v \rangle^2 - |\langle v, v^* \rangle|^2}) < 1\}. \tag{2.59}$$

Since $|\langle v, v^* \rangle| \leq \langle v, v \rangle$, one can also represent D in the form

$$D = \{v \in V : 1 - \langle v, v \rangle + \frac{1}{2}|\langle v, v^* \rangle|^2 > 0, \ \frac{1}{2}|\langle v, v^* \rangle| < 1\}.$$

Remark 2.5.2. Under the condition (2.53), $\mathbb{C}$ is not a spin factor in the usual triple product $\{a, b, c\} = a\bar{b}c$, with involution $z \mapsto \bar{z}$ and the standard inner product. However, in some literature, the factor $1/2$ is not included in the defining condition (2.53) for a spin factor (e.g. [64]), in which case $\mathbb{C}$ becomes a 'spin factor' in the usual triple product. If one adopts this definition, some dimension restriction and straightforward scaling would have to be made to our results above. For instance, the fraction $1/2$ should be dropped from the description of the Lie ball in (2.59). Let $n > 2$ and equip $\mathbb{C}^n$ with the standard inner product and involution $z = (z_1, \ldots, z_n) \in \mathbb{C}^n \mapsto \bar{z} = (\bar{z}_1, \ldots, \bar{z}_n) \in \mathbb{C}^n$. If one defines the spin triple product without $1/2$:

$$\{z, z, z\} = 2\langle z, z \rangle z - \langle z, \bar{z} \rangle \bar{z}$$

then $\mathbb{C}^n$ is a spin factor in this alternative definition and the open unit ball D has the form

$$D = \{z \in \mathbb{C}^n : 1 - 2\langle z, z \rangle + |\langle z, \bar{z} \rangle|^2 > 0, \ |\langle z, \bar{z} \rangle| < 1\}$$

which is another common representation of a finite dimensional Type IV domain (cf. [90, 108]).

According to (2.57), a maximal tripotent v has the form

$$v = e + \alpha e^*$$

where e is a minimal tripotent and $|\alpha| = 1$. It follows that $v^* = \bar{\alpha} v$ and $\|v\|_h = \sqrt{2}$. We note that the above representation of a maximal tripotent is not unique.

Lemma 2.5.3. *Let $a \in V \backslash \{0\}$ satisfy $a + \lambda a^* = 0$ for some $|\lambda| = 1$. Then $a/\|a\|$ is a maximal tripotent.*

Proof. Let a have the spectral decomposition

$$a = \alpha_1 e + \alpha_2 e^*$$

as in (2.57). Then we have

$$0 = a + \lambda a^* = (\alpha_1 + \bar{\alpha}_2 \lambda)e + (\alpha_2 + \alpha_1 \lambda)e^*$$

which implies $\alpha_2 = -\alpha_1 \lambda$ and $a = \alpha_1(e - \lambda e^*)$ where $e - \lambda e^*$ is a maximal tripotent and $\alpha_1 = \|a\|$. $\qquad\square$

Lemma 2.5.4. *Let $v \in V$ be a maximal tripotent. Then $v \square v : V \longrightarrow V$ is the identity map and v is a unitary tripotent.*

Proof. As noted earlier, we have $v^* = \lambda v$ for some $|\lambda| = 1$. For each $a \in V$, we have

$$
\begin{aligned}
v \square v(a) &= \{v, v, a\} \\
&= \frac{1}{2}\langle v, v \rangle a + \frac{1}{2}\langle a, v \rangle v - \frac{1}{2}\langle v, a^* \rangle v^* \\
&= a + \frac{1}{2}\langle a, v \rangle v - \frac{\lambda}{2}\langle a, v^* \rangle v = a.
\end{aligned}
$$

Hence $P_1(v) = 4(v \square v - (v \square v)^2) = 0$ and v is unitary. $\square$

Using the spectral decomposition of $v = \alpha_1 e + \alpha_2 e^*$, we have a useful expression of the Bergman operator $B(v, v)$. First, the triple orthogonal minimal tripotents e and e^* give rise to a joint Peirce decomposition of V:

$$V = \bigoplus_{0 \leq i \leq j \leq 2} V_{ij}$$

where each Peirce space V_{ij} is the range of a joint Peirce projection $P_{ij} : V \longrightarrow V$. In fact, we have

$$P_{00} = P_0(e)P_0(e^*); \quad P_{01} = P_0(e^*)P_1(e); \quad P_{02} = P_0(e)P_1(e^*)$$
$$P_{11} = P_2(e); \quad P_{12} = P_1(e)P_1(e^*); \quad P_{22} = P_2(e^*)$$

(cf. [37, p. 38]). For a minimal tripotent $e \in V$, the Peirce projections are given by

$$P_0(e) = \langle \cdot, e^* \rangle e^*; \quad P_2(e) = \langle \cdot, e \rangle e$$
$$P_1(e) = I - \langle \cdot, e \rangle e - \langle \cdot, e^* \rangle e^* = P_1(e^*)$$

and they are self-adjoint operators on the Hilbert space V_h. Hence for any $z \in V$, we have

$$|\langle z, e \rangle| = \|P_2(e)z\| \leq \|z\|$$

in the spin norm. We note that $P_2(e^*) = P_0(e)$ and

$$(P_1(e)v)^* = P_1(e)(v^*) \qquad (v \in V). \tag{2.60}$$

Since $\langle e, e^* \rangle = 0$, a simple computation gives $P_{00} = P_{01} = P_{02} = 0$ and $P_{12} = P_1(e)$. Hence the Bergman operator $B(v, v)$ for $v = \alpha_1 e + \alpha_2 e^*$ has the simple form

$$B(v, v) = \sum_{1 \leq i \leq j \leq 2} (1 - |\alpha_i|^2)(1 - |\alpha_j|^2)P_{ij}$$
$$= (1 - |\alpha_1|^2)^2 P_2(e) + (1 - |\alpha_1|^2)(1 - |\alpha_2|^2)P_1(e) + (1 - |\alpha_2|^2)^2 P_0(e).$$

In particular, we have

$$B(v,v)^{1/2}(v) = (1 - |\alpha_1|^2)\alpha_1 e + (1 - |\alpha_2|^2)\alpha_2 e^* \qquad (2.61)$$

$$B(v,v)^{-1/2}(v) = \frac{\alpha_1 e}{1 - |\alpha_1|^2} + \frac{\alpha_2 e^*}{1 - |\alpha_2|^2}. \qquad (2.62)$$

Example 2.5.5. Let A be a JB-algebra with identity $\mathbf{1}$ such that its complexification $\mathcal{A} = A + iA$ is a spin factor. Then A is a real spin factor $H \oplus \mathbb{R}\mathbf{1}$ for some real Hilbert space H. To see this, we note that $\mathcal{A}$ is a JB*-algebra with involution $a + ib \mapsto (a + ib)^- = a - ib$, and a spin factor with inner product $\langle \cdot, \cdot \rangle$ and conjugation $*$. Since $\mathbf{1}$ is a maximal tripotent, we have $\mathbf{1}^* = \lambda\mathbf{1}$ for some $\lambda \in \mathbb{C}$ with $|\lambda| = 1$, and (2.54) implies $\langle \mathbf{1}, \mathbf{1} \rangle = 2$. We have the orthogonal decomposition

$$A = H \oplus \mathbb{R}\mathbf{1}$$

where $H = \{a \in A : \langle a, \mathbf{1} \rangle = 0\}$. For each $a \in H$, we have

$$a = \{\mathbf{1}, a, \mathbf{1}\} = \frac{1}{2}(\langle \mathbf{1}, a \rangle \mathbf{1} + \langle \mathbf{1}, a \rangle \mathbf{1} - \langle \mathbf{1}, \mathbf{1}^* \rangle a^* = -\lambda a^*$$

and

$$a^2 = \{a, a, \mathbf{1}\} = \frac{1}{2}\langle a, a \rangle \mathbf{1}.$$

For $a, b \in H$, we have

$$\langle a, b \rangle = \langle b^*, a^* \rangle = \langle -\overline{\lambda}b, -\overline{\lambda}a \rangle = \langle b, a \rangle.$$

It follows that H is a real Hilbert space with the inner product $\langle\langle \cdot, \cdot \rangle\rangle := \frac{1}{2}\langle \cdot, \cdot \rangle$ and the Jordan product of $A = H \oplus \mathbb{R}\mathbf{1}$ satisfies

$$(a \oplus \alpha\mathbf{1})(b \oplus \beta\mathbf{1}) = \alpha b + \beta a + \alpha\beta\mathbf{1} + ab$$
$$= (\alpha b + \beta a) \oplus (\alpha\beta + \langle\langle a, b \rangle\rangle)\mathbf{1}.$$

Hence A is a real spin factor.

We end this section with representations of the TKK-Lie algebras of Cartan factors. It has been shown by Tits [165] that the TKK-Lie algebra of the Cartan factor $H_3(\mathcal{O})$ is the exceptional Lie algebra E_7 and therefore the TKK-Lie algebra of $M_{1,2}(\mathcal{O})$ identifies with a Lie subalgebra of E_7. For Cartan factors of Type I, II, III and IV, the TKK-Lie algebras can be represented as Lie algebras of matrices.

Given a C*-algebra $A \subset L(H)$, the tensor product $M_2(\mathbb{C}) \otimes A$ identifies with the matrix C*-algebra

$$M_2(\mathcal{A}) = \left\{ \begin{pmatrix} a & b \\ c & d \end{pmatrix} : a, b, c, d \in \mathcal{A} \right\}$$

with the usual matrix product and involution $\begin{pmatrix} a & b \\ c & d \end{pmatrix}^* = \begin{pmatrix} a^* & c^* \\ b^* & d^* \end{pmatrix}$ (cf. [161, p. 192]). It is a Banach Lie algebra in the Lie brackets

$$[A, B] = AB - BA \qquad (A, B \in M_2(\mathcal{A})).$$

In particular, $M_2(L(H))$ is a Banach Lie algebra. Given a subset $\mathcal{E} \subset M_2(L(H))$, we define $\mathcal{E}^* = \{A^* : A \in \mathcal{E}\}$.

Theorem 2.5.6. *Let $V \subset L(H)$ be a Cartan factor of Type I, II, III or IV. Then its TKK Lie algebra $\mathfrak{L}(V) = V \oplus V_0 \oplus \overline{V}$ is Lie isomorphic to the following Lie subalgebra of $M_2(L(H))$:*

$$\mathfrak{B}(V) = \left\{ \begin{pmatrix} \sum_{j=1}^{n} a_j b_j^* & x \\ y^* & -\sum_{j=1}^{n} b_j^* a_j \end{pmatrix} : x, y, a_1, \ldots, a_n, b_1, \ldots, b_n \in V \right\}$$

which is a TKK Lie algebra with gradings

$$\mathfrak{B}(V)_0 = \left\{ \begin{pmatrix} \sum_{j=1}^{n} a_j b_j^* & 0 \\ 0 & -\sum_{j=1}^{n} b_j^* a_j \end{pmatrix} : a_j, b_j \in V \right\}$$

$$\mathfrak{B}(V)_{-1} = \left\{ \begin{pmatrix} 0 & x \\ 0 & 0 \end{pmatrix} : x \in V \right\}, \quad \mathfrak{B}(V)_1 = \mathfrak{B}(V)_{-1}^*. \qquad (2.63)$$

The TKK involution is given by

$$\widetilde{\theta}: \begin{pmatrix} \sum_j a_j b_j^* & x \\ y^* & -\sum_j b_j^* a_j \end{pmatrix} \mapsto \begin{pmatrix} -\sum_j b_j a_j^* & \frac{y}{2} \\ 2x^* & \sum_j a_j^* b_j \end{pmatrix}.$$

Proof. Let $\mathfrak{L}(V) = V \oplus V_0 \oplus \overline{V}$ be the TKK Lie algebra of V, with TKK-involution θ, where V is contained in the JBW*-triple $L(H)$ of bounded linear operators on a Hilbert space H.

Given $a_1, \ldots, a_n$ and $b_1, \ldots, b_n$ in V, it has been proved in [48, Theorem 4.4] that the following two conditions are equivalent:

$$\text{(i)} \ \sum_{j=1}^n a_j \,\square\, b_j = 0 \qquad \text{(ii)} \ \sum_{j=1}^n a_j^* b_j = \sum_{j=1}^n b_j^* a_j = 0.$$

This enables us to define a map $\Phi : \mathfrak{L}(V) \to \mathfrak{B}(V)$ by

$$\Phi(x, \sum_j a_j \,\square\, b_j, \, y) = \begin{pmatrix} \frac{1}{2}\sum_j a_j b_j^* & \frac{x}{2} \\ y^* & -\frac{1}{2}\sum_j b_j^* a_j \end{pmatrix}.$$

Evidently, Φ is a linear bijection. Further, one can verify readily that Φ preserves the Lie multiplication in (2.45) and (2.46).

Hence Φ induces the gradings $\mathfrak{B}(V)_{-1} = \Phi(V)$ and $\mathfrak{B}(V)_0 = \Phi(V_0)$ in (2.63), as well as $\mathfrak{B}(V)_1 = \Phi(\overline{V}) = \mathfrak{B}(V)_{-1}^*$. Also, the TKK involution $\widetilde{\theta}$ of $\mathfrak{B}(V)$ is induced from the involution θ of $\mathfrak{L}(V)$ via Φ:

$$\widetilde{\theta}(\Phi(x, \sum_j a_j \,\square\, b_j, y)) := \Phi(\theta(x, \sum_j a_j \,\square\, b_j, y)) = \Phi(y, -\sum_j b_j \,\square\, a_j, x).$$

This completes the proof. $\qquad\qquad\qquad\qquad\qquad\qquad\qquad\qquad\quad\square$

Using Theorem 2.5.6, one can derive a matrix representation of TKK-Lie algebras of JB*-triples, shown in [48, Corollary 4.13], where the TKK-Lie algebras of JB*-triples have also been characterised among complex Lie algebras.

Chapter 3

Bounded symmetric domains

3.1 Algebraic structures of symmetric manifolds

We begin this Chapter by discussing the Jordan and Lie structures of symmetric Banach manifolds which are of fundamental importance. Our goal is to use these structures to show that bounded symmetric domains in Banach spaces can be realised as the open unit balls of JB*-triples.

Let M be symmetric Banach manifold modelled on a complex Banach space V and equipped with a compatible tangent norm ν. Then the group $\text{Aut}(M, \nu)$ of ν-isometries is a real Banach Lie group, by Theorem 1.3.11. In the case of a bounded symmetric domain D in a Banach space, we have $\text{Aut}(D, \nu) = \text{Aut}\, D$, by Lemma 1.3.8, where ν is the Carathéodory tangent norm and the topology of the real Banach Lie group $\text{Aut}\, D$ is that of locally uniform convergence.

Write $G = \text{Aut}(M, \nu)$ and let $s_p : M \longrightarrow M$ be a symmetry at some point $p \in M$. Then s_p is an element of the ν-isometry group $G = \text{Aut}(M, \nu)$, which is a real Banach Lie group in a topology finer

145

than the topology of locally uniform convergence, by Corollary 1.3.12. The real Banach Lie algebra $\mathfrak{g}$ of G is given by

$$\mathfrak{g} = \mathrm{aut}(M, \nu) = \{X \in \mathrm{aut}\, M : \exp tX \in \mathrm{Aut}(M, \nu), \forall t \in \mathbb{R}\}$$

where $\mathrm{aut}\, M$ is the set of complete holomorphic vector fields on M. If M is a bounded symmetric domain, then $\mathfrak{g} = \mathrm{aut}\, M$. The adjoint representation

$$\theta := Ad(s_p) : \mathfrak{g} \longrightarrow \mathfrak{g} \tag{3.1}$$

is the differential $d\sigma_e$ at the identity $e \in G$, where

$$\sigma(g) = s_p g s_p \qquad (g \in G).$$

Since $s_p^2 = e$, we have

$$(d\sigma^2)_e = (d\sigma)_{\sigma(e)} \circ d\sigma_e = d\sigma_e \circ d\sigma_e = \theta^2$$

where $(d\sigma^2)_e$ is the identity map $I : \mathfrak{g} \longrightarrow \mathfrak{g}$ and hence θ is an involution. It follows that $\mathfrak{g}$ decomposes into a direct sum, called the *canonical decomposition* of $\mathfrak{g}$,

$$\mathfrak{g} = \mathfrak{k} \oplus \mathfrak{p} \tag{3.2}$$

of ± 1-eigenspaces of θ:

$$\mathfrak{k} = \{X \in \mathfrak{g} : \theta(X) = X\}, \quad \mathfrak{p} = \{X \in \mathfrak{g} : \theta(X) = -X\} \tag{3.3}$$

satisfying

$$[\mathfrak{k}, \mathfrak{k}] \subset \mathfrak{k}, \quad [\mathfrak{k}, \mathfrak{p}] \subset \mathfrak{p}, \quad [\mathfrak{p}, \mathfrak{p}] \subset \mathfrak{k}.$$

Since θ is an automorphism of the Lie algebra $\mathfrak{g}$, we see that $\mathfrak{k}$ is a Lie subalgebra of $\mathfrak{g}$. Also, $[\mathfrak{p}, \mathfrak{p}]$ is an ideal of $\mathfrak{k}$.

Lemma 3.1.1. *In the canonical decomposition* (3.2) *induced by the symmetry* s_p *at* $p \in M$, *we have*

$$\mathfrak{k} = \{X \in \mathfrak{g} : X(p) = 0\} = \{X \in \mathfrak{g} : \exp tX(p) = p, \forall t \in \mathbb{R}\}.$$

Proof. In the notation of (1.7), we have $\theta = (s_p)_*$ and $\exp t\theta X = s_p(\exp tX)s_p$ for all $t \in \mathbb{R}$. If $\theta X = X$, then $\exp tX = s_p \exp tX s_p$ and $\exp tX(p) = s_p(\exp tX(p))$ implies $\exp tX(p) = p$ for sufficiently small t since p is an isolated fixed point of s_p. It follows that

$$X_p = \frac{d}{dt}\Big|_{t=0} \exp tX(p) = 0.$$

On the other hand, given $X_p = 0$, we have

$$\begin{aligned}
\frac{d}{dt}\Big|_{t=s} \exp tX(p) &= \frac{d}{dt}\Big|_{t=0} \exp(s+t)X(p) \\
&= \frac{d}{dt}\Big|_{t=0} \exp sX \circ \exp tX(p) \\
&= d(\exp sX)_p(X_p) = 0
\end{aligned}$$

for all $s \in \mathbb{R}$. Therefore $\exp tX$ is constant in t and we have $\exp tX(p) = p$ for all $t \in \mathbb{R}$.

In turn, the last condition implies $\theta X = X$. Indeed, by Lemma 1.3.14, the differential $(ds_p)_p$ is minus the identity map on T_pM which gives

$$d(s_p(\exp tX)s_p)_p = (ds_p)_p \circ d(\exp tX)_p \circ (ds_p)_p = d(\exp tX)_p$$

and it follows from Cartan's uniqueness theorem that

$$\exp tX = s_p(\exp tX)s_p$$

for all $t \in \mathbb{R}$. Hence

$$\begin{aligned}
\theta X(\cdot) &= \frac{d}{dt}\Big|_{t=0} \exp t\theta X(\cdot) = \frac{d}{dt}\Big|_{t=0} s_p(\exp tX)s_p(\cdot) \\
&= \frac{d}{dt}\Big|_{t=s} \exp tX(\cdot) = X(\cdot).
\end{aligned}$$

$\square$

It follows from the preceding lemma that the evaluation map

$$X \in \mathfrak{p} \mapsto X(p) \in V \tag{3.4}$$

is a real linear injection. As noted in (1.16), the map $X \in \mathfrak{g} \mapsto X(p)$ is surjective and hence the map in (3.4) is a real linear isomorphism.

Now we are going to manufacture a Jordan structure in V using the Tits-Kantor-Koecher construction and the evaluation map in (3.4). This reveals the close connection between the Jordan and Lie structures of the symmetric manifold M.

First, we define a complex structure $J : \mathfrak{p} \longrightarrow \mathfrak{p}$. Given $X \in \mathfrak{p}$, we have $iX(p) \in V$ and by (3.4), $iX(p) = JX(p)$ for a unique $JX \in \mathfrak{p}$. This defines J which is real linear and $-J^2$ is the identity map on $\mathfrak{p}$. We note that $X \in \mathfrak{g}$ does not imply $iX \in \mathfrak{g}$ and in fact, $\mathfrak{k}$ is *purely real* in the sense of the following lemma.

Lemma 3.1.2. *In the canonical decomposition (3.2), we have* $\mathfrak{k} \cap i\mathfrak{k} = \{0\}$.

Proof. Let $Y \in \mathfrak{k}$ and $Y = iZ \in i\mathfrak{k}$. We show $Y = 0$. For $s + it \in \mathbb{C}$, the map $(\exp sY)(\exp tZ) = \exp(s - it)Y : M \longrightarrow M$ is a biholomorphic isometry and therefore the map

$$F : s + it \in \mathbb{C} \longrightarrow (d\exp(s - it)Y)_p \in L(T_p M)$$

is a bounded holomorphic map, where $L(T_p M)$ is the complex Banach space of continuous linear self-maps on $T_p M$. By Liouville theorem, $\varphi \circ F : \mathbb{C} \longrightarrow \mathbb{C}$ is constant for each continuous linear functional φ on $L(T_p M)$. Hence F is constant and $(d\exp Y)_p$ is the identity map. Since $\exp Y(p) = p$, Cartan's uniqueness theorem implies that $\exp Y$ itself is an identity map on M, that is, $Y = 0$. $\square$

Nevertheless, we have

$$J\theta = \theta J, \quad [JX, Y](p) = dY_p(JX(p)) - d(JX)_p(Y(p)) = i[X, Y](p)$$

for $X \in \mathfrak{p}, Y \in \mathfrak{k}$ since $Y(p) = 0$. Hence

$$J[X, Y] = [JX, Y] \quad (X \in \mathfrak{p}, Y \in \mathfrak{k}).$$

Further, J satisfies the condition in the following lemma. A seemingly short derivation of this condition in [37, Lemma 2.5.4] requires some details, which are given below.

Lemma 3.1.3. *For $X, Y \in \mathfrak{p}$, we have $[JX, JY] = [X, Y]$.*

Proof. Let $s = s_p$ be the symmetry at $p \in M$. By Lemma 1.3.14, one can find a local chart $(\mathcal{V}, \varphi, V)$ at p with $\varphi(p) = 0 \in V$ such that the following diagram commutes:

$$
\begin{array}{ccc}
\mathcal{V} & \xrightarrow{s} & \mathcal{V} \\
\downarrow{\varphi} & & \downarrow{\varphi} \\
V & \xrightarrow{-id} & V
\end{array}
$$

where $s(\mathcal{V}) = \mathcal{V}$ and we may assume $D = \varphi(\mathcal{V})$ is a bounded domain in V.

In the local chart, write $X = h_X \frac{\partial}{\partial z}$, $Y = h_Y \frac{\partial}{\partial z}$, $JX = h_{JX} \frac{\partial}{\partial z}$ and $JY = h_{JY} \frac{\partial}{\partial z}$, where h_X, h_Y, h_{JX} and h_{JY} are holomorphic functions from $\mathcal{V}$ to V.

Consider a holomorphic vector field $Z = h_Z \frac{\partial}{\partial z}$ on M as the vector field $\underline{Z} = \underline{h}_Z \frac{\partial}{\partial z}$ on $D = \varphi(\mathcal{V})$, where the holomorphic map $\underline{h}_Z : D \longrightarrow V$ is given by $\underline{h}_Z = h_Z \circ \varphi^{-1}$.

If $Z \in \mathfrak{p}$, then we have $\theta Z = -Z$ and

$$s \circ \exp tZ \circ s = \exp ts_* Z = \exp t\theta Z = \exp t(-Z) \quad (t \in \mathbb{R})$$

which gives

$$\varphi \circ s \circ \exp tZ \circ s \circ \varphi^{-1} = \varphi \circ \exp t(-Z) \circ \varphi^{-1} \quad (t \in \mathbb{R})$$

from the above commutative diagram, where $\varphi s = -\varphi$ and $s\varphi^{-1}(\cdot) = \varphi^{-1}(-\cdot)$. It follows that

$$-\exp tZ \circ s \circ \varphi^{-1} = \exp t(-Z) \circ \varphi^{-1} \qquad (t \in \mathbb{R})$$

and hence

$$Z((\varphi^{-1}(-v)) = Z(\varphi^{-1}(v)) \qquad (v \in D).$$

Therefore

$$\underline{h}_Z(-v) = h_Z \circ \varphi^{-1}(-v) = h_Z(\varphi^{-1}(v)) = \underline{h}_Z(v) \qquad (v \in D)$$

and so $\underline{h}'_Z(0) = 0$.

In particular, for $[X,Y] = h_{[X,Y]}\frac{\partial}{\partial z}$ with $\underline{h}_{[X,Y]} = h_{[X,Y]} \circ \varphi^{-1}$, we have, as in Example 1.2.14,

$$\underline{h}_{[X,Y]}(v) = \underline{h}'_Y(v)\underline{h}_X(v) - \underline{h}'_X(v)\underline{h}_Y(v) \qquad (v \in D)$$

which entails $\underline{h}_{[X,Y]}(0) = 0$ as well as

$$\begin{aligned}
\underline{h}'_{[X,Y]}(0) &= \underline{h}''_Y(0)\underline{h}_X(0) + \underline{h}'_Y(0)\underline{h}'_X(0) - \underline{h}''_X(0)\underline{h}_Y(0) - \underline{h}'_X(0)\underline{h}'_Y(0) \\
&= \underline{h}''_Y(0)\underline{h}_X(0) - \underline{h}''_X(0)\underline{h}_Y(0).
\end{aligned}$$

Let

$$Z = [JX,Y] + [X,JY] - i[X,Y] + i[JX,JY].$$

Then we have

$$\underline{h}_Z(0) = \underline{h}_{[JX,Y]}(0) + \underline{h}_{[X,JY]}(0) - i\underline{h}_{[X,Y]}(0) + i\underline{h}_{[JX,JY]}(0) = 0$$

and repeating the preceding computation renders

$$\underline{h}'_Z(0) = \underline{h}'_{[JX,Y]}(0) + \underline{h}'_{[X,JY]}(0) - i\underline{h}'_{[X,Y]}(0) + i\underline{h}'_{[JX,JY]}(0) = 0.$$

By Lemma 1.1.7, we have $\underline{h}_Z = 0$ and hence $h_Z = 0$ (by the principle of analytic continuation for M (cf. [168, Theorem 3.1])). Hence $Z = 0$, which implies

$$[JX,Y] + [X,JY] = i[X,Y] - i[JX,JY] \in \mathfrak{k} \cap i\mathfrak{k}.$$

By Lemma 3.1.2, we conclude $[X,Y] - [JX,JY] = 0$. $\qquad\qquad\square$

We note from the previous proof that

$$[JX, Y] = -[X, JY] \qquad (X, Y \in \mathfrak{p}).$$

Let $\mathfrak{p}_c$ be the complexification of $\mathfrak{p}$ and extend J naturally to a complex linear map on $\mathfrak{p}_c$. Let

$$\mathfrak{p}_+ = \{X \in \mathfrak{p}_c : JX = iX\}, \quad \mathfrak{p}_- = \{X \in \mathfrak{p}_c : JX = -iX\} \qquad (3.5)$$

so that $\mathfrak{p}_c = \mathfrak{p}_+ \oplus \mathfrak{p}_-$ and the complexification $\mathfrak{g}_c$ of $\mathfrak{g}$ is given by

$$\mathfrak{g}_c = \mathfrak{p}_+ \oplus \mathfrak{k}_c \oplus \mathfrak{p}_- \qquad (3.6)$$

where $\mathfrak{k}_c$ is the complexification of $\mathfrak{k}$. By Lemma 3.1.3, we have $[\mathfrak{p}_+, \mathfrak{p}_+] = [\mathfrak{p}_-, \mathfrak{p}_-] = 0$. It follows that $[\mathfrak{p}_c, \mathfrak{p}_c] = [\mathfrak{p}_+, \mathfrak{p}_-]$ and the complexification of $[\mathfrak{p}, \mathfrak{p}] \oplus \mathfrak{p}$ is given by

$$\mathfrak{p}_+ \oplus [\mathfrak{p}_+, \mathfrak{p}_-] \oplus \mathfrak{p}_-$$

which is the canonical part of $\mathfrak{g}_c$.

With the complex structure, $\mathfrak{p}$ is complex linear isomorphic to $\mathfrak{p}_+$ via the map

$$\psi : X \in \mathfrak{p} \mapsto X - iJX \in \mathfrak{p}_+$$

and hence $\mathfrak{p}_+$ is complex linear isomorphic to V via the map

$$\varphi : X - iJX \in \mathfrak{p}_+ \mapsto X(p) \in V. \qquad (3.7)$$

The automorphism θ induces an involutive conjugate linear Lie automorphism σ on $\mathfrak{g}_c$, given by

$$\sigma(X + iY) = \theta X - i\theta Y \qquad (X + iY \in \mathfrak{g} + i\mathfrak{g}).$$

We have $\sigma(\mathfrak{p}_+) = \mathfrak{p}_-$ and

$$[\mathfrak{k}_c, \mathfrak{p}_\pm] \subset \mathfrak{p}_\pm.$$

Hence $\mathfrak{g}_c$ is a TKK Lie algebra with gradation in (3.6) and TKK involution σ. It follows that $\mathfrak{p}_+$ has the structure of a Jordan triple, which induces such on V via the linear isomorphism φ in (3.7). The Jordan triple product $\{\cdot, \cdot, \cdot\}_{\mathfrak{p}_+}$ of $\mathfrak{p}_+$ is given by

$$\{X, Y, Z\}_{\mathfrak{p}_+} = [[X, \sigma(Y)], Z] \qquad (X, Y, Z \in \mathfrak{p}_+).$$

Given $a, b, c \in V$ with $a = X(p)$, $b = Y(p)$ and $c = Z(p)$ for $X, Y, Z \in \mathfrak{p}$, we define the triple product on V by

$$\{a, b, c\} = \{X(p), Y(p), Z(p)\} := \varphi\{\psi(X), \psi(Y), \psi(Z)\}_{\mathfrak{p}_+}. \qquad (3.8)$$

The tangent space $T_p M$ is naturally identified with the model space V of M and we have proved the main assertion of the following far-reaching connection between symmetric manifolds and Jordan triples.

Theorem 3.1.4. *Let M be a symmetric Banach manifold modelled on a complex Banach space V. Then the tangent space $T_p M$ at each $p \in M$ carries the structure of a Jordan triple, induced by the symmetry s_p. The Jordan triple structures of the tangent spaces obtained from (3.8) by different symmetries are all mutually isomorphic.*

Proof. We need only show that the Jordan triple structures of any two tangent spaces $T_p M$ and $T_a M$, induced by the symmetries s_p and s_a in (3.8) respectively, are isomorphic. By transitivity (cf. Lemma 1.3.16), there is a biholomorphic map $\psi : M \longrightarrow M$ such that $\psi(a) = p$. By uniqueness of the symmetry, we have $s_a = \psi^{-1} s_p \psi$. As before, let $\theta = Ad(s_p)$ and $\theta_a = Ad(s_a)$ be the adjoint representations, with corresponding canonical decompositions

$$\operatorname{aut} M = \mathfrak{k} \oplus \mathfrak{p} = \mathfrak{k}_a \oplus \mathfrak{p}_a.$$

Then we have

$$\theta \psi_* = (s_p)_* \psi_* = (s_p \psi)_* = (\psi s_a)_* = \psi_* \theta_a$$

and also, ψ_* commutes with the complex structures $J : \mathfrak{p} \longrightarrow \mathfrak{p}$ and $J_a : \mathfrak{p}_a \longrightarrow \mathfrak{p}_a$ since

$$J(\psi_* X)(p) = -i\psi_* X(p) = -id\psi_a(X_a) = d\psi_a((J_a X)_a) = \psi_*(J_a X)(p)$$

for $X \in \mathfrak{p}_a$. It follows that ψ_* is a graded Lie isomorphism between the TKK Lie algebras

$$\mathfrak{g}_c = \mathfrak{p}_+ \oplus \mathfrak{k}_c \oplus \mathfrak{p}_- \quad \text{and} \quad \mathfrak{g}'_c = (\mathfrak{p}_a)_+ \oplus (\mathfrak{k}_a)_c \oplus (\mathfrak{p}_a)_-$$

and hence the Jordan triple $(\mathfrak{p}_a)_+$ is triple isomorphic to $\mathfrak{p}_+$, via $\psi_*|_{(\mathfrak{p}_a)_+}$.
□

Notes. Theorem 3.1.4 is due to Kaup [98]. One can show further in this theorem that for each $a \in V$, the box operator $a \,\square\, a : V \longrightarrow V$ is hermitian, that is, it has real numerical range. A complex Jordan triple with this property is called a *Hermitian Jordan triple system* in [98]. Conversely, given such a Hermitian Jordan triple system V, there corresponds a symmetric Banach manifold M, as constructed in [98]. We will not pursue this and its ramifications here although it is a valuable avenue to explore new grounds, but will focus on the special case of bounded symmetric domains in this correspondence. To avoid confusion, the reader should be reminded again that a complex Jordan triple in this book is called a Hermitian Jordan triple in [37].

3.2   Realisation of bounded symmetric domains

In the preceding section, we have shown that the tangent space of a symmetric Banach manifold at each point admits a Jordan triple structure, induced by the symmetry at the point and the Lie structure of the complete holomorphic vector fields on the manifold.

Now we proceed to consider a bounded symmetric domain D in a complex Banach space V, as a symmetric Banach manifold modelled on V. By Theorem 3.1.4, we can equip V with a Jordan triple structure. We aim to show that, under a suitable norm $\| \cdot \|_{sp}$, V becomes a JB*-triple and D is biholomorphic to the open unit ball of V in this norm.

To simplify our task, we make use of a result of Vigué in [172, Théorème 3.4.1] which asserts that every bounded symmetric domain is biholomorphic to a circular balanced bounded symmetric domain. We may and will therefore assume in what follows that D is circular and balanced, and pick the symmetry s_0 at the origin $0 \in D$.

Fix D and the symmetry s_0 throughout this section, and recall, in this case, the real Banach Lie algebra $\mathfrak{g} := \operatorname{aut} D$ of complete holomorphic vector fields on D decomposes into (± 1)-eigenspaces of $\theta = Ad(s_0) : \operatorname{aut} D \longrightarrow \operatorname{aut} D$,

$$\mathfrak{g} = \operatorname{aut} D = \mathfrak{k} \oplus \mathfrak{p} \,,$$

and as before, the evaluation map $X \in \mathfrak{p} \mapsto X(0) \in V$ is a real linear isomorphism.

Remark 3.2.1. A biholomorphic map on D of the form $s_p \circ s_0$, where s_p is the symmetry at $p \in D$, is called a *transvection*. Let

$$P = \{ s_p \circ s_0 : p \in D \}$$

be the set of all transvections. It can be shown that $P = \exp \mathfrak{p} = \{ \exp X : X \in \mathfrak{p} \}$ (cf. [99, Proposition 4.6]) although we shall not make use of this fact in the sequel.

There is a complex structure J on $\mathfrak{p}$ satisfying $JX(0) = iX(0)$ for $X \in \mathfrak{p}$ and the complexification

$$\mathfrak{g}_c = \mathfrak{p}_+ \oplus \mathfrak{k}_c \oplus \mathfrak{p}_-$$

given in (3.6), is a TKK Lie algebra with involution

$$\sigma(X + iY) := \theta X - i\theta Y$$

for $X, Y \in \mathfrak{g}$. Hence $\mathfrak{p}_+$ is a Jordan triple, which induces a Jordan triple structure on V via the complex linear isomorphism

$$X(0) \in V \mapsto X - iJX \in \mathfrak{p}_+ \qquad (X \in \mathfrak{p}).$$

To compute the triple product on V, we adjust (3.8) by adding a factor $\frac{1}{8}$ so that, for $X, Y, Z \in \mathfrak{p}$,

$$\{X(0), Y(0), Z(0)\} = \frac{1}{8}\varphi\{X - iJX, Y - iJ, Z - iJ\}_{\mathfrak{p}_+}$$

where triple product in $\mathfrak{p}_+$ is given by

$$
\begin{aligned}
&\frac{1}{8}\{X - iJX,\, Z - iJY,\, X - iJX\}_{\mathfrak{p}_+} \\
={}& \frac{1}{8}[\,[X - iJX,\, \sigma(Y - iJY)],\, X - iJX\,] \\
={}& \frac{1}{8}[\,[X - iJX,\, \theta(Y + iJY)],\, X - iJX\,] \\
={}& \frac{1}{8}[\,[X - iJX,\, -Y - iJY],\, X - iJX\,] \\
={}& -\frac{1}{4}[\,[X,Y],\, X\,] + \frac{1}{4}[\,[JX,Y],\, JX\,] \\
&+ i(\frac{1}{4}[\,[X,Y],\, JX\,] + \frac{1}{4}[\,[JX,Y],\, X\,]) \\
={}& -\frac{1}{4}[\,[X,Y],\, X\,] + \frac{1}{4}[\,[JX,Y],\, JX\,] \\
&- iJ(-\frac{1}{4}[\,[X,Y],\, X\,] + \frac{1}{4}[\,[JX,Y],\, JX\,]).
\end{aligned}
$$

Hence we have

$$\{X(0), Y(0), X(0)\} = -\frac{1}{4}[\,[X,Y],\, X\,](0) + \frac{1}{4}[\,[JX,Y],\, JX\,](0) \quad (3.9)$$

for $X, Y \in \mathfrak{p}$. This defines a Jordan triple product on V by polarization.

Let $X = h\frac{\partial}{\partial z} \in \mathfrak{k}$ where $h : D \longrightarrow V$ is a holomorphic map. By Lemma 3.1.1, the biholomorphic map $\exp tX : D \longrightarrow D$ satisfies $\exp tX(0) = 0$ and hence Lemma 1.1.8 implies that $\exp tX$ is (the restriction of) a continuous linear map on D, for all $t \in \mathbb{R}$. It follows from

$$X(\cdot) = \frac{d}{dt}\bigg|_{t=0} \exp tX(\cdot)$$

that h is (the restriction of) a continuous linear map on D, and by a slight abuse of notation, we may consider h as a continuous linear map on V.

We call a vector field $Z = f\frac{\partial}{\partial z}$ on D a *linear vector field* (respectively, *polynomial vector field*) if $f : D \longrightarrow V$ is (the restriction of) a linear map (respectively, a polynomial).

Let $\mathrm{co}D$ be the convex hull of D and $B = \overline{\mathrm{co}}D$ its closed convex hull. Since D contains $0 \in V$ and is bounded, open, circular and balanced, the Minkowski functional $|\cdot| : V \longrightarrow [0, \infty)$ defined by

$$|x| = \inf\{\lambda > 0 : x \in \lambda(\mathrm{co}D)\} \qquad (x \in V) \qquad (3.10)$$

is a norm, which is equivalent to the original norm $\|\cdot\|$ of V, and we have

$$B = \{x \in V : |x| \leq 1\}$$

as well as $D \subset \{x \in V : |v| < 1\}$.

We are going to show that there is another equivalent norm $\|\cdot\|_{sp}$ on V such that $(V, \|\cdot\|_{sp})$ is a JB*-triple and $D = \{v \in V : \|v\|_{sp} < 1\}$, thereby achieving our goal in this section. A proof of this has been given in [37, Theorem 2.5.27] for the case when D itself is the open unit ball of V in the norm $\|\cdot\|$. In our case, the strategy is to apply the same approach of this proof to the Jordan triple $(V, |\cdot|)$ instead. To avoid undue repetition in what follows, we shall suppress but refer to [37] for some verbatim arguments.

Returning to the preceding vector field $X = h\frac{\partial}{\partial z} \in \mathfrak{k}$ where h is linear, we have, from (1.6),

$$\exp tX(v) = \exp th(v) = v + th(v) + \frac{t^2}{2!}h^2(v) + \frac{t^3}{3!}h^3(v) + \cdots \qquad (t \in \mathbb{R})$$

for $v \in D$. By a slight abuse of notation, we write occasionally $\exp tX(v)$ for $\exp th(v)$, given $v \in V$.

Since $\exp th(D) \subset D$, we have $\exp th(B) \subset B$ by linearity and continuity. In other words, $|\exp th| \le 1$ for all $t \in \mathbb{R}$ which, by [20, p. 46], is equivalent to saying that $ih : V \longrightarrow V$ is a *hermitian* linear operator on the Banach space $(V, |\cdot|)$. We recall that a linear operator T in the Banach algebra $L(V)$ of continuous linear self-maps on V is called *hermitian* if it has real numerical range.

We have shown that each vector field in $\mathfrak{k}$ is linear. Let us now consider the vector fields in $\mathfrak{p}$. For each $\alpha \in D$, there is a unique $X_\alpha \in \mathfrak{p}$ such that $X_\alpha(0) = \alpha$. Let

$$Y_\alpha = \frac{1}{2}(X_\alpha - iJX_\alpha) \in \mathfrak{p}_+$$

where $J : \mathfrak{p} \longrightarrow \mathfrak{p}$ is the complex structure defined earlier. Then we have $Y_\alpha(0) = \alpha$.

Define a holomorphic map $F : D \longrightarrow D$ by

$$F(z) = \exp Y_z(0) \qquad (z \in D).$$

Evidently $F(0) = 0$. For $v \in V$ and sufficiently small $t \in \mathbb{R}$, we have $tv \in D$ and

$$F'(0)(v) = \frac{d}{dt}\bigg|_{t=0} F(tv) = \frac{d}{dt}\bigg|_{t=0} \exp tY_v(0) = Y_v(0) = v.$$

Again Cartan's uniqueness theorem implies that F is the identity map, that is,

$$\exp Y_z(0) = z \qquad (z \in D).$$

Since $[\mathfrak{p}_+, \mathfrak{p}_+] = 0$, we have

$$[Y_\alpha, Y_\beta] = \frac{1}{4}[X_\alpha - iJX_\alpha, X_\beta - iJX_\beta] = 0$$

for all $\alpha, \beta \in D$. Therefore the Campbell-Baker-Hausdorff series (cf. [16]) gives

$$(\exp tY_\beta)(z) = (\exp Y_{t\beta})(\exp Y_z)(0)$$
$$= \exp(Y_{t\beta} + Y_z + \frac{1}{2}[Y_{t\beta}, Y_z] + \frac{1}{12}([Y_{t\beta}, [Y_{t\beta}, Y_z]] - [Y_z, [Y_{t\beta}, Y_z]]) + \cdots)(0)$$
$$= \quad t\beta + z.$$

It follows that

$$Y_\beta(z) = \left.\frac{d}{dt}\right|_{t=0} \exp tY_\beta(z) = \beta$$

is a constant vector field.

Given $X = h\frac{\partial}{\partial z} \in \mathfrak{p}$, the vector field $[X, Y_\beta] \in \mathfrak{k}$ is linear for all Y_β which, together with the fact that Y_β has zero derivative, implies that the derivative of h is linear. Hence we must have

$$h(\cdot) = h(0) + p(\cdot)$$

where p is a homogeneous polynomial on V of degree 2. On D, we can write

$$X(\cdot) = X(0) + p(\cdot). \tag{3.11}$$

Remark 3.2.2. We have shown that each vector field $Z = f\frac{\partial}{\partial z} \in \text{aut} D = \mathfrak{k} \oplus \mathfrak{p}$ is a polynomial vector field in which we consider f as a polynomial on V of degree at most 2. To simplify notation, we shall write $Z(v)$ for $f(v)$ and $Z'(v)$ for $f'(v)$, given $v \in V$, in the next two lemmas.

Next, we compute the polynomial p in (3.11).

Lemma 3.2.3. *Let $(V, |\cdot|)$ be equipped with the Jordan triple product $\{\cdot, \cdot, \cdot\}$ in (3.9). Then in the eigenspace decomposition $\mathfrak{g} = \text{aut} D = \mathfrak{k} \oplus \mathfrak{p}$, each $X \in \mathfrak{p}$ is of the form*

$$X(z) = X(0) - \{z, X(0), z\} \qquad (z \in D).$$

Proof. The Jordan triple product of $a = X(0), b = Y(0)$ and $c = Z(0)$ in V is given by

$$\{a, b, c\} = \{X, Y, Z\}_{\mathfrak{p}}(0).$$

In the proof of Lemma 3.1.3, we have shown that $Z'(0) = 0$ for all $Z \in \mathfrak{p}$.

Let $X \in \mathfrak{p}$. By (3.11), we have

$$X(\cdot) = a - p_a(\cdot)$$

where $a = X(0)$ and $p_a(v) = P_a(v, v)$ is a homogeneous polynomial on V of degree 2, P_a being the polar form of p_a.

Pick any $z \in D$ with $z = Y(0)$ and $Y(\cdot) = z - p_z(\cdot)$, where $p_z(v) = P_z(v, v)$ is a homogeneous polynomial of degree 2. We have

$$[Y, X](v) = X'(v)(Y(v)) - Y'(v)(X(v)) = 2P_a(v, Y(v)) - 2P_z(v, X(v)).$$
$$(3.12)$$

Hence

$$[[Y, X], Y](0) = -[Y, X]'(0)(Y(0))$$
$$= -2P_a'(0, Y(0))(Y(0)) + 2P_z'(0, X(0))(Y(0))$$
$$= -2P_a(0, Y'(0)(Y(0))) - 2P_a(Y(0), Y(0))$$
$$+ 2P_z(0, X'(0)(Y(0))) + 2P_z(Y(0), X(0))$$
$$= -2P_a(z, z) + 2P_z(z, a).$$

Likewise, we have

$$[[JY, X], JY](0) = 2P_a(z, z) + 2P_z(z, a)$$

which gives

$$\{Y, X, Y\}_{\mathfrak{p}}(0) = -\frac{1}{4}[[Y, X], Y](0) + \frac{1}{4}[[JY, X], JY](0) = P_a(z, z).$$

Therefore

$$X(z) = X(0) - p_a(z) = a - P_a(z, z) = a - \{z, a, z\}.$$

$\square$

Remark 3.2.4. Since both P_a and $\{\cdot, a, \cdot\}$ above are symmetric bilinear maps, we have $P_a(z, w) = \{z, a, w\}$. Also, $P_{ia} = -iP_a$ since the triple product is conjugate linear in the middle variable.

Lemma 3.2.5. *On the Banach space* $(V, |\cdot|)$, *the operator* $a \square b + b \square a :$ $V \longrightarrow V$ *is a hermitian operator for* $a, b \in V$. *In particular,* $a \square a$ *is hermitian.*

Proof. Let $X, Y \in \mathfrak{p}$ be such that $X(0) = a$ and $Y(0) = ib$. By (3.12), we have

$$
\begin{aligned}
[Y, X](v) &= 2P_a(v, Y(v)) - 2P_{ib}(v, X(v)) \\
&= 2\{v, a, (ib - \{v, ib, v\})\} - 2\{v, ia, (a - \{v, a, v\})\} \\
&= 2\{v, a, ib\} - 2\{v, ib, a\} = 2((ib) \square a - a \square (ib))(v). \quad (3.13)
\end{aligned}
$$

Since $[Y, X] \in \mathfrak{k}$, we have shown before that $i[Y, X]$ is a hermitian linear map on V. In other words,

$$b \square a + a \square b = -\frac{i}{2}(2((ia) \square a - a \square (ia))) = -\frac{i}{2}[Y, X]$$

is hermitian. $\square$

Remark 3.2.6. Given a complex symmetric Banach manifold M, the *associated* Jordan triple constructed in [98] is the tangent space T_aM at a point $a \in M$, with a *unique* Jordan triple product $\{\cdot, \cdot, \cdot\}_a$ satisfying

$$\mathfrak{p}_a = \{(\alpha - \{z, \alpha, z\}_a)\frac{\partial}{\partial z} : \alpha \in T_aM\}$$

in the eigenspace decomposition aut $M = \mathfrak{k}_a \oplus \mathfrak{p}_a$ induced by $\theta = Ad(s_a)$ [98, (2.9)]. The triple product $\{\cdot, \cdot, \cdot\}_a$ is constructed by identifying a with $0 \in T_a M$ via a local chart. In view of Lemma 3.2.3 for the bounded symmetric domain D, the above triple product $\{\cdot, \cdot, \cdot\}_a$ associated to a point $a \in D$ coincides with the one constructed in (3.9), using the symmetry s_0 at $0 \in V = T_0 D$, by uniqueness.

The preceding lemma reveals that $a \,\square\, a$ is a bounded linear operator on $(V, |\cdot|)$. Let us estimate the norm of $a \,\square\, a = -\dfrac{i}{4}[Y, X]$. We recall from (1.15) that the norm $\|[Y, X]\|_{\text{aut } D}$ of the vector field $[Y, X]$ in the Banach Lie algebra aut D is given by

$$\sup\{|[Y, X](z)| : z \in B_0\}$$

at some neighbourhood B_0 of $0 \in D$ and hence $rD \subset B_0$ for some $r > 0$. Since, by definition of the norm in (1.15), $|Z(0)| \leq \|Z\|$ for $Z \in \text{aut } D$, the evaluation map $Z \in \mathfrak{p} \mapsto Z(0) \in V$ is a real continuous linear isomorphism. By the open mapping theorem, there is a constant $k > 0$ such that

$$\|Z\|_{\text{aut } D} \leq k|Z(0)| \qquad (Z \in \mathfrak{p}). \tag{3.14}$$

It follows that

$$
\begin{aligned}
\|a \,\square\, a\| &= \sup_{z \in D} \|a \,\square\, a(z)\| = \sup_{z \in D} \left\| -\frac{i}{4}[Y, X](z) \right\| \\
&\leq \frac{1}{4r} \|[Y, X]\|_{\text{aut } D} \leq \frac{K}{4r} \|Y\|_{\text{aut } D} \|X\|_{\text{aut } D} \\
&\leq \frac{k^2 K}{4r} |Y(0)| |X(0)| = \frac{k^2 K}{4r} |a|^2. \tag{3.15}
\end{aligned}
$$

Given $a = X(0)$, $b = Y(0)$ and $c = Z(0)$, using similar arguments as above and the formula for the triple product in (3.9), we see that the triple product $\{\cdot, \cdot, \cdot\}$ is continuous:

$$|\{a, b, c\}| \leq C|a||b||c| \tag{3.16}$$

where the constant $C > 0$ is independent of a, b and c.

Lemma 3.2.7. *Let D be the preceding symmetric domain in the Jordan triple V endowed with the norm $|\cdot|$. Given $a \in \overline{D}\backslash D$, the operator $I - a\,\square\,a : V \longrightarrow V$ is not invertible, where I is the identity operator on V.*

Proof. Assume that $I - a\,\square\,a$ is invertible. We deduce a contradiction.

For each $v \in V$, there is a unique $X_v \in \mathfrak{p}$ such that $X_v(0) = v$. Let

$$X^v = -\frac{1}{2}[X_v, X_a] + X_v \in \mathfrak{k} \oplus \mathfrak{p} = \mathrm{aut}\,D.$$

By Remark 3.2.2 and (3.13), we have

$$X^v(a) = (-\frac{1}{2}[X_v, X_a] + X_v)(a) = (I - a\,\square\,a)(v).$$

Let $F : V \longrightarrow V$ be the holomorphic map

$$F(v) = \exp X^v(a).$$

Then the differential dF_0 is the evaluation map $dF_0(v) = X^v(a) = (I - a\,\square\,a)(v)$ which, by assumption, is a linear isomorphism. Hence, by the inverse function theorem, F maps an open neighbourhood of 0 homeomorphically onto an open subset of the image

$$F(V) = \{\exp X^v(a) : v \in V\}.$$

This would contradict the fact that the image $F(V)$ is contained in the boundary $\partial D = \overline{D}\backslash D$ of D.

To see the latter, we make use of the Carathéodory distance c_D on D, introduced in Remark 1.3.10, which is invariant under biholomorphisms. In particular, the biholomorphic map $\exp X^v : D \longrightarrow D$ is an isometry with respect c_D. By continuity, we have $\exp X^v(\overline{D}) \subset \overline{D}$. If

$\exp X^v(a) \in D$, then we have $\exp X^v(a) = \exp X^v(z)$ for some $z \in D$. Let (a_n) be a sequence in D such that $\lim_n |a_n - a| = 0$. Then we have

$$\exp X^v(a_n) \to \exp X^v(a) = \exp X^v(z)$$

in D, as $n \to \infty$. By Lemma 3.5.3, shown later, there exists $\alpha > 0$ such that

$$\alpha |a_n - z| \leq c_D(a_n, z) \qquad (n = 1, 2, \ldots).$$

It follows from continuity of c_D, also shown in Proposition 3.5.5 later, that

$$\begin{aligned} 0 \ \leq \ & \lim_{n \to \infty} \alpha |a_n - z| \leq \lim_{n \to \infty} c_D(a_n, z) \\ & = \lim_{n \to \infty} c_D(\exp X^v(a_n), \exp X^v(z)) = 0 \end{aligned}$$

which gives $a = z \in D$ and is impossible. Hence $F(V) \subset \partial D$. This completes the contradiction and the proof. $\qquad\square$

We have already observed that the box operator $a \,\square\, a$ is hermitian and therefore its spectrum $\sigma(a \,\square\, a)$ must lie in $\mathbb{R}$. We will write $\sigma(a \,\square\, a) < t \in \mathbb{R}$ to mean that $\lambda < t$ for all $\lambda \in \sigma(a \,\square\, a)$.

Lemma 3.2.8. *Let $D \subset V$ be as in the preceding lemma. We have*

$$\{a \in V : \sigma(a \,\square\, a) < 1\} \subset D.$$

Consequently $\sigma(a \,\square\, a) \leq 0$ implies $a = 0$.

Proof. Let $\sigma(a \,\square\, a) < 1$. Then the operator $I - a \,\square\, a$ is invertible. We may assume $a \neq 0$ since $0 \in D$. Let

$$\mu = \inf\{\lambda > 0 : a \in \lambda D\}. \tag{3.17}$$

Since D is open, we have $\mu > 0$. Since D is balanced, we would have $a \in D$ if $\mu < 1$. Suppose, for contradiction, that $\mu \geq 1$. Then $I - \frac{a}{\mu} \,\square\, \frac{a}{\mu}$

is invertible. Since $\frac{a}{\mu} \in \overline{D}$ by definition of μ, we must have $\frac{a}{\mu} \in D$ by Lemma 3.2.7. As D is open, we can find $0 < \mu' < \mu$ so that $\frac{a}{\mu'} \in D$, contradicting (3.17). This proves $\mu < 1$.

If $\sigma(a \square a) \leq 0$, then $ta \in D$ for all $t \in \mathbb{R}$ which implies $a = 0$ because D is bounded. $\square$

We shall strengthen Lemma 3.2.5 by showing that the box operator $a \square a$ has non-negative spectrum. Let $V(a)$ be the closed subtriple in V generated by a, which is the closed complex linear span of odd powers of a. We first compare the two spectra $\sigma(a \square a)$ and $\sigma(a \square a|_{V(a)})$.

The results in the next two lemmas for arbitrary Jordan triples endowed with a complete norm and continuous triple product have been proved in detail in [37, Lemma 2.5.20] and [37, Lemma 2.5.21]. We will not repeat the proof here.

Lemma 3.2.9. *Let $b \in V(a)$ and let $b \square b : V \longrightarrow V$ have real spectrum $\sigma(b \square b)$. Then the spectrum $\sigma(b \square b|_{V(a)})$ of the operator $b \square b|_{V(a)} : V(a) \longrightarrow V(a)$ is contained in $\sigma(b \square b)$.*

Lemma 3.2.10. *Let V be a Jordan triple equipped with a complete norm and continuous triple product such that $a \square a : V \longrightarrow V$ is hermitian for each $a \in V$. Let $V(a)$ be the closed subtriple generated by an element $a \in V$. Then we have*

$$\sigma(a \square a) \subset \frac{1}{2}(S + S)$$

where $S = \sigma(a \square a|_{V(a)}) \cup \{0\}$. Also, the spectrum $\sigma(B(a, a))$ of the Bergman operator $B(a, a) : V \longrightarrow V$ is contained in $(1 - S)(1 - S)$.

Remark 3.2.11. Given $a \in V$ with $\sigma(a \square a) < 1$, the preceding lemma implies that the Bergman operator $B(a, a)$ is invertible and has non-negative spectrum. Therefore the square roots $B(a, a)^{\pm 1/2}$ are well-defined.

Now we resume the discussion of the Jordan triple $(V, |\cdot|)$.

It follows from Lemma 3.2.5 that the operator $b \square b|_{V(a)}$ is hermitian in the Banach algebra $L(V(a))$ of bounded complex linear operators on $V(a)$. Let $\overline{V(a)}_0$ be the closed linear span of $V(a) \square V(a)|_{V(a)}$ in $L(V(a))$. Since $V(a)$ is an abelian Jordan triple system by power associativity, $\overline{V(a)}_0$ is an abelian closed subalgebra of $L(V(a))$. Moreover, polarization gives

$$4x \square y = (x+y) \square (x+y) - (x-y) \square (x-y)$$
$$+ i((x+iy) \square (x+iy) - (x-iy) \square (x-iy)) \qquad (x, y \in V(a)).$$

Hence $\overline{V(a)}_0$ is the complexification of the hermitian elements in $\overline{V(a)}_0$, and by the Vidav-Palmer theorem [20, p. 65], $\overline{V(a)}_0$ is an abelian C*-algebra with involution

$$(h + ik)^* = h - ik$$

where h and k are hermitian elements in $\overline{V(a)}_0$.

Proposition 3.2.12. *For each* $a \in (V, |\cdot|)$, *the spectrum* $\sigma(a \square a)$ *of the box operator*

$$a \square a : V \longrightarrow V$$

is contained in $[0, \infty)$.

Proof. We will follow the proof given in [37, Proposition 2.5.22]. However, there is an issue concerning the notion of spectrum in this proof which seems somewhat unclear. We will make it clearer here.

Let $V(a)$ be the closed subtriple in V generated by a. By the preceding lemma, it suffices to show that $\sigma(a \square a|_{V(a)}) \subset [0, \infty)$. As remarked before Lemma 3.2.10, the closed linear span $\overline{V(a)}_0$ of

$$V(a) \square V(a)|_{V(a)} \subset L(V(a))$$

is an abelian C*-algebra.

The (quasi) spectrum σ of the hermitian element $a\,\square\,a|_{V(a)}$ in the C*algebra $\overline{V(a)}_0$ is defined to be the spectrum of $a\,\square\,a|_{V(a)}$ in the unit extension $\overline{V(a)}_0 \oplus \mathbb{C} \subset L(V(a))$. We note that the spectrum $\sigma(a\,\square\,a|_{V(a)})$ of $a\,\square\,a|_{V(a)}$ in $L(V(a))$ is contained in σ.

The C*-subalgebra $\mathcal{A}(a\,\square\,a)$ generated by $a\,\square\,a|_{V(a)}$ in $\overline{V_0(a)}$ is isometrically isomorphic to the C*-algebra $C_0(\sigma\backslash\{0\})$ of complex continuous functions on $\sigma\backslash\{0\}$, vanishing at infinity, and $a\,\square\,a|_{V(a)}$ identifies with the identity function on $\sigma\backslash\{0\}$.

If $(\sigma\backslash\{0\}) \cap (-\infty, 0) \neq \emptyset$, then there is a non-zero function $f \in C_0(\sigma\backslash\{0\})$ vanishing on $\sigma\cap[0,\infty)$. Identify f as an element in $\mathcal{A}(a\,\square\,a) \subset L(V(a))$, which is the limit of a sequence of polynomials $p_n \in \mathcal{A}(a\,\square\,a)$. By power associativity, we have $p_n(a)\,\square\,p_n(a)|_{V(a)} = (p_n^2)(a\,\square\,a)|_{V(a)}$, where $p_n \in L(V(a))$ and the latter is the product of p_n^2 and $(a\,\square\,a)|_{V(a)}$ in $\mathcal{A}(a\,\square\,a)$. It follows that $f(a)\,\square\,f(a)|_{V(a)} = (f^2)(a\,\square\,a)|_{V(a)}$, where the latter identifies as a product of two functions in $C_0(\sigma\backslash\{0\})$, with $(a\,\square\,a)|_{V(a)}$ being the identity function on $\sigma\backslash\{0\}$. As an element in $C_0(\sigma\backslash\{0\})$, the function $(f^2)(a\,\square\,a)|_{V(a)}$ has non-positive spectrum. Hence $\sigma(f(a)\,\square\,f(a)|_{V(a)}) \leq 0$ and also $\sigma(f(a)\,\square\,f(a)) \leq 0$. By Lemma 3.2.8, we have $f(a) = 0$, contradicting $(f^2)(a\,\square\,a)|_{V(a)} \neq 0$.

Hence we have $\sigma(a\,\square\,a|_{V(a)}) \subset \sigma \subset [0,\infty)$. $\qquad\qquad\square$

Remark 3.2.13. Considering $a\,\square\,a$ as a hermitian element in $L(V)$ and its closed linear span as before, we see that $a\,\square\,a$ is contained in an abelian a C*-algebra and in particular, we have $\|(a\,\square\,a)^3\| = \|a\,\square\,a\|^3$.

Corollary 3.2.14. *For the Jordan triple* $(V, |\cdot|)$, *we have*

$$\{a \in V : \sigma(a\,\square\,a|_{V(a)}) < 1\} = \{a \in V : \sigma(a\,\square\,a) < 1\}$$

where $V(a)$ *denotes the closed subtriple generated by* a, *and the following*

$$\|a\|_{sp} := \|a\,\square\,a\|^{1/2} \qquad (a \in V)$$

defines a norm on V.

Proof. The first assertion follows from Lemmas 3.2.9 and 3.2.10.

Since $a \square a$ is an hermitian operator on V, the norm $\|a \square a\|$ equals the spectral radius of $a \square a$. If $\|a \square a\| = 0$, then Lemma 3.2.8 implies $a = 0$. It remains to show that $\| \cdot \|_{sp}$ satisfies the triangle inequality.

Let $\|a\|_{sp}, \|b\|_{sp} \leq 1$. We show $\|a + b\|_{sp} \leq 2$. Observe that

$$4 - (a+b) \square (a+b) = 2(I - a \square a) + 2(I - b \square b) + (a - b) \square (a - b)$$

where I is the identity operator on V and the operator on the right-hand side has non-negative spectrum. Hence $\sigma((a+b) \square (a+b)) \subset [0, 4]$ concludes the proof. $\qquad\square$

By Lemma 3.2.8, the ball $D_\infty := \{a \in V : \|a\|_{sp} < 1\}$ is contained in the open unit ball of V, equivalently, $|\cdot| \leq \|\cdot\|_{sp}$. Hence these two norms on V are equivalent by (3.15).

We are almost ready to prove the main theorem in this section. Let $\mathfrak{g} = \mathfrak{k} \oplus \mathfrak{p}$ be the decomposition in (3.2), which is induced by the symmetry s_0 at $0 \in D$. Following the method of integrating quadratic polynomial vector fields in [99], one can find an explicit description of the biholomorphic maps $\exp X : D \longrightarrow D \subset V$ for $X \in \mathfrak{p}$. In finite dimensions, this has also been obtained by Loos in [125, Proposition 9.8]. We omit the derivation, but refer to [99, 125] for details.

Let V be a Jordan triple, which is endowed with a complete norm so that the triple product is continuous and the box operator $a \square a :$ $V \longrightarrow V$ is hermitian for each $a \in V$.

For each $a \in V$, let X^a be the constant analytic vector field on V taking value a and write $X^a = a \frac{\partial}{\partial z}$. Then it is easy to see that $\exp X^a : V \longrightarrow V$ is the translation $z \in V \mapsto a + z$ since the flow

$$\alpha(t, z) := ta + z \qquad (t \in \mathbb{R}, z \in V)$$

solves the initial value problem

$$\frac{d}{dt}\alpha(t,z) = X(\alpha(t,z)), \quad \alpha(0,z) = z.$$

For each $a \in V$, we define a homogeneous quadratic polynomial $q_a \in \mathcal{P}^2(V,V)$ by

$$q_a(v) = \{v,a,v\} \qquad (v \in V).$$

We wish to integrate the polynomial vector field $X_a = (a - q_a)\frac{\partial}{\partial z}$ on V. Consider the Banach space $\mathcal{P}$ of quadratic polynomials on V, which is the ℓ_∞-sum

$$\mathcal{P} = \mathcal{P}^0(V,V) \oplus \mathcal{P}^1(V,V) \oplus \mathcal{P}^2(V,W)$$

where $\mathcal{P}^1(V,V) = L(V)$. For each quadratic polynomial vector field $q\frac{\partial}{\partial z}$ on V, its norm $\|q\frac{\partial}{\partial z}\|$ is defined to be that of $q \in \mathcal{P}$. Denote by

$$Q = \overline{\{q_a : a \in V\}} \subset \mathcal{P}^2(V,V)$$

the closure of $\{q_a : a \in V\}$. Then it has been shown in [98, §3] that the direct sum

$$\mathfrak{l} = \left\{ a\frac{\partial}{\partial z} : a \in V \right\} \oplus \left\{ f\frac{\partial}{\partial z} : f \in L(V), [f\frac{\partial}{\partial z}, q\frac{\partial}{\partial z}] = p\frac{\partial}{\partial z} : q, p \in Q \right\}$$

$$\oplus \left\{ q\frac{\partial}{\partial z} : q \in Q \right\}$$

forms a graded complex Banach Lie algebra of a Lie group L, which acts holomorphically (on a manifold M containing V, and also) on V. In particular, we can consider $\exp X : V \longrightarrow V$ for each $X \in \mathfrak{l}$.

Lemma 3.2.15. *Let V be a Jordan triple, which is endowed with a complete norm so that the triple product is continuous and the box operator $a \,\square\, a : V \longrightarrow V$ is hermitian for each $a \in V$. Let*

$$\Omega = \{v \in V : \sigma(v \,\square\, v) < 1\}.$$

Then for each $a \in \Omega$, there is an element $c \in \Omega$ such that

$$\exp X_a = \exp(a - q_a)\frac{\partial}{\partial z} = (\exp X^c) \circ B(c, c)^{1/2} \circ \exp(X_c - X^c),$$

which maps Ω onto Ω.

Conversely, given $c \in \Omega$, there exists $a \in V$ such that

$$(\exp X^c) \circ B(c, c)^{1/2} \circ \exp(X_c - X^c) = \exp X_a.$$

Proof. Cf. [99, (2.25), (4.2), (4.3)]. We note that the square root $B(c, c)^{1/2}$ of the Bergman operator $B(c, c)$ is well-defined by Remark 3.2.11. □

Returning to the preceding symmetric domain D in the Jordan triple $(V, |\cdot|)$ and the decomposition $\mathfrak{g} = \operatorname{aut} D = \mathfrak{k} \oplus \mathfrak{p}$, we have shown that

$$\mathfrak{p} = \{X \in \mathfrak{g} : X = (a - q_a)\frac{\partial}{\partial z} \text{ where } a = X(0)\}$$

in which case $\exp X_a \in \operatorname{Aut} D$ for each $a \in V = \{X(0) : X \in \mathfrak{p}\}$ and by Lemma 3.2.15, $a \in D_\infty$ implies

$$\exp X_a = (\exp X^c) \circ B(c, c)^{1/2} \circ \exp(X_c - X^c)$$

for some $c \in D_\infty$. The biholomorphic map $\exp X_a$ plays an important role in the geometry of D and we give it a name.

Definition 3.2.16. For each $c \in D_\infty$, the biholomorphic map

$$g_c := (\exp X^c) \circ B(c, c)^{1/2} \circ \exp(X_c - X^c) : D \longrightarrow D$$

is called the *Möbius transformation* on D induced by c.

The Möbius transformation g_c appears somewhat formidable to compute. To facilitate computation, let us express it in terms of the Jordan triple product of V, which will be more manageable. We have already noted that $\exp X^c(z) = c + z$ for $z \in V$. The vector field $X_c - X^c$

is the quadratic polynomial vector field $-\{z, c, z\}\frac{\partial}{\partial z}$, for which we need to solve

$$\frac{d}{dt}\alpha(t, z) = -\{\alpha(t, z), c, \alpha(t, z)\}, \quad \alpha(0, z) = z.$$

We define a local flow

$$\alpha : \left(-\frac{1}{\sqrt{\delta}}, \frac{1}{\sqrt{\delta}}\right) \times \{z \in V : \sigma(z \square z) < \delta < 1\} \longrightarrow V$$

by $\alpha(t, z) = (I + tz \square c)^{-1}(z)$, where $I : V \longrightarrow V$ is the identity map. For this to be well-defined, we need to show that the linear operator $I + tz \square c$ is invertible. Indeed, we have $\sigma(tz \square tz) < 1$ and by polarization, we can write

$$tz \square c = \sum_{\varepsilon^4 = 1} \varepsilon \left(\frac{tz + \varepsilon c}{2}\right) \square \left(\frac{tz + \varepsilon c}{2}\right)$$

which implies that the numerical range $N(tz \square c)$ of $tz \square c$ is contained in the square $(-1, 1)^2 \subset \mathbb{C}$ by Lemma 3.2.10. Replacing tz by λtz in the above argument for all complex numbers λ of unit modulus, we see that

$$N(tz \square a) \subset \bigcap_{|\lambda|=1} \lambda(-1, 1)^2 = \{z \in \mathbb{C} : |z| < 1\} \tag{3.18}$$

and $1 \notin \sigma(tz \square c)$.

For α defined above, we have $\alpha(0, z) = z$ and

$$\frac{d}{dt}\alpha(t, z) = -(I + tz \square c)^{-2}(z \square c)(z)$$

which, by the triple identities (2.14) and (2.15), can be written as

$$-\{(I + tz \square c)^{-1}(z), c, (I + tz \square c)^{-1}(z)\}$$

by observing that

$$(I + tz \square c)^{-1}(z) = z - t\{z, c, z\} + t^2\{z, c, \{z, c, z\}\} - \cdots$$

and

$$(I + tz \square c)^{-2}(z) = z - 2t\{z, c, z\} + 3t^2\{z, c, \{z, c, z\}\} - \cdots .$$

This shows that $\alpha(t, z)$ is a local flow for $X_c - X^c$ and we have

$$\exp(X_c - X^c)(z) = (I + z \square c)^{-1}(z) \quad \text{for} \quad \sigma(z \square z) < 1.$$

It follows that, for $\sigma(c \square c) < 1$ and $\sigma(z \square z) < 1$, we have the formula

$$g_c(z) = c + B(c, c)^{1/2}(I + z \square c)^{-1}(z) \in D_\infty. \qquad (3.19)$$

In particular, we have

$$g_c(0) = c \quad \text{and} \quad g_c'(0) = B(c, c)^{1/2} \qquad (3.20)$$

by observing the derivatives

$$d\{z, c, z\} = 2z \square c, \quad d\{z, c, \{z, c, z\}\} = 2(z \square c)^2 + Q_z(c) \square c$$

and so on.

Example 3.2.17. Consider $\mathbb{C}$ as a Jordan triple in the triple product $\{x, y, z\} = x\bar{y}z$, we have $(c \square c)(z) = |c|^2 z$ and

$$B(c, c)(z) = z - 2|c|^2 z + |c|^4 z = (1 - |c|^2)^2 z \qquad (z \in \mathbb{C}).$$

The formula in (3.19) reduces to

$$g_c(z) = \frac{c + z}{1 + \bar{c}z} \qquad (z \in \mathbb{D} = \{z \in \mathbb{C} : \sigma(z \square z) < 1\})$$

which is the usual Möbius transformation on $\mathbb{D}$.

After some considerable effort and lengthy arguments in the preceding exposition, we are now ready to complete the proof of the main theorem in this section, which realises bounded symmetric domains as open unit balls of JB*-triples and can be regarded as a generalisation of the Riemann mapping theorem.

Theorem 3.2.18. *Let D be a bounded symmetric domain in a complex Banach space V. Then there is an equivalent norm $\|\cdot\|_{sp}$ on V such that $(V, \|\cdot\|_{sp})$ is a JB*-triple and D is biholomorphic to the open unit ball D_∞ of $(V, \|\cdot\|_{sp})$.*

Proof. Given a bounded symmetric domain D_0 in a Banach space V with a norm $\|\cdot\|$, we have noted earlier that, by a result in [172], D_0 is biholomorphic to a circular and balanced bounded symmetric domain D containing the origin $0 \in V$. By (3.10), V can be endowed with an equivalent norm $|\cdot|$ so that $D \subset \{v \in V : |v| < 1\}$.

The automorphism group $\operatorname{Aut} D$ of D is a real Banach Lie group and the symmetry s_0 at $0 \in D$ induces an involution θ on the Lie algebra $\operatorname{aut} D$ of complete holomorphic vector fields on D. Using the Tits-Kantor-Koecher construction, one can define a Jordan triple product on V, shown in (3.9), and by (3.16), this triple product is continuous in the norm $|\cdot|$. By Lemma 3.2.14, we can define a norm on V by

$$\|a\|_{sp}^2 = \|a \square a\| \qquad (a \in V). \tag{3.21}$$

It follows from Lemma 3.2.8 and a previous remark that $\|\cdot\|_{sp}$ is equivalent to the norm $|\cdot|$, and hence to the original norm of V as well, and the open unit ball in the norm $\|\cdot\|_{sp}$ satisfies

$$D_\infty = \{v \in V : \|v\|_{sp} < 1\} \subset D.$$

Further, $(V, \|\cdot\|_{sp})$ is a JB*-triple by Lemma 3.2.5, Proposition 3.2.12 and (3.21).

To complete the proof, it suffices to show $D_\infty = D$. Indeed, if D *properly* contains D_∞, then one can find a point $v \in D \cap \partial D_\infty$ as D is balanced. We deduce a contradiction to conclude the proof.

We have $|v| < 1$ as $v \in D$. Also, $\|v \square v\| = \|v\|_{sp}^2 = 1$ since v resides in the boundary ∂D_∞. Hence $\sigma(tv \square tv) < 1$, that is, $tv \in D_\infty$, for $0 < t < 1$. Let $\gamma : (0,1) \longrightarrow D_\infty$ be the curve $\gamma(t) = tv$. We show that γ has infinite length in D_∞, giving a contradiction.

At each point $tv \in D_\infty$, the tangent space $T_{tv}D_\infty = V$ is endowed with the Carathéodory tangent norm (cf. Example 1.3.6)

$$\|u\|_{tv} = \sup\{|f'(tv)(u)| : f \in H(D_\infty, U), f(tv) = 0\} \qquad (u \in V).$$

Let $g_{tv} : D \longrightarrow D$ be the Möbius transformation in Definition 3.2.16. By (3.19), the restriction $g_{tv} : D_\infty \longrightarrow D_\infty$ has the form

$$g_{tv}(z) = tv + B(tv, tv)^{1/2}(I + z \,\square\, tv)^{-1}(z) \qquad (z \in D_\infty).$$

We have $g_{tv}(0) = tv$ and $g'_{tv}(0) = B(tv, tv)^{1/2}$. When restricted to the closed subtriple $V(v)$ generated by v in V, the Bergman operator $B(tv, tv)$ has the form

$$B(tv, tv) = id_v - 2t^2 v \,\square\, v + t^4 (v \,\square\, v)^2 = (id_v - t^2(v \,\square\, v))^2 \qquad (3.22)$$

by power associativity, where id_v is the identity map on $V(v)$. Therefore $B(tv, tv)^{1/2}v = (id_v - t^2(v \,\square\, v))(v)$. Let $f = g_{tv}^{-1}|_{D_\infty}$ so that $f(tv) = 0$ and

$$\begin{aligned} f'(tv)(v) &= g'_{tv}(0)^{-1}(v) = (id_v - t^2(v \,\square\, v))^{-1}(v) \\ &= v + t^2(v \,\square\, v)(v) + t^4(v \,\square\, v)^2(v) + \cdots. \qquad (3.23) \end{aligned}$$

Let $R(v)$ be the closed real linear span of odd powers of v in V and let $\mathcal{R} \subset L(V)$ be the closed real linear span of the box operators $R(v) \,\square\, R(v)$. Since $R(v)$ is flat, each box operator $a \,\square\, b = \frac{1}{2}(a \,\square\, b + b \,\square\, a)$ is hermitian for $a, b \in R(v)$, by Lemma 3.2.5. Hence we have $\|a \,\square\, b\| \leq \|a\|_{sp}\|b\|_{sp}$ since, using analogous arguments for (3.18), $\|a\|_{sp}, \|b\|_{sp} < 1$ implies $\sigma(a \,\square\, b) \subset \mathbb{D}$. Equip the direct sum

$$\mathcal{A} = R(v) \oplus \mathcal{R}|_{V(v)}$$

with the ℓ_1-norm

$$\|(a, h|_{V(v)})\|_1 = \|a\|_{sp} + \|h|_{V(v)}\| \qquad (a \in R(v), h \in \mathcal{R})$$

and define a product in $\mathcal{A}$ by

$$(a \oplus h)(b \oplus k) = (h(b) + k(a)) \oplus (a \square b|_{V(v)} + hk).$$

Then $\mathcal{A}$ is a real abelian Banach algebra. Since $\|v \square v|_{V(v)}\| = 1$, there exists a real linear character $\chi : \mathcal{A} \longrightarrow \mathbb{C}$ such that $\chi(v \square v|_{V(v)}) = 1$ which implies $\chi(v) \neq 0$.

From (3.23), we have

$$\|f'(tv)(v)\| \geq |\chi(f'(tv)(v))| = |\chi(v) + t^2\chi(v) + t^4\chi(v) + \cdots| = \frac{|\chi(v)|}{1 - t^2}.$$

It follows that

$$\int_0^s \|\gamma'(t)\|_{tv} dt = \int_0^s \|v\|_{tv} dt \geq \int_0^s \frac{|\chi(v)|}{1 - t^2} = \frac{|\chi(v)|}{2} \log \frac{1 + s}{1 - s}$$

for $s < 1$. Letting $s \to 1$, we get the contradiction that γ has infinite length.

Hence $D_\infty = D$ and D_0 is biholomorphic to D_∞. $\qquad\square$

Remark 3.2.19. We have already noted, as in the proof of Theorem 3.2.18, that for all a, b in a JB*-triple, (3.18) implies $\|a \square b\| \leq \|a\|\|b\|$ if $a \square b$ is hermitian. Moreover, since $a \square b + b \square a$ is hermitian, (3.18) also implies $\|a \square b + b \square a\| \leq 2\|a\|\|b\|$, as well as $\|a \square b - b \square a\| = \|ia \square b + b \square ia\| \leq 2\|a\|\|b\|$. It follows that

$$\|a \square b\| \leq 2\|a\|\|b\|. \tag{3.24}$$

In fact, the constant 2 can be dropped in the above inequality. This follows from the following fundamental inequality

$$\|\{a, b, c\}\| \leq \|a\|\|b\|\|c\| \qquad (a, b, c \in V) \tag{3.25}$$

which will be shown in Corollary 3.8.19. By Remark 3.2.13, we also have

$$\|\{a, a, a\}\|^2 = \|\{a, a, a\} \square \{a, a, a\}\| = \|(a \square a)^3\| = \|a \square a\|^3 = \|a\|^6.$$

In particular, $\|a\| = 1$ if a is a non-zero tripotent.

The preceding theorem, together with the one below, shows that bounded symmetric domains are exactly (biholomorphically equivalent to) the open unit balls of JB*-triples. This identification is a valuable device for the geometric function theory of these domains.

Theorem 3.2.20. *A complex Banach space V is a JB*-triple if, and only if, its open unit ball D is a symmetric domain.*

Proof. Let V a JB*-triple. By definition, its open unit ball is just

$$D = \{a \in V : \sigma(a \square a) < 1\}$$

which has a symmetry $s_0(v) = -v$ at the origin 0. Given $a \in V$, the Möbius transformation $g_a : D \longrightarrow D$ satisfies $g_a(0) = a$ and, analogous to (3.22), by considering the restriction of the Bergman operator $B(a, a)$ to the closed subtriple $V(a)$ generated by a, we have

$$
\begin{aligned}
g_{-a}(a) &= -a + B(a, a)^{1/2}(I - a \square a)^{-1}(a) \\
&= -a + (I - a \square a)(I - a \square a)^{-1}(a) = 0.
\end{aligned}
$$

Hence the automorphism group $\operatorname{Aut} D$ acts transitively on D and it follows that D is a symmetric domain (cf. Example 1.3.22).

Conversely, if the open unit ball D of a complex Banach space $(V, \|\cdot\|)$ is symmetric, then the proof of Theorem 3.2.18 shows that there is an equivalent norm $\|\cdot\|_{sp}$ such that $(V, \|\cdot\|_{sp})$ is a JB*-triple and $D = D_\infty = \{v \in V : \|v\|_{sp} < 1\}$, which implies $\|\cdot\| = \|\cdot\|_{sp}$ and completes the proof. $\square$

Remark 3.2.21. As remarked before, the seminal work of É. Cartan [32] and Harish-Chandra [78] identifies finite dimensional Hermtian symmetric spaces of non-compact type as bounded symmetric domains in complex Euclidean spaces $\mathbb{C}^n$, using Lie theory. Further, Hermann's convexity theorem [81] reveals that the domain in Harish-Chandra's realisation is actually the open unit ball of a finite dimensional complex Banach space. One can therefore view Theorem 3.2.18 as an infinite dimensional extension of this result, via Jordan theory. This Jordan approach in finite dimensions has also been shown by Loos [125]. Interestingly, it has been shown by Mok and Tsai [131, 132] that any realisation of a finite dimensional non-compact irreducible Hermitian symmetric space of (rank ≥ 2) as a bounded convex domain is, up to a complex affine transformation, the Harish-Chandra realisation.

Concerning the JB*-triple realisation in Theorem 3.2.18, it is also unique in the following sense. Let V be the Jordan triple in the theorem. Then $\|\cdot\|_{sp}$ is the only norm turning V into a JB*-triple. For if $(V, \|\|\cdot\|\|)$ is a JB*-triple, then we must have

$$\|a\|_{sp}^2 = \sup\{|\lambda| : \lambda \in \sigma(a \,\square\, a)\} = \|\|a\|\|^2 \qquad (a \in V).$$

Moreover, if D is biholomorphic to the open unit ball B of another JB*-triple W, then V and W must be linearly isometric and triple isomorphic. Indeed, let $\varphi : D \longrightarrow B$ be a biholomorphic map. Then the Möbius transformation $g := g_{-\varphi(0)} \in \operatorname{Aut} B$ satisfies $g(\varphi(0)) = 0$ and Cartan's uniqueness theorem (cf. Lemma 1.1.8) implies that the derivative $(g \circ \varphi)'(0) : V \longrightarrow W$ is a linear isometry. By the following result, V and W are triple isomorphic.

Theorem 3.2.22. *Let* $\psi : V \longrightarrow W$ *be a surjective linear isometry between two JB*-triples V and W. Then ψ is a triple isomorphism.*

Proof. Let D_V and D_W be the open unit balls of V and W respectively and let

$$\text{aut}\, D_W = \mathfrak{k} \oplus \mathfrak{p}$$

be the eigenspace decomposition induced by the symmetry $s_0(w) = -w$ at $0 \in D_W$.

Let $a \in V$. Then there is a unique vector field $Y_{\psi(a)} \in \mathfrak{p}$ such that $Y_{\psi(a)}(0) = \psi(a)$. Likewise, let X_a be the unique complete holomorphic vector field on D_V such that $X_a(0) = a$. Since $\psi : D_V \longrightarrow D_W$ is a surjective linear isometry, $\psi X_a \psi^{-1}$ is a complete holomorphic vector filed on D_W satisfying

$$(\psi X_a \psi^{-1})(0) = \psi(a).$$

Hence we have $\psi X_a \psi^{-1} = Y_{\psi(a)}$. This gives, by Lemma 3.2.3,

$$
\begin{aligned}
\psi(a) - \{\psi(z), \psi(a), \psi(z)\} &= Y_{\psi(a)}(\psi(z)) = (\psi X_a \psi^{-1})(\psi(z)) \\
&= \psi(a) - \psi\{z, a, z\} \qquad (z \in D_V)
\end{aligned}
$$

and so

$$\psi\{z, a, z\} = \{\psi(z), \psi(a), \psi(z)\} \qquad (z \in D_V),$$

proving that ψ is a triple isomorphism by polarization. $\qquad\square$

Remark 3.2.23. As a converse to Theorem 3.2.22, it is easy to see that a continuous triple isomorphism $\psi : V \longrightarrow W$ between two JB*-triples V and W must be an isometry since $\psi(a) \square \psi(a) = \psi \circ (a \square a) \circ \psi^{-1}$ implies $\sigma(\psi(a) \square \psi(a)) = \sigma(a \square a)$ for all $a \in V$. In fact, it can be shown that a triple homomorphism between JB*-triples is already continuous, in fact, contractive [37, Corollary 3.1.19].

Corollary 3.2.24. *Let V_1 and V_2 be two closed subtriples of a JB*-triple V. If $V_1 \square V_2 = \{0\}$, then $V_1 + V_2$ is a closed subtriple of V, linearly isometric to the ℓ_∞-sum $V_1 \oplus V_2$.*

Given $x, y \in V$ with $x \square y = 0$, we have $\|x + y\| = \max\{\|x\|, \|y\|\}$.

Proof. By orthogonality, the closure $\overline{V_1 + V_2}$ is a closed subtriple of V. The linear map

$$(x, y) \in V_1 \oplus V_2 \mapsto x + y \in \overline{V_1 + V_2} \subset V$$

is a triple monomorphism and, as remarked above, contractive, that is,

$$\max\{\|x\|, \|y\|\} = \|(x, y)\| \leq \|x + y\| \qquad (x \in V_1, y \in V_2).$$

It follows that $V_1 + V_2$ is norm-complete and hence closed in V. Moreover, $V_1 + V_2$ is triple isomorphic to $V_1 \oplus V_2$ and therefore these two spaces are isometric by the preceding remark.

The last assertion follows from the fact that the closed subtriples $V(x)$ and $V(y)$ in V generated by x and y, respectively, satisfy

$$V(x) \,\square\, V(y) = \{0\}.$$

$\square$

In earlier arguments, we have made use of the Möbius transformations g_c, which will play an important role later in the function theory of bounded symmetric domains. The following facts will be useful.

Lemma 3.2.25. *Let $g_c : D \longrightarrow D$ be the Möbius transformation induced by c in a bounded symmetric domain D, realised as the open unit ball of a JB*-triple V. Its derivative*

$$g_c'(z) : V \longrightarrow V$$

is given by

$$g_c'(z) = B(c, c)^{1/2} B(z, -c)^{-1}$$

for $z \in D$, where the Bergman operator $B(z, -c) = I + 2z \,\square\, c + Q_z Q_c$ on V is invertible and I denotes the identity map. The Möbius transformation g_{-c} is the inverse of g_c.

Proof. We have, for $z \in D$,

$$g_c(z) \; = \; c + B(c,c)^{1/2}(I + z \,\square\, c)^{-1}(z)$$
$$= c + B(c,c)^{1/2}(z - \{z,c,z\} + \{z,c,\{z,c,z\}\} - \{z,c,\{z,c,\{z,c,z\}\}\} + \cdots)$$

and hence

$$g_c'(z) = B(c,c)^{1/2}[I - 2z \,\square\, c + (\{z,c,z\} \,\square\, c + 2(z \,\square\, c)^2)$$
$$- (\underline{\{z,c,\{z,c,z\}\} \,\square\, c} + (z \,\square\, c)(\{z,c,z\} \,\square\, c) + 2(z \,\square\, c)^3) + \cdots]. \quad (3.26)$$

Using the main triple identity and the identity $Q_x(y \,\square\, x) = (x \,\square\, y)Q_x$, the expression in the above square brackets can be written as

$$f(z) := I - 2z \,\square\, c + (4(z \,\square\, c)^2 - Q_z Q_c)$$
$$- (\underline{4(z \,\square\, c)^3 - 3Q_z Q_c(z \,\square\, c)} + 2(z \,\square\, c)^3 - Q_z Q_c(z \,\square\, c) + 2(z \,\square\, c)^3) + \cdots$$
$$= \; I - 2z \,\square\, c + (4(z \,\square\, c)^2 - Q_z Q_c) - (8(z \,\square\, c)^3 - 4Q_z Q_c(z \,\square\, c))$$
$$+ (16(z \,\square\, c)^4 - 12Q_z Q_c(z \,\square\, c)^2 + (Q_z Q_c)^2) - \cdots$$

which gives

$$f(z)(I + 2z \,\square\, c + Q_z Q_c) = I.$$

This proves that $B(z, -c)$ is invertible and $B(z, -c)^{-1} = f(z)$. It follows that

$$g_c'(z) = B(c,c)^{1/2}B(z, -c)^{-1}.$$

In particular, we have $g_c'(-c) = B(c,c)^{1/2}B(-c, -c)^{-1} = B(c,c)^{-1/2}$ and

$$(g_c \circ g_{-c})'(0) = g_c'(-c)g_{-c}'(0) = B(c,c)^{-1/2}B(c,c)^{1/2} = I$$

where $(g_c \circ g_{-c})(0) = 0$. It follows from Cartan's uniqueness theorem that $g_c \circ g_{-c} = I$. $\square$

Remark 3.2.26. We can also extend the Möbius transformation

$$z \in D \mapsto g_c(z) = c + B(c,c)^{1/2}(I + z \square c)^{-1}(z) \in D \subset V$$

to a neighbourhood of $\overline{D}$. Indeed, let $U = \{z \in V : \|z \square c\| < 1\}$, then $\overline{D} \subset U$ (cf. Remark 3.2.19) and the inverse $(I + z \square c)^{-1} : V \longrightarrow V$ exists, which enables us to extend g_c to a holomorphic map on U. The extension of g_c maps $\overline{D}$ to $\overline{D}$ biholomorphically. This extension on $\overline{D}$ will still be denoted by g_c.

On a JB*-triple V, the Bergman operator

$$B(a,b) : x \in V \mapsto x - 2(a \square b)(x) + Q_a Q_b(x) \in V$$

is a continuous linear operator with norm

$$\begin{aligned}
\|B(a,b)x\| &\leq \|x\| + 2\|a\|\|b\|\|x\| + \|a\|^2\|b\|^2\|x\| \\
&= (1 + \|a\|\|b\|)^2\|x\| \qquad (x \in V) \tag{3.27}
\end{aligned}$$

where, by (3.25), we have

$$\|Q_{a,b}(x)\| = \|\{a, x, b\}\| \leq \|a\|\|x\|\|b\|.$$

For $V = \mathbb{C}$, we have $B(a,b)x = (1 - a\overline{b})^2 x$ and $\|B(a,b)\| = |1 - a\overline{b}|^2$.

Example 3.2.27. Let $V = \overset{\ell_\infty}{\underset{\alpha \in \Lambda}{\bigoplus}} V_\alpha$ be an ℓ_∞-sum of JB*-triples V_α. For $\mathbf{a} = (a_\alpha), \mathbf{b} = (b_\alpha)$ and $\mathbf{x} = (x_\alpha)$ in V, the Bergman operator $B(\mathbf{a}, \mathbf{b})\mathbf{x}$ is given by

$$B(\mathbf{a}, \mathbf{b})\mathbf{x} = (B(a_\alpha, b_\alpha)x_\alpha).$$

For each $\mathbf{c} = (c_\alpha)$ in the open unit ball D of V, the Möbius transformation $g_{\mathbf{c}}$ is given by

$$g_{\mathbf{c}}(\mathbf{x}) = (g_{c_\alpha}(x_\alpha)) \qquad (\mathbf{x} \in D).$$

In applications, we often need to estimate the norm of the Bergman operator. The following lemma will be useful.

Lemma 3.2.28. *Let D be the open unit ball of a JB*-triple V. For $a, b \in D$, we have*

(i) $\|B(a,a)^{-1/2}\| = \dfrac{1}{1 - \|a\|^2}$,

(ii) $\dfrac{1}{1 - \|g_{-b}(a)\|^2} = \|B(a,a)^{-1/2} B(a,b) B(b,b)^{-1/2}\|$.

Proof. For $V = \mathbb{C}$, (i) is obvious since $B(a,a)z = (1 - |a|^2)z$ for $z \in \mathbb{C}$. A complete proof of (i) for all JB*-triples has already been given in [37, Proposition 3.2.13]. The identity in (ii) has also been proved in [129] and [37, Lemma 3.2.17]. We repeat the proof as a simple consequence of (i) and Lemma 3.2.25.

Indeed, by Cartan uniqueness theorem, $g_{g_{-b}(a)}(0) = g_{-b} \circ g_a(0)$ implies that

$$g_{g_{-b}(a)} = \varphi \circ g_{-b} \circ g_a$$

for some linear isometry φ on D. By Lemma 3.2.25, we have

$$
\begin{aligned}
B(g_{-b}(a), g_{-b}(a))^{1/2} &= g'_{g_{-b}(a)}(0) \\
&= \varphi \circ g'_{-b}(a) \circ g'_a(0) \\
&= \varphi \circ g'_{-b}(a) \circ B(a,a)^{1/2}.
\end{aligned}
$$

Hence we deduce from (i) that

$$
\begin{aligned}
\frac{1}{1 - \|g_{-b}(a)\|^2} &= \|B(g_{-b}(a), g_{-b}(a))^{-1/2}\| = \|B(a,a)^{-1/2} g'_{-b}(a)^{-1}\| \\
&= \|B(a,a)^{-1/2} B(a,b) B(b,b)^{-1/2}\|.
\end{aligned}
$$

$\square$

Example 3.2.29. Let D be the open unit ball of a Hilbert space, with inner product $\langle \cdot, \cdot \rangle$. The Bergman operator $B(a,b)$ has the form

$$
\begin{aligned}
B(a,b)(v) &= v - 2(a \square b)(v) - \{a\{b,v,b\},a\} \\
&= v - \langle a,b\rangle v - \langle v,b\rangle a + \langle v,b\rangle\langle a,b\rangle a.
\end{aligned}
$$

In particular, $B(a,b)a = (1 - \langle a,b\rangle)^2 a$ and $B(a,b) = 0$ if and only if $\langle a,b\rangle = 1$, equivalent to $a = b \in \partial D$. We also have

$$
B(a,a)(v) = (1 - \|a\|^2)(v - \langle v,a\rangle a) \qquad (v \in V)
$$

and

$$
1 - \|g_{-b}(a)\|^2 = \frac{(1 - \|a\|^2)(1 - \|b\|^2)}{|1 - \langle a,b\rangle|^2} \qquad (a,b \in D). \tag{3.28}
$$

The Möbius transformation g_c for $c \in D$ can be expressed as

$$
g_c(z) = c + \frac{\sqrt{1 - \|c\|^2}}{1 + \langle z,c\rangle}\left(z + (\sqrt{1 - \|c\|^2} - 1)\langle z,c\rangle\frac{c}{\|c\|^2}\right).
$$

Example 3.2.30. Let D be the open unit ball of a closed subtriple V of $L(H)$ over a Hilbert space H. For $a \in V$ and $c \in D$, we have

$$
B(a,a)(x) = x - 2\{a,a,x\} + \{a,\{a,x,a\},a\} = (1 - aa^*)x(1 - a^*a)
$$

for $x \in V$, and the Möbius transformation g_c has the form

$$
\begin{aligned}
g_c(x) &= c + B(c,c)^{1/2}(1 + x \square c)^{-1}(x) \\
&= c + (1 - cc^*)^{1/2}(1 + x \square c)^{-1}x(1 - c^*c)^{1/2} \\
&= c + (1 - cc^*)^{1/2}(1 - x \square c + (x \square c)^2 - (x \square c)^3 + \cdots)x(1 - c^*c)^{1/2} \\
&= c + (1 - cc^*)^{1/2}(x - xc^*x + xc^*xc^*x - xc^*xc^*xc^*x + \cdots)(1 - c^*c)^{1/2} \\
&= c + (1 - cc^*)^{1/2}(1 - xc^* + (xc^*)^2 - (xc^*)^3 + \cdots)x(1 - c^*c)^{1/2} \\
&= c + (1 - cc^*)^{1/2}(1 + xc^*)^{-1}x(1 - c^*c)^{1/2}
\end{aligned}
$$

which coincides with the one introduced in [79] and [142].

Given a tripotent u in a JB*-triple V, we have the following useful description of the linear isometry $\exp it(u \square u) : V \longrightarrow V$ in terms of the Peirce projections of u.

Proposition 3.2.31. *Let u be a tripotent in a JB*-triple V. Then the Peirce projections $P_k(u)$ $(k = 0, 1, 2)$ induced by u are all contractive and we have*

$$\exp it(u \square u) = P_0(u) + e^{it/2} P_1(u) + e^{it} P_2(u).$$

Proof. Since the box operator $u \square u$ has eigenvalues $0, 1/2$ or 1, with corresponding eigenspaces $P_0(u)V, P_1(u)V$ and $P_2(u)V$, the spectral mapping theorem implies that the linear isometry $\exp it(u \square u)$ has eigenvalues $0, e^{it/2}$ or e^{it}, as well as

$$\exp it(u \square u) = P_0(u) + e^{it/2} P_1(u) + e^{it} P_2(u) \qquad (t \in \mathbb{R}).$$

For each $v \in V$, we have

$$\|v\| = \| \exp it(u \square u)v\| = \|P_0(u)v + e^{it/2} P_1(u)v + e^{it} P_2(u)v\|$$

for all $t \in \mathbb{R}$. It follows that

$$\|P_k(u)v\| \leq \|v\| \qquad (k = 0, 1, 2).$$

$\square$

Putting $t = 2\pi$ in Proposition 3.2.31 gives

$$\exp 2\pi i(u \square u) = P_0(u) - P_1(u) + P_2(u) = B(u, 2u) \qquad (3.29)$$

where $B(u, 2u) : V \longrightarrow V$ is the Bergmann operator and the last equality follows from (2.35). The isometry $\exp 2\pi i(u \square u)$ is involutive and fixes u. It is called the *Peirce symmetry* at u, and by Theorem 3.2.22, it is a triple isomorphism. We note that

$$P_0(u) + P_2(u) = \frac{1}{2}(1 + B(u, 2u))$$

which is a contractive projection on V.

As in the case of the unit disc $\mathbb{D}$, the automorphism group $\mathrm{Aut}\,D$ of a bounded symmetric domain D is determined by the Möbius transformations.

Theorem 3.2.32. *Let D be a bounded symmetric domain realised as the open unit ball of a JB*-triple V. Then we have*

$$\mathrm{Aut}\,D = \{\varphi \circ g_a : a \in D, \ \varphi \text{ is a linear isometry of } V\}.$$

Proof. Let $g \in \mathrm{Aut}\,D$ and $a = g^{-1}(0)$. Let $\varphi = g_{-a}g^{-1}$. Then $\varphi \in \mathrm{Aut}\,D$ and $\varphi(0) = 0$. By Cartan's uniqueness theorem (cf. Lemma 1.1.8), φ is a linear isometry and we have $g = \varphi^{-1}g_{-a}$. $\qquad\square$

In what follows, we will identify a bounded symmetric domain as the open unit ball of a (unique) JB*-triple and discuss diverse applications of this realisation in geometry and analysis of these domains.

Notes. The fundamental correspondence between bounded symmetric domains and JB*-triples in Theorems 3.2.18 and 3.2.20 has been established in the seminal paper of Kaup [99]. Prior to this, Loos showed in [125, Theorem 4.1] the correspondence between finite dimensional circled bounded symmetric domains and Jordan pairs with involution. A proof of the special case of Theorem 3.2.18 for open unit balls of Banach spaces has been presented in [37, Theorem 2.5.26].

3.3   Rank of a bounded symmetric domain

The realisation of a bounded symmetric domain as the open unit ball of a JB*-triple provides a useful Jordan description of the rank of a bounded symmetric domain of any dimension. For a finite dimensional bounded symmetric domain D in a complex Euclidean space $V = \mathbb{C}^n$, with the Cartan decomposition

$$\mathfrak{g} = \mathfrak{k} \oplus \mathfrak{p}$$

of the real Banach Lie algebra $\mathfrak{g}$ of complete holomorphic vector fields on D, the *rank* of D is defined to be the common dimension of a maximal abelian subspace $\mathfrak{a} \subset \mathfrak{p}$, where $\mathfrak{p}$ is real linear isomorphic to V.

As shown previously, V becomes a JB*-triple in an equivalent norm and D is biholomorphic to (and identified with) the open unit ball of V in this norm. Let

$$\mathfrak{g}_c = \mathfrak{p}_- \oplus \mathfrak{k}_c \oplus \mathfrak{p}_+$$

be the complexification in (3.6) with the complex linear isomorphism

$$\psi : X \in \mathfrak{p} \mapsto X - iJX \in \mathfrak{p}_+$$

where $\mathfrak{p}$ is equipped with the complex structure J, in which case $\mathfrak{p}$ is complex linear isomorphic to V via the map $X \in \mathfrak{p} \mapsto X(0)$.

To formulate the rank of D in terms of the Jordan structures of V, we first show that the (real) abelian subspaces of $\mathfrak{p}$ correspond to real flat subspaces of $\mathfrak{p}_+$.

Lemma 3.3.1. *Given* $X, Y \in \mathfrak{p}$, *the following conditions are equivalent.*

(i) $[X, Y] = 0$.

(ii) $[X - iJX, \sigma(Y - iJY)] = [Y - iJY, \sigma(X - iJX)]$.

In particular, a real subspace $\mathfrak{a}$ *of* $\mathfrak{p}$ *is abelian if and only if its isomorphic image*

$$\{X(0) : X \in \mathfrak{a}\} \subset V$$

is flat.

Proof. The first assertion follows from the direct computation

$$[X - iJX, \sigma(Y - iJY)] = [X - iJX, -Y - iJY] = -2[X,Y] + 2i[JX,Y]$$

and $[Y - iJY, \sigma(X - iJX)] = 2[X,Y] + 2i[JX,Y]$.

To conclude the proof, we note that $X(0) \,\square\, Y(0) = Y(0) \,\square\, X(0)$ if and only if condition (ii) above holds, by the definition of the triple product in V. $\square$

By [125, Proposition 5.2, Theorem 5.3], all (real) maximal flat subspaces of a finite dimensional JB*-triple have the same dimension and are of the form

$$\mathbb{R}e_1 \oplus \cdots \oplus \mathbb{R}e_r \qquad\qquad (3.30)$$

for some mutually orthogonal minimal tripotents $e_1, \ldots, e_r$.

It follows from the preceding lemma that the rank of a finite dimensional bounded symmetric domain D, realised as the open unit ball of a JB*-triple V, is the common dimension of a real maximal flat subspace of V, which is also used as the definition of the *rank* of V in this case. We note from (3.30) that the rank is exactly the maximal number of mutually orthogonal minimal tripotents in V.

We now consider JB*-triples and domains of any dimension. Let $a = e_1 + 2e_2 + \cdots + re_r$ say, in (3.30). Then it can be seen that the smallest real closed subtriple $R(a)$ of V containing a, being the real linear span of odd powers of a, is just the given maximal flat subspace itself and hence the dimension of $R(a)$ is the rank of V. For arbitrary $a \in V$, we define $R(a)$ as before. Let $V(a)$ be the smallest closed subtriple of V containing a. Then we have

$$V(a) = R(a) + iR(a)$$

where $V(a)$ is an abelian subtriple and $R(a)$ is flat.

Definition 3.3.2. Let D be a bounded symmetric domain, realised as the open unit ball of a JB*-triple $V \neq \{0\}$. The *rank* of D, and of V also, is defined to be

$$\operatorname{rank} D = \operatorname{rank} V := \sup\{\dim V(a) : a \in V\} \in \mathbb{N} \cup \{\infty\}.$$

By the discussions following Lemma 3.3.1, this definition of the rank coincides with the one for finite dimensional bounded symmetric domains defined in terms of the Cartan decomposition. Evidently, if V is finite dimensional (and $V \neq \{0\}$), then the rank of V is a positive integer. We note that, however, a JB*-triple of finite rank, that is,

$$\sup\{\dim V(a) : a \in V\} \in \mathbb{N},$$

need not be finite dimensional.

Example 3.3.3. The rank of a Hilbert space V is 1 since for each $a \in V \backslash \{0\}$, $\dim V(a) = \dim V(a/\|a\|) = 1$, where $a/\|a\|$ is a tripotent. The JB*-triple $C[0,1]$ of complex continuous functions on $[0,1]$ has infinite rank since, for instance, $\dim V(a) = \infty$ for the function $a(x) = x$.

Let $\operatorname{rank} V < \infty$ and $a \in V$. By Theorem 2.4.21, the JB*-triple $V(a)$ is linearly isometric to the JB*-triple $C_0(S_a)$ of continuous functions vanishing on a locally compact space S_a, and if $\dim V(a) = r < \infty$, then $C_0(S_a)$ is just the JB*-triple $\mathbb{C}^r$ equipped with the ℓ_∞-norm and the standard basis $\{e_1, \ldots, e_r\}$ in $\mathbb{C}^r$ forms a maximal family of mutually orthogonal minimal tripotents. It follows that the rank of a finite-rank JB*-triple V is exactly the maximal number of mutually orthogonal tripotents in V.

Example 3.3.4. The dimension of the Cartan factor $H_3(\mathcal{O})$ is 27, whereas its rank is 3 since it has a maximal family $\{e_1, e_2, e_3\}$ of mutually orthogonal minimal tripotents, where e_j is the diagonal matrix in $H_3(\mathcal{O})$ with 1 in the jj-entry and 0 elsewhere, for $j = 1, 2, 3$. The Cartan factor $H_{1,2}(\mathcal{O})$ has rank 2.

Given a tripotent e in a JB*-triple V, the *rank of e* is defined to be the rank of the Peirce 2-space $V_2(e)$. In particular, if e is a finite sum $e = e_1 + \cdots + e_r$ of mutually orthogonal minimal tripotents $e_1, \ldots, e_r$, then we have $e_1, \ldots, e_r \in V_2(e)$ and the rank of e is r.

We now prove a structure theorem for finite-rank JB*-triples. First, we recall that a Banach space V is said to have the *Radon-Nikodym property* [56, p. 61] if for any finite measure space (Ω, Σ, μ) and any μ-continuous vector measure $L : \Sigma \to V$ of bounded total variation, there exists a Bochner integrable function $g : \Omega \to V$ such that

$$L(E) = \int_E g \, d\mu \qquad (E \in \Sigma).$$

If V possesses the Radon-Nikodym property, then so does its dual V^* and also, V has the *Krein-Milman property*, that is, every closed bounded convex subset of V is the norm closed convex hull of its extreme points [56, p. 190].

Theorem 3.3.5. *Let V be a JB*-triple. The following conditions are equivalent.*

  (i) *V is of finite rank.*

 (ii) *V is a reflexive Banach space.*

(iii) *V has the Krein-Milman property.*

(iv) *V has the Radon-Nikodym property.*

 (v) *V is a finite ℓ_∞-sum of finite-rank Cartan factors.*

Proof. The equivalence of (ii), (iii) and (iv) has been proven in [45, Theorem 6], and the implication (v) $\Rightarrow$ (i) is obvious.

(i) $\Rightarrow$ (iv). Let V be of finite rank. Then we have $\dim V(a) < \infty$ for all $a \in V$, where the JB*-triple $V(a)$ is linearly isometric to the JB*-triple $C_0(S_a)$ of complex continuous functions vanishing at infinity on a locally compact space $S_a \subset [0, \infty)$. Hence S_a is a finite set. By [28, Proposition 4.5], V has the Radon-Nikodym property.

(iv) $\Rightarrow$ (v). If V has the Radon-Nikodym property, then V is a reflexive Banach space by the equivalence of (iv) and (ii). Using the fact that the weak* closed triple ideal generated by a minimal tripotent in $V = V^{**}$ is a Cartan factor (which will be proved in Lemma 3.8.11 and Theorem 3.8.17), it has been shown in [27, Lemma 3.2, Lemma 3.4] that V contains a net $\{J_\alpha\}_{0 \leq \alpha \leq \gamma}$ of closed triple ideals J_α, indexed by the ordinals, such that $J_0 = \{0\}$, $J_\gamma = V$, $J_\alpha \subset J_\beta$ for $\alpha \leq \beta$, $J_\beta = \overline{\cup_{\alpha < \beta} J_\alpha}$ if $\beta \in [0, \gamma]$ is a limit ordinal, and $(J_{\alpha+1}/J_\alpha)^{**}$ is a Cartan factor.

Since the second dual J^{**} of a closed triple ideal J in V identifies with a weak* closed triple ideal in V^{**}, there is a weak* closed triple ideal $Z \subset V^{**}$ such that V^{**} decomposes into a triple orthogonal sum $V^{**} = J^{**} \oplus Z$, by Lemma 3.8.9 which will be proved in Section 3.8. By Corollary 3.2.24, this sum is an ℓ_∞-sum. Using the duality of subspaces and quotients, it can be seen that $(V/J)^{**}$ is isometrically isomorphic to Z and hence we have the ℓ_∞-sum

$$V^{**} = J^{**} \oplus (V/J)^{**}.$$

This argument can be applied inductively to the above triple ideals $\{J_\alpha\}$.

It follows that $V = V^{**}$ is an ℓ_∞-sum

$$\bigoplus_{\alpha < \gamma} V_\alpha$$

of Cartan factors $V_\alpha := (J_{\alpha+1}/J_\alpha)^{**}$, which are reflexive Banach spaces. Since the Banach space ℓ_∞ of bounded sequences does not have the

Radon-Nikodym property [56, p. 219] and a closed subspace of a Banach space with the Radon-Nikodym property also has this property, the above ℓ_∞-sum must be finite.

It remains to show that each reflexive Cartan factor is of finite rank. Let V_α be a reflexive Cartan factor. An examination of the list of Cartan factors in Section 2.5 reveals that V_α must be one of the following factors:

$$
\begin{array}{lll}
\text{Type I} & L(\mathbb{C}^\ell, K) \quad (\ell = 1, 2, \ldots),\ \text{rank} = \ell \leq \dim K, & \\
\text{Type II} & \{z \in L(\mathbb{C}^\ell) : z^t = -z\} \quad (\ell = 5, 6, \ldots),\ \text{rank} = \left[\dfrac{\ell}{2}\right] & \\
\text{Type III} & \{z \in L(\mathbb{C}^\ell) : z^t = z\} \quad (\ell = 2, 3, \ldots),\ \text{rank} = \ell & (3.31) \\
\text{Type IV} & \text{spin factor, rank} = 2 & \\
\text{Type V} & M_{1,2}(\mathcal{O}),\ \text{rank} = 2 & \\
\text{Type VI} & H_3(\mathcal{O}),\ \text{rank} = 3 &
\end{array}
$$

where $L(\mathbb{C}^\ell, K)$ is the JB*-triple of bounded linear operators from $\mathbb{C}^\ell$ to a Hilbert space K and z^t denotes the transpose of z in the JB*-triple $L(\mathbb{C}^\ell)$ of $\ell \times \ell$ complex matrices. The only infinite dimensional factors in the above list are the infinite dimensional spin factors and the Type I factors $L(\mathbb{C}^\ell, K)$ with $\dim K = \infty$.

If V_α is a spin factor, then by (2.56), each element $a \in V_\alpha$ has a spectral decomposition $a = \lambda e + \mu u$, where e, u are mutually orthogonal minimal tripotents and $\lambda, \mu \in \mathbb{C}$. Hence $\dim V_\alpha(a) \leq 2$. For each minimal tripotent $e \in V_\alpha$, the minimal tripotent e^* is triple orthogonal to e. Hence a spin factor has rank 2.

To complete the proof, we show that the Type I factor $L(\mathbb{C}^\ell, K)$ has rank ℓ. Let $T \in L(\mathbb{C}^\ell, K)$ and let $\{u_1, \ldots, u_d\}$ be an orthonormal basis in $T(\mathbb{C}^\ell)$, where $d \leq \ell$ and $u_j = T(e_j)$ for some $e_j \in \mathbb{C}^\ell$, for $j = 1, \ldots, d$. With the standard inner product $\langle \cdot, \cdot \rangle$ in $\mathbb{C}^\ell$, the rank-one operators $\frac{e_j}{\|e_j\|} \otimes u_j : \mathbb{C}^\ell \to K$ defined by

$$\frac{e_j}{\|e_j\|} \otimes u_j(h) = \left\langle h, \frac{e_j}{\|e_j\|} \right\rangle u_j \qquad (h \in \mathbb{C}^\ell)$$

are mutually orthogonal minimal tripotents in $L(\mathbb{C}^\ell, K)$ and we have

$$T = e_1 \otimes u_1 + \cdots + e_d \otimes u_d$$

which implies $\dim V(T) = d$. From this, we infer that $L(\mathbb{C}^\ell, K)$ has rank ℓ. $\qquad \square$

Corollary 3.3.6. *A JB-algebra is reflexive if and only if it is a finite ℓ_∞-sum $\bigoplus_j A_j$, where A_j is either a real spin factor or a finite dimensional JB-algebra.*

Proof. If A is a reflexive JB-algebra, then it is a JBW-algebra which is unital. The complexification $A + iA$ is a JBW*-algebra and by Theorem 3.3.5, it is a finite ℓ_∞-sum of reflexive Cartan factors, which are either complex spin factors or finite dimensional JB*-algebras. By Example 2.5.5, A is a finite ℓ_∞-sum of either real spin factors or finite dimensional JB-algebras, or both. $\qquad \square$

We have shown in Theorem 2.2.34 that each (non-zero) element in a finite dimensional Jordan triple has a spectral decomposition. This is also true for a JB*-triple V of finite rank p. Each element $a \in V$ has a *spectral decomposition*

$$a = \alpha_1 e_1 + \cdots + \alpha_p e_p \tag{3.32}$$

where $\|a\| = \alpha_1 \geq \cdots \geq \alpha_p \geq 0$ and $e_1, \ldots, e_p$ are mutually orthogonal minimal tripotents in V. Indeed, we see in the proof of Theorem 3.3.5 that the closed subtriple $V(a)$ generated by a is linearly isometric (and triple isomorphic) to $C(S_a)$, where S_a is a finite set, say $\{t_1, \ldots, t_k\}$ in $[0, \infty)$. Identifying a as a function in $C(S_a)$, we can write

$$a = a_1 \chi_1 + \cdots + a_k \chi_k \qquad (a_1, \ldots, a_k \in \mathbb{C}, k \leq p)$$

where χ_j is the indicator function of $\{t_j\}$ for $j = 1, \ldots, k$, and we may assume $\|a\| = |a_1|$ by a judicious choice of indices. The indicator functions $\chi_1, \ldots, \chi_k$ are mutually orthogonal minimal tripotents in the JB*-triple $C(S_a)$ and identified with mutually orthogonal minimal tripotents $c_1, \ldots, c_k$ in $V(a)$. It follows that

$$a = \alpha_1 e_1 + \cdots + \alpha_k e_k$$

where a_j has the polar form $a_j = \alpha_j e^{i\theta_j}$ and $e_j = e^{i\theta_j} c_j$ is a minimal tripotent. We may assume $k = p$ for otherwise one can complete $\{e_1, \ldots, e_k\}$ to a set of p mutually orthogonal minimal tripotents and set $\alpha_j = 0$ for $j > k$.

Theorem 3.3.5 gives a classification of finite-rank bounded symmetric domains.

Corollary 3.3.7. *Let D be a bounded symmetric domain of finite rank. Then D decomposes into a finite Cartesian product*

$$D_1 \times \cdots \times D_k$$

of bounded symmetric domains $D_1, \ldots, D_k$, each of which is biholomorphic to the open unit ball of a Cartan factor listed in (3.31).

Proof. Let D be realised as the open unit ball of a finite-rank JB*-triple V. Then V decomposes into a finite ℓ_∞-sum

$$V = V_1 \oplus \cdots \oplus V_k$$

of finite-rank Cartan factors $V_1, \ldots, V_k$, by Theorem 3.3.5 (v). These Cartan factors are among the ones listed in (3.31) and the open unit ball of V is the Cartesian product of the open unit balls $D_1, \ldots, D_k$ of $V_1, \ldots, V_k$ respectively. $\qquad\square$

Remark 3.3.8. In the proof of the preceding corollary, we have made use of Theorem 3.3.5 (v), which in turn depends on the crucial result, to be proved in Section 3.8, that the weak* closed triple ideal generated by a minimal tripotent in a JBW*-triple is a Cartan factor.

A bounded symmetric domain D is called *irreducible* if it is not (biholomorphic to) a Cartesian product of two bounded symmetric domains. Using Corollary 3.3.7, one can formulate Cartan's classification of finite dimensional bounded symmetric domains in terms of JB*-triples. Indeed, a finite dimensional irreducible bounded symmetric domain D is (biholomorphic to) the open unit ball of one of the following finite dimensional Cartan factors, which are JB*-triples of matrices:

(i) $M_{m,n}(\mathbb{C})$ (ii) $S_n(\mathbb{C})$ (iii) $H_n(\mathbb{C})$

(iv) Sp_n (v) $M_{1,2}(\mathcal{O})$ (vi) $H_3(\mathcal{O})$

where $M_{m,n}(\mathbb{C}) = L(\mathbb{C}^n, \mathbb{C}^m)$, both $S_n(\mathbb{C})$ and $H_n(\mathbb{C})$ are norm closed subspaces of $M_{n,n}(\mathbb{C})$, consisting of $n \times n$ skew-symmetric and symmetric matrices respectively, and Sp_n is a spin factor of dimension $n > 2$, which is a closed subspace of $M_{n,n}(\mathbb{C})$ such that $a \in Sp_n$ implies $a^* \in Sp_n$ and $a^2 \in \mathbb{C}1$. The open unit balls in (i) – (iv) are known as the *classical domains* or *Cartan domains*. The *exceptional domains* are the open unit balls in (v) and (vi).

We will relate in Section 3.7 the connection between the rank of a bounded symmetric domain and the extent to which a Hilbert ball can be 'squeezed' inside the domain.

Following the previous remark about the maximal flat subspace in (3.30), one can actually show that two maximal flat subspaces in a finite dimensional JB*-triple V are linear isomorphic via a triple automorphism of V (cf. [125, Theorem 5.3]). If follows that, given two maximal tripotents e and u in V, there is a triple automorphism $\psi : V \longrightarrow V$ such that $\psi(e) = u$, and in particular, we have $\mathrm{Trace}\,(e \,\square\, e) = \mathrm{Trace}\,(u \,\square\, u)$.

Definition 3.3.9. Let D be a finite dimensional bounded symmetric domain realised as the open unit ball of a JB*-triple V. The *genus p* of D, or of V, is defined to be

$$p = \frac{2}{r}\mathrm{Trace}\,(e \,\square\, e) \qquad (r = \mathrm{rank}\,D)$$

for a maximal tripotent $e \in V$.

We note that $V = V_2(e) \oplus V_1(e)$ for a maximal tripotent $e \in V$, where $V_2(e)$ and $V_1(e)$ are eigenspaces for the eigenvalues 1 and $1/2$ of $e \,\square\, e$, respectively. It follows that

$$rp = 2\mathrm{Trace}\,(e \,\square\, e) = 2(\dim V_2(e) + \frac{1}{2}\dim V_1(e)) = \dim V_2(e) + \dim V.$$
$$(3.33)$$

The genuses of the finite dimensional Cartan factors are listed below.

(i) $M_{m,n}(\mathbb{C})$, $p = m + n$

(ii) $S_n(\mathbb{C})$, $p = 2(n - 1)$

(iii) $H_n(\mathbb{C})$, $p = n + 1$

(iv) Sp_n, $p = n$

(v) $M_{1,2}(\mathcal{O})$, $p = 12$

(vi) $H_3(\mathcal{O})$, $p = 18$.

The Euclidean ball $B \subset \mathbb{C}^n \approx M_{n1}$ has genus $p = n+1$, and the polydisc

$$\mathbb{D}^n \subset \underbrace{\mathbb{C} \oplus \cdots \oplus \mathbb{C}}_{n\text{-fold } \ell_\infty\text{-sum}}$$

has genus $p = 2$.

3.4   Boundary structures

An important difference between finite and infinite dimensional bounded domains in Banach spaces is that the former is relatively compact, whereas the latter can never be. When bounded symmetric domains are realised as bounded convex domains, the closure of a finite dimensional bounded symmetric domain contains an abundance of extreme points, but an infinite dimensional one may contain none. For instance, the closed unit ball of the JB*-triple $C_0(0,1)$ of complex continuous functions vanishing at infinity on $(0,1)$ has no extreme point. This indicates the different boundary features of finite and infinite dimensional bounded symmetric domains. The contrast can also be seen in the profile of the Shilov boundary. To discuss this, we begin with some examples.

Example 3.4.1. While all continuous complex functions on the closure of a finite dimensional bounded domain are bounded, this is not the case in infinite dimension.

Let D be the open unit ball of $C_0(0,1)$. One can find a continuous functional $\ell \in C_0(0,1)^*$ such that $\|\ell\| = 1$ and $|\ell(z)| < 1$ for all $z \in \overline{D}$ (cf. [93]). The function $f : \overline{D} \longrightarrow \mathbb{C}$ defined by

$$f(z) = \frac{1}{1 - \ell(z)} \qquad (z \in \overline{D})$$

is unbounded, continuous on $\overline{D}$ and holomorphic in D.

Let D be a bounded symmetric domain realised as the open unit ball of a JB*-triple V and let $f : \overline{D} \longrightarrow \mathbb{C}$ be a continuous function which is holomorphic in D. For each $0 < r < 1$, the Cauchy integral formula (cf. [133, § II.8])

$$f(z) = \frac{1}{2\pi i} \int_0^{2\pi} f(0 + re^{i\theta}z)d\theta$$

implies

$$|f(z)| \leq \sup\{|f(w)| : \|w\| = r\} \quad \text{for all} \quad \|z\| \leq r.$$

If f is bounded, then

$$\sup\{|f(z)| : z \in \overline{D}\} = \sup\{|f(z)| : z \in \partial D\}$$

which is just a case of the maximum principle. A finer version of the principle is shown below.

Lemma 3.4.2. *Let D be a domain in a complex Banach space and B the open unit ball of a Hilbert space H, with inner product $\langle \cdot, \cdot \rangle$. Given any holomorphic function $f : D \longrightarrow \overline{B}$, we have either $f(D) \subset B$ or $f(z) = \xi \in \partial B$ for all $z \in D$.*

Proof. Suppose $f(z_0) = \xi \in \partial B$ for some $z_0 \in D$. We show that $f(z) = \xi$ for all $z \in D$. There is an open ball $B(z_0, r) \subset D$ for some $r > 0$ and it suffices to show that $f(z) = \xi$ for all $z \in B(z_0, r)$. Pick any $v \in B(z_0, r)$ and define $\varphi_v : \mathbb{D} \longrightarrow \mathbb{C}$ by

$$\varphi_v(\lambda) = \langle f(z_0 + \lambda(v - z_0)), \xi \rangle \qquad (\lambda \in \mathbb{D}).$$

Then $\varphi_v : \mathbb{D} \longrightarrow \overline{\mathbb{D}}$ and $\varphi_v(0) = 1$ imply $\varphi_v(\mathbb{D}) = \{1\}$ by the maximum modulus principle. It follows that $f(z_0 + \lambda(v - z_0)) = \xi$ for all $\lambda \in \mathbb{D}$ which gives $f(v) = \xi$ by continuity. $\qquad\qquad \square$

Example 3.4.3. Let D be the open unit ball of a (complex) Hilbert space V with inner product $\langle \cdot, \cdot \rangle$. Then its topological boundary is given by

$$\partial D = \{v \in V : \|v\| = 1\} = \{v \in V : v \text{ is an extreme point of } \overline{D}\}.$$

Let $e \in \partial D$. Define a bounded holomorphic function $f : \overline{D} \longrightarrow \mathbb{C}$ by

$$f(z) = \frac{1}{2 - \langle z, e \rangle} \qquad (z \in \overline{D}).$$

It can be seen that

$$f(e) = 1 > |f(z)| \quad \text{for all } z \in \overline{D}\backslash\{e\}$$

and hence

$$\sup\{|f(z)| : z \in \partial D\backslash\{e\}\} < \sup\{|f(z)| : z \in \partial D\}.$$

This shows that ∂D is the smallest closed set satisfying

$$\sup\{|f(z)| : z \in \partial D\} = \sup\{|f(z)| : z \in \overline{D}\}.$$

Actually, the topological boundary ∂D of a bounded domain D need not be the smallest closed set where the maximum of *all* bounded holomorphic functions on a neighbourhood of D is achieved.

Example 3.4.4. Let $D = \mathbb{D} \times \mathbb{D}$ be the bidisc. Then we have

$$\partial D = (\partial \mathbb{D} \times \overline{\mathbb{D}}) \cup (\overline{\mathbb{D}} \times \partial \mathbb{D}).$$

Applying the maximum principle, we see that $\partial \mathbb{D} \times \partial \mathbb{D}$ is the smallest closed subset of ∂D satisfying

$$\sup\{|f(z)| : z \in \partial \mathbb{D} \times \partial \mathbb{D}\} = \sup\{|f(z)| : z \in \overline{D}\}$$

for every continuous function f on $\overline{D}$, which is holomorphic in D.

The notion of Shilov boundary originated from the theory of commutative Banach algebras in relation to an analogue of the maximum principle and integral representations. Given a unital commutative Banach algebra $\mathcal{A}$, the Shilov boundary of $\mathcal{A}$ exists and is the smallest closed set S in the maximal ideal space of $\mathcal{A}$ such that the Gelfand transform $\widehat{x}$ of each $x \in \mathcal{A}$ attains its norm on S (cf. [84, Theorem 4.15.4]). The maximal ideal space, alias *spectrum* or *structure space*, identifies with the set Ω of homomorphisms of $\mathcal{A}$ *onto* $\mathbb{C}$, where Ω

is equipped with the weak* topology $w(\mathcal{A}^*, \mathcal{A})$ and is a weak* closed (hence compact) subset of the closed unit ball of the dual $\mathcal{A}^*$.

Let D be a bounded domain in a complex Banach space V and let $C_b(\overline{D})$ be the C*-algebra of bounded complex continuous functions on the norm closure $\overline{D}$, with the supremum norm. Let $A(\overline{D})$ be the closed subalgebra of $C_b(\overline{D})$, consisting of functions which are holomorphic on D. We note that each linear functional $\varphi \in V^*$ restricts to a function $\varphi|_{\overline{D}}$ in $A(\overline{D})$. The evaluation map

$$x \in \overline{D} \mapsto \varepsilon_x \in A(\overline{D})^*, \quad \varepsilon(f) := f(x) \quad (f \in A(\overline{D}))$$

is continuous and bijective onto its image $\varepsilon(\overline{D}) = \{\varepsilon_x : x \in \overline{D}\} \subset \Omega$, which is equipped with the weak* topology $w(A(\overline{D})^*, A(\overline{D}))$.

We identify $\overline{D}$ with $\varepsilon(\overline{D})$ via the evaluation map, but note that $\varepsilon(\overline{D})$ need not be weak*-closed in Ω for otherwise, $\overline{D}$ would be compact in the weak topology $w(V, V^*)$. Indeed, if $\varepsilon(\overline{D})$ is weak* compact, then for any sequence (x_n) in $\overline{D}$, the sequence (ε_{x_n}) has a subsequence (ε_{x_k}) weak* convergent to some $\varepsilon_y \in \varepsilon(\overline{D})$ and hence $(\varphi(x_k))$ converges to $\varphi(y)$ for each $\varphi \in V^*$.

Let $S \subset \Omega$ be the Shilov boundary of $A(\overline{D})$. For each $f \in A(\overline{D})$, the norm $\|\widehat{f}\|$ of the Gelfand transform $\widehat{f}$ is given by $\sup\{|\omega(f)| : \omega \in \Omega\}$ and we have

$$\begin{aligned}
\|f\| &= \sup\{|\varepsilon_x(f)| : x \in \overline{D}\} = \sup\{|\omega(f)| : \omega \in \Omega\} \\
&= \sup\{|\omega(f)| : \omega \in S\}
\end{aligned} \tag{3.34}$$

where $|\omega(f)| \leq \|\omega\|\|f\| \leq \|f\|$ for all $\omega \in \Omega$. Identifying $\overline{D}$ as a subset of Ω, not necessarily weak*-closed, (3.34) implies that the Shilov boundary S is contained in the *weak*-closure* of D in Ω.

However, if $\overline{D}$ is compact, in which case V is finite dimensional, then the evaluation map $x \in \overline{D} \mapsto \varepsilon_x \in \varepsilon(\overline{D})$ is a homeomorphism and we have $S \subset \overline{D}$, which is the smallest (norm) closed set in $\overline{D}$ where each $f \in A(\overline{D})$ attains its norm. In fact, $S \subset \partial D$ since $\|f\| = \sup\{|\varepsilon_x(f)| : x \in \partial D\}$ for each $f \in A(\overline{D})$ and ∂D is weak* closed in Ω.

Definition 3.4.5. Let D be a bounded symmetric domain in a complex Banach space V, with norm closure $\overline{D}$. The *Bergman-Shilov boundary* of D, if it exists, is the smallest (norm) closed set $\Sigma \subset \partial D$ such that

$$\sup\{|f(z)| : z \in \Sigma\} = \sup\{|f(z)| : z \in \overline{D}\}$$

for all bounded continuous functions f on $\overline{D}$, which are holomorphic in D.

The discussion leading to this definition reveals that the Bergman-Shilov boundary of a finite dimensional bounded symmetric domain always exists. The geometry of the Bergman-Shilov boundary for these domains and some other manifolds in several complex variables has been analysed in [116] and [151]. An interesting result in [116] states that a finite dimensional Hermitian symmetric space is of tube type if and only if its dimension is twice of that of its Bergman-Shilov boundary.

We have shown in the previous two examples that the Bergman-Shilov boundary of a Hilbert ball coincides with its topological boundary, whereas the Bergman-Shilov boundary of the bidisc $\mathbb{D} \times \mathbb{D}$ is $\partial \mathbb{D} \times \partial \mathbb{D}$ which is properly contained in the topological boundary. In fact, the Bergman-Shilov boundary exists in all finite-rank bounded symmetric domains and admits an appealing Jordan description.

Theorem 3.4.6. *Let D be a finite-rank bounded symmetric domain, realised as the open unit ball of a JB*-triple V. Then the Bergman-Shilov boundary of D exists and is exactly the set of maximal tripotents in V.*

Proof. We recall that a tripotent $e \in V$ is maximal if its Peirce 0-projection $P_0(e)$ vanishes. By continuity of the Jordan triple product, we see that the set of maximal tripotents in V must be closed.

We first show that every $f \in A(\overline{D})$ achieves its norm on the set of maximal tripotents, by extending the proof in [51, Theorem 10] for finite dimensional domains.

Let p be the rank of V and let $z \in \overline{D}$, with spectral decomposition

$$z = \alpha_1 e_1 + \cdots + \alpha_p e_p$$

where $e_1, \ldots, e_p$ are mutually orthogonal minimal tripotents in V and

$$1 \geq \|z\| = \alpha_1 \geq \cdots \geq \alpha_p \geq 0.$$

By orthogonality, the closed linear span W of $\{e_1, \ldots, e_p\}$ in V is the ℓ_∞-sum

$$W = \mathbb{C}e_1 \oplus_\infty \cdots \oplus_\infty \mathbb{C}e_p$$

and the open unit ball of W is the intersection $W \cap D$, which identifies with the polydisc

$$W \cap D = \underbrace{\mathbb{D}e_1 \times \cdots \times \mathbb{D}e_p}_{p\text{-times}}. \tag{3.35}$$

Analogous to Example 3.4.4, $\partial \mathbb{D}e_1 \times \cdots \times \partial \mathbb{D}e_p$ is the Shilov boundary of $\overline{\mathbb{D}}e_1 \times \cdots \times \overline{\mathbb{D}}e_p$. We have $z \in W \cap \overline{D} = \overline{\mathbb{D}}e_1 \times \cdots \times \overline{\mathbb{D}}e_p$. Hence for any $f \in A(\overline{D})$,

$$|f(z)| \leq \sup_{W \cap \overline{D}} |f| = \sup_{\partial \mathbb{D}e_1 \times \cdots \times \partial \mathbb{D}e_p} |f|$$

where $\partial \mathbb{D}e_1 \times \cdots \times \partial \mathbb{D}e_p$ is the set $\{\beta_1 e_1 + \cdots \beta_p e_p : \beta_1, \ldots, \beta_p \in \partial \mathbb{D}\}$ in the identification (3.35). Each element in this set is a maximal tripotent in V. It follows that

$$\sup\{|f(z)| : z \in \overline{D}\}$$

is just the supremum of $|f|$ over the set of maximal tripotents in V.

To show that the maximal tripotents form the Bergman-Shilov boundary, let S be a closed subset of ∂D where every $f \in A(\overline{D})$ attains its norm. We prove that S contains all maximal tripotents in V.

Pick a maximal tripotent $e \in V$. Then there are mutually orthogonal minimal tripotents $e_1, e_2, \ldots, e_p$ in V such that

$$e = e_1 + e_2 + \cdots + e_p.$$

Let $P_2(e_j)$ be the Peirce 2-projection of e_j for $j = 1, \ldots, p$. By minimality, the Peirce 2-space $V_2(e_j) = P_2(e_j)(V) = \mathbb{C}e_j$ is one-dimensional. Define a linear map $Q : V \longrightarrow V$ by $Q = \sum_{j=1}^{p} P_2(e_j)$. Then we have $Q(V) = \sum_{j=1}^{p} \mathbb{C}e_j$, which can be identified with the Euclidean space $\mathbb{C}^p \approx \bigoplus_{j=1}^{p} \mathbb{C}e_j$, with $\{e_1, \ldots, e_p\}$ as the standard basis. This induces an inner product $\langle \cdot, \cdot \rangle$ on $Q(V)$ and in particular, for $Q(v) = \lambda_1 e_1 + \cdots + \lambda_p e_p$, we have

$$\langle Q(v), e \rangle = \sum_{j=1}^{p} \lambda_j \quad \text{and} \quad \langle e, e \rangle = p. \tag{3.36}$$

For $v \in \overline{D}$, we have $|\lambda_j| \leq 1$ for all j and $\langle Q(v), Q(v) \rangle \leq p$.

Define a continuous function $f : \overline{D} \longrightarrow \mathbb{C}$ by

$$f(z) = \frac{1}{2}\left(1 + \frac{\langle Q(z), e \rangle}{\langle e, e \rangle}\right)$$

which is holomorphic on D and $f(e) = 1$. Observe that $|f(z)| = 1$ if and only if $\left|\frac{\langle Q(z), e \rangle}{\langle e, e \rangle}\right| = 1$ since $|f(z)| \leq 1$. This in turn is equivalent to $p^2 = |\langle Q(z), e \rangle|^2 \leq \langle Q(z), Q(z) \rangle \langle e, e \rangle$, and equivalent to $Q(z) = \lambda e$ for some $|\lambda| = 1$. For such λ, we have $\left|\frac{1}{2}(1 + \lambda)\right| = 1$ if and only if $\lambda = 1$. Hence we conclude that $|f(z)| = 1$ if and only if $z = e$ and e must reside in S. This completes the proof. $\qquad \square$

Remark 3.4.7. We have shown in Theorem 3.3.5 that finite-rank JB*-triples are reflexive Banach spaces. Using (3.36), one can endow a finite-rank JB*-triple with a concrete inner product.

Maximal tripotents in a JB*-triple V are contained in the topological boundary ∂D of the open unit ball D. They can be characterised by the convex structure of D. A convex subset F of $\overline{D}$ is called a *face* if for all $x, y \in \overline{D}$, the condition $tx + (1 - t)y \in F$ for some $t \in (0, 1)$ entails $x, y \in F$. A point $x \in \overline{D}$ is a (real) extreme point if and only if $\{x\}$ is a face of $\overline{D}$. A point $u \in \overline{D}$ is called a *complex extreme point* if whenever $v \in V$ satisfies $u + \lambda v \in \overline{D}$ for all $\lambda \in \overline{\mathbb{D}}$, we have $v = 0$. Evidently, a real extreme point of $\overline{D}$ is also a complex extreme point.

A point $u \in \overline{D}$ is called a *holomorphic extreme point* if for every open neighbourhood $\Omega \subset \mathbb{C}$ of 0 and holomorphic map $f : \Omega \longrightarrow \overline{D} \subset V$ satisfying $f(0) = u$, we have $f'(0) = 0$. Plainly, a holomorphic extreme point is also a complex extreme point.

Theorem 3.4.8. *Let D be the open unit ball of a JB*-triple V and let $u \in V$. The following conditions are equivalent.*

(i) *u is a (real) extreme point of $\overline{D}$.*

(ii) *u is a complex extreme point of $\overline{D}$.*

(iii) *u is a maximal tripotent.*

(iv) *u is a holomorphic extreme point of $\overline{D}$.*

Proof. (ii) $\Rightarrow$ (iii). The closed subtriple $V(u)$ generated by u is triple isomorphic to an abelian C*-algebra $\mathcal{A}$ in which the complex extreme points of the closed unit ball are partial isometries. It follows that u is a tripotent. One needs to show that $P_0(u)V = \{0\}$. We first observe that $\{z, u, z\} = 0$ for $z \in P_0(u)V$, by Peirce multiplication rule.

Let $X(\cdot) = u - \{\cdot, u, \cdot\}$ be the unique complete holomorphic vector field on D such that $X(0) = u$. One can verify that, for $z \in P_0(u)V$, the flow

$$g_t(z) = (\tanh t)u + z$$

solves the differential equation

$$\frac{dg_t(z)}{dt} = X(g_t(z)).$$

Hence

$$D \supset \exp tX(P_0(u)V \cap D) = (\tanh t)u + (P_0(u)V \cap D).$$

Letting $t \to \infty$, we see that $u + (P_0(u)V \cap D) \subset \overline{D}$ which implies $P_0(u)V = \{0\}$.

(iii) $\Rightarrow$ (iv). For $v \in V$, we write

$$v = v_1 + v_2 \in V_1(u) \oplus V_2(u)$$

for its Peirce decomposition. Let Ω be an open neighbourhood of $0 \in \mathbb{C}$ and $f : \Omega \longrightarrow \overline{D}$ a holomorphic map satisfying $f(0) = u$.

Let X be the unique vector field on D such that $X(0) = u$. For each $t \in \mathbb{R}$, write $g_t = \exp(-tX)$. We have

$$g_t(u) = u \quad \text{and} \quad g_t'(u)(v_1 + v_2) = e^t v_1 + e^{2t} v_2.$$

Observe that $\{g_t \circ f : t \in \mathbb{R}\}$ is a family of holomorphic maps from Ω to V, uniformly bounded and $g_t \circ f(0) = u$. By the Cauchy inequality (1.2), $\{g_t'(u)(f'(0)) : t \in \mathbb{R}\}$ is bounded in V and hence $f'(0) = 0$.

(iv) $\Rightarrow$ (i). By (iv), u is a complex extreme point of $\overline{D}$ and hence a maximal tripotent by (ii) $\Rightarrow$ (iii). Let $u + \lambda v \in \overline{D}$ for all $\lambda \in [-1, 1]$. We show $v = 0$ to complete the proof.

Let $v = v_1 + v_2 \in P_1(u)V \oplus P_2(u)V$ be the Peirce decomposition of v. By Proposition 3.2.31, we have

$$u + \lambda v_2 \in \overline{D} \quad \text{and} \quad u + \lambda e^{it} v_1 + \lambda v_2 \in \overline{D} \qquad (t \in \mathbb{R}).$$

Now $P_2(u)V$ is a JB*-algebra as noted in Example 2.4.18 and u is a complex extreme point of its closed unit ball. By the result [24, Lemma 4.1] for JB*-algebras, u is an extreme point of the ball and hence $v_2 = 0$. It follows that $v_1 = 0$ also.   □

It follows from the preceding result and the Krein-Milman theorem that every JBW*-triple, which is a dual Banach space, contains a maximal tripotent.

Example 3.4.9. For the open unit ball D of a Hilbert $(V, \langle \cdot, \cdot \rangle)$, the topological boundary ∂D is exactly the set of extreme points of the closed ball $\overline{D}$. Given $a \in V \backslash \{0\}$, the element $u = a/\|a\|$ is a maximal tripotent and therefore $V = V_1(u) \oplus V_2(u)$, where the Peirce spaces are mutually orthogonal by associativity of the inner product shown in (2.50):

$$\langle x, y \rangle = \langle 2\{u, u, x\}, y \rangle = 2\langle x, \{u, u, y\} \rangle = 2\langle x, y \rangle \quad (x \in V_1(u), y \in V_2(u)).$$

It follows from Example 3.2.29 that

$$B(a, a)(v) = \begin{cases} (1 - \|a\|^2)v & (v \in V_1(u)) \\ (1 - \|a\|^2)^2 v & (v \in V_2(u)). \end{cases}$$

To end this section, we discuss the boundary components of a bounded symmetric domain. In finite dimensions, a theory of boundary components has been developed in [176], where the boundary components in ∂D of a finite dimensional bounded symmetric domain D, contained in its compact dual X via the Borel embedding, are shown to be the orbits in X of the identity component $(\text{Aut } D)_0$ of the automorphism group $\text{Aut } D$ and they are classified using Lie groups.

We are going to describe the boundary components of arbitrary bounded symmetric domains in terms of the Jordan structure of the underlying JB*-triples.

Let U be a convex domain in a complex Banach space V. A *holomorphic arc* in the closure $\overline{U}$ is a holomorphic map $\gamma : \mathbb{D} \longrightarrow \overline{U}$.

Definition 3.4.10. A subset $\Gamma \subset \overline{U}$ is called a *holomorphic boundary component*, or simply, a *boundary component* of $\overline{U}$ if the following conditions are satisfied:

(i) $\Gamma \neq \emptyset$;

(ii) for each holomorphic arc γ in $\overline{U}$, either $\gamma(\mathbb{D}) \subset \Gamma$ or $\gamma(\mathbb{D}) \subset \overline{U} \backslash \Gamma$;

(iii) Γ is minimal with respect to (i) and (ii).

Two boundary components are either equal or disjoint. The interior U is the unique open boundary component of $\overline{U}$, all others are contained in the boundary ∂U [102]. Of interest are the boundary components in ∂U. For each $a \in \overline{U}$, we denote by Γ_a the boundary component containing a.

Lemma 3.4.11. *Let D be a domain in a complex Banach space and $h : D \longrightarrow \overline{U}$ a holomorphic map. Then the image $h(D)$ is contained entirely in one boundary component of $\overline{U}$.*

Proof. Let $z_0 \in D$. Then $a := h(z_0)$ is contained in the boundary $\Gamma_a \subset \overline{U}$. We show that $h(D) \subset \Gamma_a$.

Let $r > 0$ so that the closed ball $\overline{B(z_0, r)}$ centred at z_0 is contained in D. For each $v \in B(z_0, r) \backslash \{z_0\}$, the holomorphic arc $\gamma : \mathbb{D} \longrightarrow \overline{U}$ defined by

$$\gamma(\lambda) = h\left(z_0 + \frac{\lambda r(v - z_0)}{\|v - z_0\|}\right) \qquad (\lambda \in \mathbb{D})$$

satisfies $\gamma(0) = a$ and $\gamma(\frac{\|v - z_0\|}{r}) = h(v)$. Hence $h(v) \in \Gamma_a$. This proves $h(B(z_0, r)) \subset \Gamma_a$.

Given any boundary point $b \in \partial B(z_0, r)$, there is an open ball $B(b, s) \subset D$ for some $s > 0$, which intersects $B(z_0, r)$. Hence

$$h(B(b, s) \cap B(z_0, r)) \subset \Gamma_{h(b)} \cap \Gamma_a$$

and $h(b) \in \Gamma_{h(b)} = \Gamma_a$.

Let $z \in D \backslash \overline{B(z_0, r)}$. We show $h(z) \in \Gamma_a$ to conclude the proof. By connectedness, there is a continuous path $\varphi : [0, 1] \longrightarrow D$ such that $\varphi(1) = z$ and $\varphi(0) = b$ is a boundary point of $B(z_0, r)$. By compactness, there are a finite number of distinct points $t_0, t_1, \ldots, t_k$ in $[0, 1]$ such that $\varphi[0, 1]$ is covered by the open balls $B(\varphi(t_0), r_0), \ldots, B(\varphi(t_k), r_0)$ in D, with $b \in B(\varphi(t_0), r_0)$ say. As before, $h(b) \in \Gamma_a$ implies $h(B(\varphi(t_0), r_0)) \subset \Gamma_{h(\varphi(t_0))}$ and $\Gamma_{h(\varphi(t_0))} \cap \Gamma_a \neq \emptyset$, which gives $\Gamma_{h(\varphi(t_0))} = \Gamma_a$.

Since $\varphi[0, 1]$ is connected, there exists $p \in [0, 1]$ such that $\|\varphi(p) - \varphi(t_0)\| = r_0$, that is, $\varphi(p) \in \partial B(\varphi(t_0), r_0)$. We have $\varphi(p) \in B(\varphi(t_1), r_1)$, say. Again, this implies $h(B(\varphi(t_1), r_1)) \subset \Gamma_a$ since $h(\varphi(p)) \in \Gamma_a$.

Repeat the above arguments, we conclude that $h(B(\varphi(t_j), r_j)) \subset \Gamma_a$ for $j = 0, 1, \ldots, k$ and in particular $h(z) \in \Gamma_a$. $\qquad\square$

In Definition 3.4.10, we call the set Γ an *affine boundary component* of $\overline{U}$ if condition (ii) is satisfied by all *affine* maps $\alpha : \mathbb{D} \longrightarrow \overline{U}$ instead of holomorphic arcs.

Lemma 3.4.12. *For the open unit ball U of a Banach space, the holomorphic and affine boundary components coincide.*

Proof. We follow the proof in [102, 4.2]. One only needs to show that each affine boundary component Γ of $\overline{U}$ is a holomorphic boundary component. Let $\gamma : \mathbb{D} \longrightarrow \overline{U}$ be a holomorphic arc such that $\gamma(\mathbb{D}) \cap \Gamma \neq \emptyset$. We show $\gamma^{-1}(\Gamma) = \mathbb{D}$.

Fix $a \in \gamma^{-1}(\Gamma)$ and define $f : \mathbb{D} \longrightarrow \overline{U}$ by

$$f(\lambda) = \gamma \left(\frac{\lambda - a}{\overline{a}\lambda - 1} \right)$$

where $f(0) = \gamma(a)$. By [63, Lemma III.1.2], a complex number ζ satisfies $f(0) + \zeta(f(\lambda) - f(0)) \in \overline{U}$ if and only if $2|\zeta\lambda| \leq 1 - |\lambda|$. In particular, we have $f(\lambda) \in \Gamma$ if $|\lambda| \leq 1/3$. Equivalently, we have $\gamma(\zeta) = f(\frac{a-\zeta}{1-\bar{a}\zeta}) \in \Gamma$ if $|\frac{a-\zeta}{1-\bar{a}\zeta}| \leq 1/3$. This implies $\gamma^{-1}(\Gamma)$ is open and closed in $\mathbb{D}$. □

Theorem 3.4.13. *Let D be a bounded symmetric domain realised as the open unit ball of a JB*-triple V. Given a tripotent $e \in \partial D$, the boundary component containing e is the convex set $\Gamma_e = e + (V_0(e) \cap D) \subset \partial D$, where $V_0(e)$ is the Peirce 0-space of e. If D is of finite rank, then all boundary components in ∂D are of this form.*

Proof. By Corollary 3.2.24, the convex set $e + (V_0(e) \cap D)$ lies in ∂D. For each $v \in V_0(e)$ with $0 < \|v\| < 1$, we see that $e + v \in \Gamma_e$ by considering the holomorphic arc $\gamma : \lambda \in \mathbb{D} \mapsto e + \lambda v/\|v\|$. This shows $e + (V_0(e) \cap D) \subset \Gamma_e$.

To show the reverse inclusion, let $x \in \Gamma_e$ and $v = x - e$. We need to show $v \in V_0(e) \cap D$. Let $v = v_2 + v_1 + v_0 \in P_2(e)V \oplus P_1(e)V \oplus P_0(e)V$ be the Peirce decomposition induced by the tripotent e. By the preceding lemma, Γ_e is also an affine boundary component. Hence there is an open neighbourhood U of $0 \in \mathbb{C}$ such that the image of the affine map $f : \lambda \in U \mapsto e + \lambda v \in \overline{D}$ is contained in Γ_e.

For each $s \in (-1, 1)$, the extended Möbius transformation $g_{se} : \overline{D} \longrightarrow \overline{D}$ is biholomorphic (cf. Remark 3.2.26) and satisfies

$$g_{se}(e) = e \quad \text{and} \quad g_{se}(\Gamma_e) = \Gamma_e.$$

Observe that $\sup\{\|g_{se} \circ f(\lambda)\| : \lambda \in U\} \leq 1$ for all $s \in (-1, 1)$ and by Cauchy inequality (1.2), the family $\{(g_{se} \circ f)'(0) : s \in (-1, 1)\}$ is uniformly bounded. From (3.26), it can be seen that

$$(g_{se} \circ f)'(0) = g'_{se}(e)v = \left(\frac{1-s}{1+s}\right)v_2 + \left(\frac{1-s}{1+s}\right)^{1/2} v_1 + v_0 \quad (-1 < s < 1).$$

Since $\lim\limits_{s\downarrow -1}(1+s) = 0$, we must have $v_2 = v_1 = 0$ by uniform bound-edness. Hence $v = v_0 \in V_0(e)$ and $\|v\| \leq \|e + v\| = \|x\| = 1$. Since $\|v\| = 1$ is impossible, we have $v \in V_0(e) \cap D$ as required.

Finally, if D is of finite rank, then each $a \in \partial D$ has a spectral decomposition

$$a = e_1 + \alpha_1 e_2 + \cdots + \alpha_k e_k$$

where $\alpha_1 e_2 + \cdots + \alpha_k e_k \in V_0(e_1) \cap D$ and hence $\Gamma_a = e_1 + (V_0(e_1) \cap D)$.

$\square$

Remark 3.4.14. The norm closure $\overline{\Gamma}_e = e + (V_0(e) \cap \overline{D})$ is a face of $\overline{D}$. To see this, let $x, y \in \overline{D}$ with $tx + (1-t)y \in e + (V_0(e) \cap \overline{D})$ for some $t \in (0, 1)$. Consider the Peirce decompositions $x = x_2 + x_1 + x_0 \in V_2(e) \oplus V_1(e) \oplus V_0(e)$ and $y = y_2 + y_1 + y_0$. There is a point $v \in V_0(e) \cap \overline{D}$ such that

$$
\begin{aligned}
e + v &= tx + (1-t)y \\
&= (tx_2 + (1-t)y_2) + (tx_1 + (1-t)y_1) + (tx_0 + (1-t)y_0) \\
&\in V_2(e) \oplus V_1(e) \oplus V_0(e)
\end{aligned}
$$

which gives $e = tx_2 + (1-t)y_2$. Since $\|x_2\|, \|y_2\| \leq 1$ and e, being the identity of the JB*-algebra $V_2(e)$ (cf. Example 2.4.18), is an extreme point of the closed unit ball of $V_2(e)$, we have $x_2 = e = y_2$.

As $1 = \|tx + (1-t)y)\| \leq t\|x\| + (1-t)\|y\| \leq 1$, we must have $\|x\| = \|y\| = 1$. It follows from Proposition 3.2.31 that, for all $\theta \in \mathbb{R}$,

$$
\begin{aligned}
\|e + e^{i\theta/2}x_1 + e^{i\theta}x_0\| &= \|P_2(e)(x) + e^{i\theta/2}P_1(e)(x) + e^{i\theta}P_0(e)x)\| \\
&= \|\exp-i\theta(e\,\square\,e)(x)\| = \|x\| = 1
\end{aligned}
$$

and by the maximum principle, the map $\gamma : \lambda \in \mathbb{D} \mapsto e + \lambda^{1/2}x_1 + \lambda x_0 \in V$ is a holomorphic arc in $\overline{D}$. Since $\gamma(0) = e \in e + V_0(e) \cap D$, we have

$$e + \frac{1}{2}x_1 + \frac{1}{4}x_0 = \gamma\left(\frac{1}{4}\right) \in e + V_0(e) \cap D$$

which implies $x_1 = 0$. Likewise $y_1 = 0$ and hence $x, y \in e + V_0(e) \cap \overline{D}$.

Example 3.4.15. Let D be the open unit ball of a JB*-triple V. Given a tripotent $e \in \partial D$, the open unit ball D_e of the Peirce 0-space $V_0(e)$ equals $V_0(e) \cap D$. Since the Peirce projection $P_0(e)$ is contractive, we see that $D_e = P_0(e)(D)$ and $\overline{\Gamma}_e = e + \overline{D_e} = e + P_0(e)(\overline{D})$. Also, the boundary $\partial \Gamma_e$ of Γ_e equals $e + \partial D_e$.

Each tripotent c in $V_0(e)$ is orthogonal to e and its Peirce 0-space in $V_0(e)$ is the eigenspace $(V_0(e))_0(c) = \{v \in V_0(e) : (c \,\square\, c)(v) = 0\}$ and we have

$$(V_0(e))_0(c) = V_0(e + c).$$

Indeed, considering $V_0(e + c)$ in terms of the joint Peirce spaces induced by $\{e, c\}$, we have

$$\begin{aligned} V_0(e + c) &= \{v \in V : (e \,\square\, e + c \,\square\, c)(v) = 0\} \\ &= V_{00}(e, c) = V_0(e) \cap V_0(c) = (V_0(e))_0(c). \end{aligned}$$

Hence the boundary component in ∂D_e containing the tripotent $c \in V_0(e)$ is of the form $c + (V_0(e))_0(c) = c + V_0(e + c)$, and the boundary component of $\partial \Gamma_e$ containing $e + c$ is of the form $e + c + V_0(e + c) = \Gamma_{e+c}$, which is also a boundary component of ∂D.

Example 3.4.16. For a Lie ball D in a spin factor V, introduced in Section 2.5, we have a very simple description of the boundary components of $\overline{D}$. We first note that $K_e = \{e\}$ if e is a maximal tripotent of V. For a minimal tripotent e, we have shown $P_0(e) = \langle \cdot, \cdot \rangle e^*$ which gives $V_0(e) = \mathbb{C}e^*$ and hence $K_e = e + \mathbb{D}e^*$. Together with D, these are all the boundary components of $\overline{D}$.

Notes. Lemma 3.4.2 has been shown in [47]. For finite dimensional bounded symmetric domains, Theorem 3.4.6 and Theorem 3.4.8 have been proved in [125, Theorem 6.5]. A list of Shilov boundaries of finite dimensional tube type bounded symmetric domains in terms of Lie groups has been appended to [51]. Theorem 3.4.13 is due to Kaup and Sauter [102]. It has also been proved by Loos [125, Theorem 6.3] for finite dimensional bounded symmetric domains.

3.5   Invariant metrics, Schwarz lemma and dynamics

It is impossible to overestimate the fundamental importance of the Poincaré metric on the open unit disc $\mathbb{D}$, and its higher dimensional generalisation, the Bergman metric, in unifying function theory and geometry for complex domains (e.g. [71]). The fact that these metrics are invariant under biholomorphic maps (and contracted by holomorphic maps) on bounded domains means that geometric invariants are preserved by biholomorphic maps which in turn, provides a powerful tool in studying automorphisms and holomorphic maps on these domains. The initial ideas go back to the Schwarz lemma and the Schwarz-Pick lemma, which can be interpreted geometrically via the Poincaré metric.

In finite dimensions, the classical invariant metrics on complex domains are well documented in literature. Our object in this section is to study two classical invariant metrics on infinite dimensional domains and some variants of the Schwarz lemma on bounded symmetric domains as well as related results in holomorphic dynamics.

The Bergman metric is not available in infinite dimension. Instead, we consider the most common Carathéodory and Kobayashi metrics on domains in complex Banach spaces. Let D be a domain in a complex Banach space V. For $p \in D$ and $v \in V$, the *(infinitesimal) Carathéodory pseudo-metric* of v at p is actually the Carathéodory tangent semi-norm defined in Section 1.3, namely,

$$\mathcal{C}_D(p,v) = \sup\{|f'(p)(v)| : f \in H(D,\mathbb{D}) \text{ and } f(p) = 0\}$$

where $H(M,\mathbb{D})$ denotes the set of all holomorphic maps from M to $\mathbb{D}$. If D is a bounded domain, then $\mathcal{C}_D(p,v)$ is a compatible tangent norm, as shown in Example 1.3.6, in which case $\mathcal{C}_D(p,v)$ is called the *(infinitesimal) Carathéodory metric*, or *Carathéodory (differential) metric*, on D.

On a bounded domain D in a complex Banach space V, the Carathéodory differential metric is an *invariant metric*, that is, it is preserved by biholomorphic maps $g : D \longrightarrow D$ as follows:

$$\mathcal{C}_D(g(p), dg_p(v)) = \mathcal{C}_D(p,v) \qquad (p \in D, v \in V).$$

This has been shown in Example 1.3.5. Likewise, a distance function d on a domain D, i.e. a metric $d : D \times D \longrightarrow [0,\infty)$, is called an *invariant distance* if $d(g(x), g(y)) = d(x,y)$ for all biholomorphic maps $g : D \longrightarrow D$. To avoid confusion with differential metrics on tangent bundles, a pseudo-metric $d : D \times D \longrightarrow [0,\infty)$ will be called a *pseudo-distance*.

We have also shown in Example 1.3.9 that the Carathéodory metric on the complex unit disc $\mathbb{D}$ coincides with the Poincaré metric and the integrated distance $d_{\mathcal{C}_\mathbb{D}}$ of $\mathcal{C}_\mathbb{D}(p,v)$, given in (1.14), is the Poincaré distance on $\mathbb{D}$, which we shall denote by ρ in this section.

The *Carathéodory pseudo-distance* c_D on a domain D in a complex Banach space V is defined by

$$c_D(z,w) := \sup\{\rho(f(z), f(w)) : f \in H(D,\mathbb{D})\} \qquad (z,w \in D).$$

It is called the *Carathéodory distance* if $c_D(z,w) > 0$ whenever $z \neq w$, which is the case if D is bounded. We note that $c_\mathbb{D}$ is the Poincaré distance ρ. The Carathéodory distance is an invariant distance as holomorphic maps between domains contract (i.e. decrease) the Carathéodory pseudo-distance. By definition, the Carathéodory pseudo-distance is the *smallest* pseudo-distance on a domain D for which every holomorphic map $f : D \longrightarrow \mathbb{D}$ is a contraction (i.e. distance-decreasing).

Lemma 3.5.1. *Let B be the open unit ball of a complex Banach space V. Then we have $c_B(0, v) = \rho(0, \|v\|) = \tanh^{-1} \|v\|$ for all $v \in B$.*

Proof. Let $v \in B \backslash \{0\}$ and define a holomorphic map $\zeta : z \in \mathbb{D} \mapsto \frac{zv}{\|v\|} \in B$. Then for each $f \in H(B, \mathbb{D})$, we have

$$\rho(f(0), f(v)) = \rho(f \circ \zeta(0), f \circ \zeta(\|v\|)) \leq \rho(0, \|v\|).$$

On the other hand, let $\varphi \in V^*$ be a continuous linear functional such that $\|\varphi\| \leq 1$ and $\varphi(v) = \|v\|$. Then $\varphi \in H(B, \mathbb{D})$ and

$$\rho(\varphi(0), \varphi(v)) = \rho(0, \|v\|).$$

$\square$

Given that the integrated distance $d_{\mathcal{C}_\mathbb{D}}$ of the Carathéodory metric $\mathcal{C}_\mathbb{D}(p, v)$ on $\mathbb{D}$ coincides with the Poincaré distance, and also identical with the Carathéodory distance $c_\mathbb{D}$ defined above, one may ask if this is also true for all bounded domains D. This need not be the case even in finite dimensions [14] although it can be seen readily that $d_{\mathcal{C}_D} \geq c_D$.

We now introduce another invariant distance on complex domains. Let D be a domain in a complex Banach space V. Given two points $z, w \in D$, choose

◇ $z_0, z_1, \ldots, z_k \in D$ such that $z_0 = z$ and $z_k = w$;

◇ $a_1, \ldots, a_k$ and $b_1, \ldots, b_k$ in $\mathbb{D}$ and

◇ holomorphic maps $f_j : \mathbb{D} \longrightarrow D$ satisfying $f_j(a_j) = z_{j-1}$ and $f_j(b_j) = z_j$ for $j = 1, \ldots, k$.

Define

$$k_D(z,w) = \inf\{\rho(a_1, b_1) + \cdots + \rho(a_k, b_k) : a_1, \ldots, a_k, b_1, \ldots, b_k \in \mathbb{D}\}$$

where the infimum is taken over all possible choices above. It can be verified readily that k_D is a pseudo-distance on D. We call k_D the *Kobayashi pseudo-distance* on D. It is called the *Kobayashi distance* on D if $k_D(z,w) = 0$ implies $z = w$. Evidently, holomorphic maps between domains are contractions with respect to the Kobayashi pseudo-distances on these domains and hence Kobayashi distance is an invariant distance. Further, the Kobayashi pseudo-distance is the *largest* pseudo-distance on D for which every holomorphic map $f : \mathbb{D} \longrightarrow D$ is a contraction. In particular, we have

$$c_D(z,w) \le k_D(z,w) \qquad (z, w \in D).$$

Given two domains D and D' with $D \subset D'$, we have $c_{D'} \le c_D$ and $k_{D'} \le k_D$.

Lemma 3.5.2. *On the complex open unit disc $\mathbb{D}$, the Kobayashi distance $k_{\mathbb{D}}$ coincides with the Poincaré distance ρ.*

Proof. This follows from the fact that a holomorphic map $f : \mathbb{D} \longrightarrow \mathbb{D}$ is distance-decreasing with respect to ρ. $\qquad\square$

Analogous to Lemma 3.5.1, we also have

$$k_B(0, v) = \rho(0, \|v\|) \tag{3.37}$$

for all v in the open unit ball B of a complex Banach space. Let $B(p, r)$ be the open ball centred at $p \in V$ of radius $r > 0$. It follows from invariance that

$$c_{B(p,r)}(p, x) = k_{B(p,r)}(p, x) = \rho(0, \|x - p\|/r)$$

for all $x \in B(p, r)$.

Lemma 3.5.3. *Let D be a bounded domain in a complex Banach space V. Then for each $p \in D$, there exists $\alpha > 0$ such that*

$$\alpha \|x - p\| \leq c_D(x, p) \qquad (x \in D).$$

Proof. Let $B(p, R)$ is the open ball centred at p of radius $R > 0$ such that $D \subset B(p, R)$. Then we have

$$c_D(x, p) \geq c_{B(p,R)}(x, p) = \rho(0, \|x - p\|/R) \geq \|x - p\|/R$$

for all $x \in D$. $\qquad\qquad\square$

Lemma 3.5.4. *Let D be a domain in a complex Banach space V. Let $p \in D$ with $B(p, r) \subset D$, where $B(p, r)$ is the open ball centred at p of radius $r > 0$. Then for $0 < s < r$, there exists $\beta > 0$ such that*

$$k_D(x, p) \leq \beta \|x - p\| \qquad (x \in B(p, s)).$$

Proof. We have

$$k_D(x, p) \leq k_{B(p,r)}(x, p) = \rho(0, \|x - p\|/r) = \tanh^{-1}(\|x - p\|/r)$$

for $x \in D$. Let $0 < s < r$ so that $s/r < 1$. Since $\tanh^{-1}(0) = 0$ and the derivative of $\tanh^{-1}$ is bounded on $[0, s/r]$, one can find a constant $c > 0$ such that $\tanh^{-1} t \leq ct$ for $t \in [0, s/r]$. It follows that

$$\tanh^{-1}(\|x - p\|/r) \leq \frac{c}{r} \|x - p\|$$

for all $x \in B(p, s)$. $\qquad\qquad\square$

Both pseudo-distances c_D and k_D are continuous functions.

Proposition 3.5.5. *Let D be a domain in a complex Banach space V. Then both functions $c_D : D \times D \longrightarrow [0, \infty)$ and $k_D : D \times D \longrightarrow [0, \infty)$ are continuous.*

Proof. By Lemma 3.5.4, the functions $x \in D \mapsto c_D(x,q) \in [0,\infty)$ and $x \in D \mapsto k_D(x,q) \in [0,\infty)$ are continuous, for each $q \in D$. The joint continuity of c_D and k_D on $D \times D$ follows from the inequalities

$$|c_D(x,y) - c_D(a,b)| \le c_D(x,a) + c_D(y,b)$$

and $|k_D(x,y) - k_D(a,b)| \le k_D(x,a) + k_D(y,b)$. □

Theorem 3.5.6. *Let D be a bounded domain in a complex Banach space V. Then the Carathéodory pseudo-distance c_D and the Kobayashi pseudo-distance k_D are equivalent to the norm-distance on any closed ball strictly contained in D. In particular, c_D and k_D are distances on D.*

Proof. This follows from Lemmas 3.5.3 and 3.5.4. □

A finite dimensional domain is called *hyperbolic* in [108] if its Kobayashi pseudo-distance is a distance, in which case it can be shown that the relative topology of the domain is equivalent to the one induced by the Kobayashi distance (cf. [63, Proposition IV.2.3]). According to the following definition, all bounded domains in Banach spaces are hyperbolic.

Definition 3.5.7. A domain D in a complex Banach space V is called *hyperbolic* if the Kobayashi pseudo-distance k_D is a distance and the topology defined by k_D is equivalent to the relative topology of D in V.

Example 3.5.8. Let $\mathbb{D}^k = \mathbb{D} \times \cdots \times \mathbb{D}$ be the k-dimensional polydisc. Then the Carathéodory and Kobayashi distances on $\mathbb{D}^k$ are given by

$$c_D((z_1,\ldots,z_k),(w_1,\ldots,w_k)) = k_D((z_1,\ldots,z_k),(w_1,\ldots,w_k))$$
$$= \max\{\rho(z_1,w_1),\ldots,\rho(z_k,w_k)\}$$

which differs from the integrated distance of the Bergman metric for $k > 1$. We refer to [108] for a derivation of this formula, from which one can deduce that $c_M = k_M$ for a finite dimensional bounded symmetric domain M (see [108, Chap. IV, Example 2]). In fact, we have the following general result.

Theorem 3.5.9. *Let D be a bounded symmetric domain. Then we have*

$$c_D(x, y) = k_D(x, y) = \tanh^{-1} \|g_{-x}(y)\| \qquad (x, y \in D)$$

where g_{-x} is the Möbius transformation induced by $-x$. Given a Cartesian product $D = D_1 \times \cdots \times D_k$ of bounded symmetric domains, we have

$$k_D((x_1, \ldots, x_k), (y_1, \ldots, y_k)) = \max\{k_{D_1}(x_1, y_1), \ldots, k_{D_k}(x_k, y_k)\}.$$

Proof. The domain D can be realized as the open unit ball of a JB*-triple. Given $x, y \in D$, the Möbius transformation $g_{-x} : D \longrightarrow D$ satisfies $g_{-x}(x) = 0$. Hence by Lemma 3.5.1 and (3.37), we have

$$\begin{aligned} c_D(x, y) &= c_D(g_{-x}(x), g_{-x}(y)) = c_D(0, g_{-x}(y)) \\ &= k_D(0, g_{-x}(y)) = k_D(x, y) \end{aligned}$$

where

$$k_D(0, g_{-x}(y)) = \rho(0, \|g_{-x}(y))\|) = \tanh^{-1} \|g_{-x}(y))\|.$$

For $j = 1, \ldots, k$, the bounded symmetric domain D_j identifies with the open unit ball of a JB*-triple V_j and the Cartesian product $D = D_1 \times \cdots \times D_k$ identifies with the open unit ball of the ℓ_∞-sum $V_1 \oplus^{\ell_\infty} \cdots \oplus^{\ell_\infty} V_k$. By the preceding result just proved and Example 3.2.27, we have

$$k_D((x_1, \ldots, x_k), (y_1, \ldots, y_k)) = \max\{k_{D_1}(x_1, y_1), \ldots, k_{D_k}(x_k, y_k)\}.$$

$\square$

Corollary 3.5.10. *On a bounded symmetric D, realised as the open unit ball of a JB*-triple, we have*

$$c_D(\alpha x + (1 - \alpha)y, \alpha z + (1 - \alpha)w) \leq \max\{c_D(x, z), c_D(y, w)\}$$

for $0 < \alpha < 1$ and $x, y, z, w \in D$. The same inequality holds for k_D.

Proof. Both the Carathéodory and Kobayashi distances are contracted by the holomorphic map

$$h : (x, y) \in D^2 \mapsto \alpha x + (1 - \alpha)y \in D.$$

$\square$

Although the Carathéodory distance c_D need not be the integrated distance of the Carathéodory differential metric, the Kobayashi distance k_D, on the other hand, is indeed the integrated form of the Kobayashi differential metric. Let D be a domain in a complex Banach space V. The *(infinitesimal) Kobayashi pseudo-metric* is defined by

$$\mathcal{K}_D(p, v) = \inf\{|\alpha| : \exists \text{ holomorphic } f : \mathbb{D} \to D, \ f(0) = p, f'(0)(\alpha) = v\}$$

for $p \in D$ and $v \in V$. It is clear that $\mathcal{K}_D(p, v)$ is a tangent semi-norm on the tangent bundle TD, as defined in Section 1.3. If $\mathcal{K}_D(p, v)$ is a compatible tangent norm, it will be called the *(infinitesimal) Kobayashi metric*, or *Kobayashi (differential) metric*, on D. One can verify readily that the Kobayashi pseudo-metric is an invariant metric and

$$\mathcal{C}_D(p, v) \leq \mathcal{K}_D(p, v).$$

Lemma 3.5.11. *Let D be the open unit ball of a Banach space V. Then we have*

$$\mathcal{C}_D(0, v) = \mathcal{K}_D(0, v) = \|v\| \qquad (v \in V).$$

Moreover, if V is a JB*-triple, then

$$\mathcal{C}_D(p,v) = \mathcal{K}_D(p,v) = \|g'_{-p}(p)(v)\| = \|B(p,p)^{-1/2}(v)\|$$

for $(p,v) \in D \times V$, where g_{-p} is the Möbius transformation induced by p and $B(p,p)$ is the Bergman operator.

Proof. The first assertion has been proved in [63, Lemma V.1.5]. Let V be a JB*-triple. Then we have

$$\mathcal{K}_D(p,v) = \mathcal{K}_D(g_{-p}(p), g'_{-p}(p)v) = \mathcal{K}_D(0, g'_{-p}(p)v) = \|g'_{-p}(p)v\|$$

which is equal to $\|B(p,p)^{-1/2}(v)\|$ by Lemma 3.2.25. $\square$

Remark 3.5.12. We see from the preceding result that $\mathcal{K}_D(p,v)$ is a compatible tangent norm on a bounded symmetric domain D.

Example 3.5.13. Let D be a finite dimensional bounded domain and let $h_p(u,v)$ be the Bergman metric on D. It has been shown in [74] that $\mathcal{C}_D(p,v)^2 \leq h_p(v,v)$ for $p \in D$. If further, D is a symmetric domain realised as the open unit ball of a JB*-triple V, then we have

$$h_0(u,v) = 2\mathrm{Trace}(u \,\square\, v) \qquad (u,v \in V)$$

by [125, Theorem 2.10]. Let g_{-p} be the Möbius transformation induced by $-p \in D$. By Lemma 1.3.4, we have

$$h_p(v,v) = h_{g_{-p(p)}}(g'_{-p}(p)v, g'_{-p}(p)v) = h_0(g'_{-p}(p)v, g'_{-p}(p)v)$$

and hence

$$h_p(v,v) \leq 2\mathrm{Trace}(g'_{-p}(p)v \,\square\, g'_{-p}(p)v) \leq 2n\|g'_{-p}(p)v \,\square\, g'_{-p}(p)v\|.$$

Therefore we have

$$\mathcal{K}_D(p,v)^2 = \mathcal{C}_D(p,v)^2 \leq h_p(v,v) \leq 2n\|g'_{-p}(p)v\|^2 = 2n\mathcal{K}_D(p,v)^2.$$

Let D be a domain in a complex Banach space V and let $d_{\mathcal{K}_D}(a,b)$ be the integrated pseudo-distance of $\mathcal{K}_D$, as defined in (1.13):

$$d_{\mathcal{K}_D}(a,b) = \inf_{\gamma}\{\ell(\gamma) : \gamma(0) = a, \gamma(1) = b\}$$

where $\gamma : [0,1] \longrightarrow D$ is a piecewise smooth curve. It has been shown by Royden [152] that $d_{\mathcal{K}_D}$ coincides with the Kobayashi pseudo-distance k_D in the case of $\dim V < \infty$. This is also true in infinite dimension, which has been proved in [63, Theorem V.4.1].

Theorem 3.5.14. *Let D be a domain in a complex Banach space V and let $d_{\mathcal{K}_D}$ be the integrated form of the Kobayashi pseudo-metric $\mathcal{K}_D$ on D. Then we have $d_{\mathcal{K}_D} = k_D$.*

We now turn to the question of completeness of the distances c_D and k_D.

Lemma 3.5.15. *Let D be a bounded domain in a complex Banach space V. If the Carathéodory distance c_D is complete on D, then so is the Kobayashi distance k_D. The converse is false.*

Proof. Let (x_n) be a Cauchy sequence with respect to k_D. Then it is also a c_D-Cauchy sequence since $c_D \leq k_D$. By assumption, (x_n) converges to some point $a \in D$ with respect to the distance c_D, and hence converges to a in the norm topology of D, by Theorem 3.5.6. Now continuity of k_D gives $\lim_{n\to\infty} k_D(x_n, a) = 0$.

For the converse, we note that the Kobayashi distance $d_{\mathbb{D}\backslash\{0\}}$ on the punctured disc $\mathbb{D}\backslash\{0\}$ is complete while the Carathéodory distance $c_{\mathbb{D}\backslash\{0\}}$ is not (cf. [108, p. 56]). $\square$

In finite dimensions, it has been shown in [89] that a bounded domain D is a domain of holomorphy if (D, c_D) is complete. This useful result can be extended to infinite dimension. We first recall that an open set D in a complex Banach space V is called a *domain of holomorphy* [133, 10.4] if there are no open sets U and W in V satisfying the following conditions:

  (i) U is connected and $U \not\subset D$.

 (ii) $\emptyset \neq W \subset D \cap U$.

 (iii) For each holomorphic function $f \colon D \to \mathbb{C}$, there is a holomorphic function $\tilde{f} \colon U \to \mathbb{C}$ such that $\tilde{f}(z) = f(z)$ for each $z \in W$.

Proposition 3.5.16. *Let D be a bounded domain in a complex Banach space V. If D is complete with respect to the Carathéodory distance c_D, then it is a domain of holomorphy.*

Proof. Suppose that D is not a domain of holomorphy. Then by definition, there are open subsets U, W of V satisfying the following conditions:

  (i) U is connected and $U \not\subset D$.

 (ii) $\emptyset \neq W \subset D \cap U$.

 (iii) For each holomorphic function $f \colon D \to \mathbb{C}$, there is a holomorphic function $\tilde{f} \colon U \to \mathbb{C}$ such that $\tilde{f}(z) = f(z)$ for each $z \in W$.

We deduce a contradiction. Without loss of generality, we may assume that U is bounded. Let W_0 be a connected component of $U \cap D$ with $W_0 \cap W \neq \emptyset$. Then we have

$$\partial W_0 \cap \partial D \cap U \neq \emptyset.$$

Indeed, if this is not the case, then for each $p \in U \backslash W_0 \neq \emptyset$, either $p \notin \partial W_0$ or $p \notin \partial D$. If $p \notin \partial W_0$, then there is a norm-open ball $B_p \subset U$ containing p such that either $B_p \cap W_0 = \emptyset$ or $B_p \cap W_0^c = \emptyset$. Since

$p \notin W_0$, we must have $B_p \cap W_0 = \emptyset$. On the other hand, if $p \notin \partial D$, then there is an open ball B_p containing p such that $B_p \cap D = \emptyset$ or $B_p \cap D^c = \emptyset$. In either case, we have $B_p \cap W_0 = \emptyset$ since, if $B_p \subset D$, then the connected ball B_p resides in a connected component W_1 of $U \cap D$ and we must have $W_1 \neq W_0$ as $p \notin W_0$. Now the disconnection

$$U = W_0 \cup \left(\bigcup_{p \in U \setminus W_0} B_p \right)$$

contradicts the connectedness of U.

Pick a point $p \in \partial W_0 \cap \partial D \cap U$ and let (z_n) be a sequence in W_0 norm-converging to p. By omitting the first few terms of the sequence if necessary, we may assume that (z_n) and p are contained in a closed ball strictly contained in U. It follows that (z_n) also converges to p with respect to the Carathéodory distance c_U.

By condition (iii) above, each holomorphic function $f: D \to \mathbb{C}$ with $|f(z)| < 1$ extends to a holomorphic function $\tilde{f} : U \to \mathbb{C}$, which coincides with f on the connected component W_0 by the identity principle. Moreover, if $|\tilde{f}(u)| > 1$ for some $u \in U$, then we deduce a contradiction by considering the extension to U of the function $\frac{1}{f - \tilde{f}(u)}$ on D. Hence we must have $|\tilde{f}(u)| \leq 1$ for all $u \in U$ and, by the maximum principle, $|\tilde{f}(u)| < 1$ for all $u \in U$. It follows that

$$c_D(z_n, z_m) \leq c_U(z_n, z_m)$$

for $n, m = 1, 2, \ldots$, where $c_U(z_n, z_m)$ converges to $c_U(p, p) = 0$ as $n, m \to \infty$. Hence (z_n) is a Cauchy sequence in D with respect to c_D. However, (z_n) does not converge in D, with respect to c_D. Indeed, if (z_n) c_D-converges to some point $z \in D$ say, then by Lemma 3.5.3, there is a constant $\alpha > 0$ such that

$$\alpha \| z_n - z \| \leq c_D(z_n, z) \to 0 \quad \text{as} \quad n \to \infty$$

which is impossible since (z_n) does not converge in D with respect to the norm-distance. This shows that (D, c_D) fails to be complete, which is a contradiction. We therefore conclude that D is a domain of holomorphy. □

Proposition 3.5.17. *A bounded homogeneous domain in a complex Banach space is complete in the distances c_D and k_D. In particular, it is a domain of holomorphy.*

Proof. Let D be a bounded homogeneous domain in a Banach space V. In view of Lemma 3.5.15 and Proposition 3.5.16, it suffices to show that (D, c_D) is complete.

Fix $p \in D$ with $\overline{B}(p, r) \subset D$, where $\overline{B}(p, r)$ is the norm closed ball centred at p of radius $r > 0$. By Lemma 3.5.3, there exists $s > 0$ such that $B_c(p, s) \subset B(p, r)$, where $B_c(p, s) := \{x \in V : c_D(x, p) < s\}$. Now let (x_n) be a c_D-Cauchy sequence in D. Pick x_k such that $c_D(x_n, x_k) < s$ for all $n \geq k$. Since D is homogeneous, we can find $g \in \operatorname{Aut} D$ such that $g(x_k) = p$. Then the sequence $(g(x_n))$ is a c_D-Cauchy sequence and

$$c_D(g(x_n), p) = c_D(g(x_n), g(x_k)) = c_D(x_n, x_k) < s$$

for all $n \geq k$. Hence $g(x_n) \in \overline{B}(p, r)$ for all $n \geq k$. By Theorem 3.5.6, $(g(x_n))$ norm converges, as well as c_D-converges, to some point $q \in \overline{B}(p, r)$. It follows that (x_n) c_D-converges to $g^{-1}(q) \in D$, which proves completeness of c_D. □

Now the Schwarz lemma. A familiar variant (of its main assertion) is the Schwarz-Pick lemma for holomorphic functions $h : \mathbb{D} \longrightarrow \mathbb{D}$ which, as noted at the outset of the section, can be interpreted geometrically in terms of the Poincaré metric. Indeed, the following assertion of the Schwarz-Pick lemma

$$\left| \frac{h(z) - h(w)}{1 - \overline{h(z)}h(w)} \right| \leq \left| \frac{z - w}{1 - \overline{z}w} \right| \qquad (z, w \in \mathbb{D})$$

is just saying that the holomorphic map h contracts the Poincaré distance ρ:

$$\rho(h(z), h(w)) \leq \rho(z, w) = \tanh^{-1} \left| \frac{z - w}{1 - \bar{z}w} \right|.$$

The Schwarz-Pick lemma for $\mathbb{D}$ can be extended to bounded symmetric domains via the Hahn-Banach theorem.

Lemma 3.5.18. *Let D be the open unit ball of a complex Banach space V and let $h : D \longrightarrow D$ be a holomorphic map satisfying $h(0) = 0$. Then we have $\|h(z)\| \leq \|z\|$ for all $z \in D$.*

If V is a JB*-triple, then for any holomorphic self-map f on D, we have

$$\|g_{-f(w)}(f(z))\| \leq \|g_{-w}(z)\| \qquad (z, w \in D)$$

where $g_\bullet$ denotes the Möbius transformation.

Proof. Fix $z \in D \backslash \{0\}$ and define a holomorphic map $\zeta : \mathbb{D} \longrightarrow D$ by

$$\zeta(\alpha) = \frac{\alpha z}{\|z\|} \qquad (\alpha \in \mathbb{D}).$$

For the given holomorphic map h, and for a continuous linear map $\varphi \in V^*$ with $\|\varphi\| \leq 1$, apply the Schwarz lemma to the composite map $\varphi \circ h \circ \zeta : \mathbb{D} \longrightarrow \mathbb{D}$, which gives

$$|\varphi \circ h \circ \zeta(\alpha)| \leq |\alpha| \quad \text{for all} \quad \alpha \in \mathbb{D}.$$

In particular, for $\alpha = \|z\|$, we have $|\varphi(h(z))| \leq \|z\|$ and hence $\|h(z)\| \leq \|z\|$ since φ was arbitrary.

Now, if V is a JB*-triple and $f : D \longrightarrow D$ is holomorphic, then we have

$$g_{-f(w)} \circ f \circ g_w(0) = 0 \qquad (z, w \in D)$$

by composing f with Möbius transformations. Hence the preceding result implies

$$\|g_{-f(w)} \circ f \circ g_w(g_{-w}(z))\| \leq \|g_{-w}(z)\| \qquad (z, w \in D)$$

which gives

$$\|g_{-f(w)}(f(z))\| \le \|g_{-w}(z)\|.$$

$$\square$$

There is another assertion of the Schwarz lemma, which says that a holomorphic map $h : \mathbb{D} \longrightarrow \mathbb{D}$ with $h(0) = 0$ must satisfy $|h'(0)| \le 1$ (and it must be a rotation $h(z) = e^{i\theta}z$ if either $|h'(0)| = 1$ or $|h(z)| = |z|$ for some $z \ne 0$). Without the assumption of $h(0) = 0$, the Schwarz-Pick version states that $|h'(z)| \le \frac{1-|h(z)|^2}{1-|z|^2}$. This can also be extended to bounded symmetric domains using the Hahn-Banach theorem.

Proposition 3.5.19. *Let D_j be the open unit ball of a JB*-triple V_j, for $j = 1, 2$. Given a holomorphic map $f : D_1 \longrightarrow D_2$, we have*

$$\|f'(z)\| \le \frac{\|B(f(z), f(z))^{1/2}\|}{1 - \|z\|^2} \qquad (z \in D_1)$$

where $B(\cdot, \cdot)$ denotes the Bergman operator.

In particular, if $V_2 = \mathbb{C}$, we have $\|f'(z)\| \le \frac{1-|f(z)|^2}{1-\|z\|^2}$ for all $z \in D_1$, and if V_2 is a Hilbert space of dimension at least 2, then $\|f'(z)\| \le \frac{\sqrt{1-\|f(z)\|^2}}{1-\|z\|^2}$.

Proof. As before, we denote by $g.$ the Möbius transformation. First, assume $f(0) = 0$. Let $z \in D_1 \backslash \{0\}$ and as in the proof of Lemma 3.5.18, define a holomorphic map $\zeta : \mathbb{D} \longrightarrow D$ by $\zeta(\alpha) = \alpha z / \|z\|$. For each $\varphi \in V_2$ with $\|\varphi\| \le 1$, apply the Schwarz lemma to the holomorphic function $\varphi \circ f \circ \zeta$ on $\mathbb{D}$, we have $|(\varphi \circ f \circ \zeta)'(0)| \le 1$ which gives

$$|\varphi(f'(0)z)| \le \|z\|$$

and it follows that $\|f'(0)\| \le 1$.

Now, without the assumption of $f(0) = 0$, but if $f(w) = 0$ for some $w \in D_1$, then we have $f \circ g_w(0) = 0$ and the preceding arguments imply

$$
\begin{aligned}
\|f'(w)\| &= \|(f \circ g_w \circ g_{-w})'(w)\| = \|(f \circ g_w)'(0)g'_{-w}(w)\| \\
&= \|(f \circ g_w)'(0)B(w,w)^{-1/2}\| \leq \frac{1}{1 - \|w\|^2}.
\end{aligned}
$$

Finally, for each $z \in D_1$, we have $(g_{-f(z)} \circ f)(z) = 0$ and therefore, as in the previous case, we deduce

$$
\|(g_{-f(z)} \circ f)'(z)\| \leq \frac{1}{1 - \|z\|^2}.
$$

Observe that

$$
(g_{-f(z)} \circ f)'(z) = (g'_{-f(z)}(f(z))f'(z) = B(f(z), f(z))^{-1/2}f'(z).
$$

Hence we have

$$
\|f'(z)\| = \|B(f(z), f(z))^{1/2}(g_{-f(z)} \circ f)'(z)\| \leq \frac{\|B(f(z), f(z))^{1/2}\|}{1 - \|z\|^2}.
$$

The last assertion follows from Examples 3.2.17 and 3.4.9.   □

We consider another variant of the Schwarz lemma, which is given by the classical theorem of Wolff [177]. One can view the Schwarz lemma as an invariant theorem in the following sense. Given a holomorphic self-map h on $\mathbb{D}$ with $h(0) = 0$, the Schwarz lemma asserts that every open Euclidean disc $\mathbb{D}(0, r) := \{z \in \mathbb{D} : |z| < r\}$ is invariant under h. More generally, if h fixes some $z \in \mathbb{D}$ instead of 0, then we have

$$
h(\Delta(z, r)) \subset \Delta(z, r) \qquad (0 < r < 1)
$$

for each *Poincaré disc*

$$
\Delta(z, r) := \{x \in \mathbb{D} : \rho(x, z) < \tanh^{-1} r\}
$$

(with radius $\tanh^{-1} r$) by the Schwarz-Pick lemma. If h has no fixed-point in $\mathbb{D}$, then we have $h(\Delta(z, r)) \subset \Delta(h(z), r)$. By considering the Poincaré discs $\Delta(z, r)$ for which z tends to a boundary point $\xi \in \partial \mathbb{D}$ and $r \to 1$, one can show that the *'limit'* of these discs is invariant under h, and is in fact a horodisc with horocentre at ξ. This can be viewed as a *boundary* analogue of the Schwarz lemma. Actually, it can be shown that ξ is a *boundary fixed-point* of h, meaning $\xi = \lim_{r \to 1^-} h(r\xi)$.

We now make precise Wolff's theorem and extend it to bounded symmetric domains. Given a holomorphic self-map f on $\mathbb{D}$, without fixed-point, Wolff has shown in [177] that there is a (unique) boundary point $\xi \in \partial \mathbb{D}$ so that every Euclidean disc internally tangent to $\mathbb{D}$ at ξ is invariant under f. Such a disc is called a *horodisc* with *horocentre* at ξ, and is of the form

$$H(\xi, \lambda) = \frac{\lambda}{1 + \lambda} \xi + \frac{1}{1 + \lambda} \mathbb{D}$$

where $\lambda > 0$ and the radius of the disc $H(\xi, \lambda)$ is written $1/(1 + \lambda)$ so that the horodiscs can be parametrised by positive real numbers.

To show that $H(\xi, \lambda)$ as the *'limit'* of a sequence $\Delta(z_k, r_k)$ of Poincaré discs as mentioned earlier, where $z_k \to \xi$ and $r_k \to 1$, we first need to choose the sequence (z_k). For this, we pick a sequence $\alpha_k \in (0, 1)$ with $\alpha_k \uparrow 1$. By comparing the identity function id_k on the closed disc $\overline{\mathbb{D}}(0, \alpha_k) \subset \mathbb{D}$ with the function $id_k - \alpha_k f$, one finds a point $z_k \in \mathbb{D}(0, \alpha_k)$ satisfying $\alpha_k f(z_k) = z_k$, using Rouché's theorem. Further, choosing a subsequence if necessary, we may assume (z_k) converges to some point $\xi \in \overline{\mathbb{D}}$. Since f has no fixed-point in $\mathbb{D}$, we must have $\xi \in \partial \mathbb{D}$.

Given $\lambda > 0$, pick $r_k \in (0, 1)$ so that $1 - r_k^2 = \lambda(1 - |z_k|^2)$, for sufficiently large k. The Poincaré disc

$$P_k(\lambda) := \Delta(z_k, r_k) = \{z \in \mathbb{D} : \rho(z, z_k) < \tanh^{-1} r_k\} \tag{3.38}$$

is the image $g_{z_k}(\mathbb{D}(0, r_k))$ of the Euclidean disc $\mathbb{D}(0, r_k)$ under the Möbius transformation

$$g_{z_k}(z) = \frac{z + z_k}{1 + \overline{z}_k z} \qquad (z \in \mathbb{D}).$$

We define the *limit* of the Poincaré discs $P_k(\lambda)$ to be the set

$$S(\xi, \lambda) := \{z \in \overline{\mathbb{D}} : z = \lim_k x_k, \ x_k \in P_k(\lambda)\}.$$

Then the horodisc $H(\xi, \lambda)$ is exactly the interior $S_0(\xi, \lambda)$ of $S(\xi, \lambda)$. In fact, we have

$$H(\xi, \lambda) = S_0(\xi, \lambda) = \left\{z \in \mathbb{D} : \frac{|1 - z\overline{\xi}|^2}{1 - |z|^2} < \frac{1}{\lambda}\right\} = \frac{\lambda}{1 + \lambda}\xi + \frac{1}{1 + \lambda}\mathbb{D}.$$
$$(3.39)$$

Further, the f-invariance of these horodiscs implies that ξ is a boundary fixed-point of f. Indeed, given $\varepsilon > 0$, the intersection $\mathbb{D}(\xi, \varepsilon) \cap \mathbb{D}$ contains a horodisc $H(\xi, \lambda)$ of sufficiently small radius $1/(1 + \lambda)$ (Draw a picture!). Hence for $\lambda/(1 + \lambda) < r < 1$, we have $r\xi \in H(\xi, \lambda)$ and $f(r\xi) \in H(\xi, \lambda) \subset \mathbb{D}(\xi, \varepsilon)$.

One may observe that our presentation of Wolff's theorem here bears little resemblance to the original work in [177]. We formulate it as such so that the preceding construction can be extended naturally to bounded symmetric domains, which is what we are going to do presently.

There is another more direct approach to the construction of $H(\xi, \lambda)$, however, also applicable to bounded symmetric domains. We discuss this first. This approach depends on the crucial observation that the horodisc $H(\xi, \lambda)$ is identical with the following sets:

$$
\begin{aligned}
H(\xi, \lambda) &= \left\{z \in \mathbb{D} : \lim_k \frac{|1 - z\overline{z}_k|^2}{1 - |z|^2} < \frac{1}{\lambda}\right\} \\
&= \left\{z \in \mathbb{D} : \lim_k \frac{1 - |z_k|^2}{1 - |g_{-z_k}(z)|^2} < \frac{1}{\lambda}\right\}
\end{aligned}
\qquad (3.40)
$$

which can be formulated for bounded symmetric domains. Hinted by this, we proved the following lemma.

Lemma 3.5.20. *Let D be a bounded symmetric domain realised as the open unit ball of a JB*-triple V. Let (z_k) be a sequence in D (norm) converging to a boundary point $\xi \in \partial D$. The function $F : D \to [0, \infty)$ given by*

$$F(x) = \limsup_{k \to \infty} \frac{1 - \|z_k\|^2}{1 - \|g_{-z_k}(x)\|^2} \qquad (x \in D)$$

is well-defined and continuous.

Proof. For each $x \in D$, we have, by Lemma 3.2.28 and (3.27),

$$
\begin{aligned}
\frac{1 - \|z_k\|^2}{1 - \|g_{-z_k}(x)\|^2} &= \|B(x,x)^{-1/2}B(x,z_k)B(z_k,z_k)^{-1/2}\|(1 - \|z_k\|^2) \\
&\leq \|B(x,x)^{-1/2}B(x,z_k)\|\|B(z_k,z_k)^{-1/2}\|(1 - \|z_k\|^2) \\
&= \|B(x,x)^{-1/2}B(x,z_k)\| \leq \frac{(1 + \|x\|\|z_k\|)^2}{1 - \|x\|^2} \leq \frac{1 + \|x\|}{1 - \|x\|}
\end{aligned}
$$

and hence the defining sequence for $F(x)$ is bounded. Therefore F is well-defined.

For continuity, let $x, y \in D$ and write $x_k = \dfrac{1 - \|z_k\|^2}{1 - \|g_{-z_k}(x)\|^2}$, also $y_k = \dfrac{1 - \|z_k\|^2}{1 - \|g_{-z_k}(y)\|^2}$. Then we have

$$|x_k - y_k| \leq \|B(x,x)^{-1/2}B(x,z_k) - B(y,y)^{-1/2}B(y,z_k)\|$$

which gives

$$
\begin{aligned}
|F(x) - F(y)| &= |\limsup_k x_k - \limsup_k y_k| \\
&\leq \limsup_k \|B(x,x)^{-1/2}B(x,z_k) - B(y,y)^{-1/2}B(y,z_k)\| \\
&= \|B(x,x)^{-1/2}B(x,\xi) - B(y,y)^{-1/2}B(y,\xi)\|
\end{aligned}
$$

since (z_k) norm converges to ξ. Now continuity of F follows from that of the function $h(x) = \|B(x,x)^{-1/2}B(x,\xi)\|$. $\qquad\square$

A computation similar to the previous proof yields the following result.

Lemma 3.5.21. *Let (x_k) be a sequence in D norm converging to $x \in D$. Then we have*

$$\limsup_{k \to \infty} \frac{1 - \|z_k\|^2}{1 - \|g_{-z_k}(x_k)\|^2} = \limsup_{k \to \infty} \frac{1 - \|z_k\|^2}{1 - \|g_{-z_k}(x)\|^2}.$$

The sequence (z_k) in Wolff's theorem for $\mathbb{D}$ has a limit point ξ by relative compactness of $\mathbb{D}$. An infinite dimensional domain D need not be relatively compact and the existence of limit points is not guaranteed. For this reason, we consider *compact* maps $f : D \to D$, which are the ones having relatively compact image $f(D)$, that is, the norm closure $\overline{f(D)}$ is compact in V. All continuous self-maps on a finite dimensional bounded domain are necessarily compact.

Now let $f : D \longrightarrow D$ be a *compact* holomorphic map without fixed-point. To prove an analogue of Wolff's theorem, the first step is to construct a sequence (z_k) in D converging to the boundary, as in the case of $\mathbb{D}$. As before, pick a sequence $\alpha_k \in (0, 1)$ with $\alpha_k \uparrow 1$, and consider the function $\alpha_k f : D \longrightarrow D$. To find a fixed-point for $\alpha_k f$, one can make use of the fixed-point theorem of Earle and Hamilton [59], instead of Rouché's theorem, which states that a holomorphic self-map h on a bounded domain D in a complex Banach space must have a fixed-point if the image $h(D)$ is strictly contained in D. Using this result, we obtain $\alpha_k f(z_k) = z_k$ for some $z_k \in D$ and for each k. By compactness of f, we may assume (z_k) norm converges to some point $\xi \in \overline{D}$, via a subsequence if necessary. Since f has no fixed-point in D, we must have $\xi \in \partial D$.

We are now ready to extend Wolff's theorem to all bounded symmetric domains, following the observation in (3.40).

Theorem 3.5.22. *Let f be a fixed-point free compact holomorphic self-map on a bounded symmetric domain D. Then there is a sequence (z_k) in D converging to a boundary point $\xi \in \overline{D}$ such that, for each $\lambda > 0$, the set*

$$H(\xi, \lambda) = \left\{ x \in D : \limsup_{k \to \infty} \frac{1 - \|z_k\|^2}{1 - \|g_{-z_k}(x)\|^2} < \frac{1}{\lambda} \right\}$$

is a convex domain and f-invariant, that is, $f(H(\xi, \lambda)) \subset H(\xi, \lambda)$. Moreover, we have $D = \bigcup_{\lambda > 0} H(\xi, \lambda)$ and $0 \in \bigcap_{\lambda < 1} H(\xi, \lambda)$.

Proof. Let (z_k) be the sequence in D constructed previously, which converges to a boundary point $\xi \in D$. By Lemma 3.5.20, the set $H(\xi, \lambda)$ is open. To see that $H(\xi, \lambda)$ is convex, let $x, y \in H(\xi, \lambda)$ and $0 < \alpha < 1$. Then there exists k_0 such that $k \geq k_0$ implies

$$\frac{1 - \|z_k\|^2}{1 - \|g_{-z_k}(x)\|^2}, \quad \frac{1 - \|z_k\|^2}{1 - \|g_{-z_k}(y)\|^2} < \frac{1}{\beta} < \frac{1}{\lambda}$$

for some β satisfying $\beta(1 - \|z_k\|^2) = 1 - s_k^2$ with $s_k \in (0, 1)$. This gives $\|g_{-z_k}(x)\| < s_k$ and $\|g_{-z_k}(y)\| < s_k$, that is, $k_D(z_k, x) < \tanh^{-1} s_k$ and $k_D(z_k, y) < \tanh^{-1} s_k$. By Corollary 3.5.10, the element $w := \alpha x + (1 - \alpha)y$ satisfies

$$k_D(z_k, w) < \tanh^{-1} s_k \qquad (k \geq k_0).$$

Therefore

$$\limsup_{k \to \infty} \frac{1 - \|z_k\|^2}{1 - \|g_{-z_k}(w)\|^2} \leq \limsup_{k \to \infty} \frac{1 - \|z_k\|^2}{1 - s_k^2} = \frac{1}{\beta} < \frac{1}{\lambda}$$

and $w \in H(\xi, \lambda)$.

For f-invariance, let $x \in H(\xi, \lambda)$. We need to show $f(x) \in H(\xi, \lambda)$. Let $f_k = \alpha_k f$. Then by the Schwarz-Pick lemma, we have

$$\|g_{-z_k}(f_k(x))\| = \|g_{-f_k(z_k)}(f_k(x)) \leq \|g_{-z_k}(x)\|$$

and

$$\frac{1 - \|z_k\|^2}{1 - \|g_{-z_k}(f_k(x))\|^2} \leq \frac{1 - \|z_k\|^2}{1 - \|g_{-z_k}(x)\|^2}.$$

Hence Lemma 3.5.21 implies

$$\limsup_{k \to \infty} \frac{1 - \|z_k\|^2}{1 - \|g_{-z_k}(f(x))\|^2} = \limsup_{k \to \infty} \frac{1 - \|z_k\|^2}{1 - \|g_{-z_k}(f_k(x))\|^2}$$

$$\leq \limsup_{k \to \infty} \frac{1 - \|z_k\|^2}{1 - \|g_{-z_k}(x)\|^2} < \frac{1}{\lambda}$$

which gives $f(x) \in H(\xi, \lambda)$.

Finally, for each $y \in D$, we have $y \in H(\xi, \lambda)$ whenever $F(y) < 1/\lambda$. Since $F(0) = 1$, we have $0 \in H(\xi, \lambda)$ for $\lambda < 1$. This proves the last assertion. $\qquad \square$

Remark 3.5.23. We see from (3.40) that the preceding theorem reduces to Wolff's theorem when $D = \mathbb{D}$, in which case ξ is unique for if there is another boundary point ξ' satisfying the theorem, one can choose horodiscs $H(\xi, \lambda)$ and $H(\xi', \lambda')$ which are tangent to each other at a point $p \in \mathbb{D}$. By f-invariance, we find $f(p)$ in the intersection of the closures $\overline{H}(\xi, \lambda) \cap \overline{H}(\xi', \lambda') = \{p\}$, contradicting fixed-point freeness of f. In higher dimensions, such argument may not be available.

Remark 3.5.24. Evidently, the family $\{H(\xi, \lambda)\}_{\lambda > 0}$ in Theorem 3.5.22 is decreasing in the sense that $\lambda > \lambda'$ implies $H(\xi, \lambda) \subset H(\xi, \lambda')$. Hence, given $H(\xi, \lambda) \neq \emptyset$, then $H(\xi, \lambda') \neq \emptyset$ for all $\lambda' \leq \lambda$. In fact, we have $H(\xi, \lambda) \neq \emptyset$ for all $\lambda > 0$, by Theorem 3.5.27 below.

In view of the preceding discussion, we shall call $H(\xi, \lambda)$ a *horoball* at ξ (of '*radius*' $\frac{1}{1+\lambda}$) in higher dimensional bounded symmetric domains.

Example 3.5.25. Let D be the open unit ball of the JB*-triple $C(\Omega)$ of complex continuous functions on a compact Hausdorff space Ω, with Jordan triple product $\{x, y, z\} = x\overline{y}z$ where $\overline{y}$ denotes the complex conjugate of the function $y \in C(\Omega)$. For $a, b \in D$, the Bergmann operator $B(a, b)$ is given by a product of functions:

$$B(a, b)(z) = (\mathbf{1} - a\bar{b})^2 z \qquad (z \in C(\Omega))$$

where $\mathbf{1}$ denotes the constant function with value 1 and we have

$$\|B(a, b)\| = \|(\mathbf{1} - a\bar{b})^2\| = \sup\{|1 - a(\omega)\bar{b}(\omega)|^2 : \omega \in \Omega\}.$$

Let (z_k) be a sequence in D converging to $\xi \in \partial D$ as in Theorem 3.5.22. We may assume $0 \notin z_k(\Omega)$ by omitting the first few terms of the sequence. Then

$$
\begin{aligned}
&\|B(x, x)^{-1/2} B(x, z_k) B(z_k, z_k)^{-1/2}\|(1 - \|z_k\|^2) \\
&= \left\| \frac{(\mathbf{1} - x\bar{z}_k)^2 (1 - \|z_k\|^2)}{(1 - |x|^2)(1 - |z_k|^2)} \right\|.
\end{aligned}
$$

Since $\left\| \frac{1 - \|z_k\|^2}{1 - |z_k|^2} \right\| \le 1$ and the sequence $(\mathbf{1} - x\bar{z}_k)$ converges to $\mathbf{1} - x\bar{\xi}$ in $C(\Omega)$, we have

$$
\begin{aligned}
&\limsup_k \|B(x, x)^{-1/2} B(x, z_k) B(z_k, z_k)^{-1/2}\|(1 - \|z_k\|^2) \\
&= \limsup_k \left\| \frac{(\mathbf{1} - x\bar{\xi})^2}{1 - |x|^2} \frac{1 - \|z_k\|^2}{1 - |z_k|^2} \right\|
\end{aligned}
$$

and for $\lambda > 0$,

$$H(\xi, \lambda) = \left\{ x \in D : \limsup_k \left\| \frac{(\mathbf{1} - x\bar{\xi})^2}{1 - |x|^2} \left(\frac{1 - \|z_k\|^2}{1 - |z_k|^2} \right) \right\| < \frac{1}{\lambda} \right\}$$

which reduces to the horodisc in (3.39) if Ω is a singleton in which case the function $\frac{1 - \|z_k\|^2}{1 - |z_k|^2} = \mathbf{1}$.

Another construction of the horodisc $H(\xi, \lambda)$ in $\mathbb{D}$ is by taking the limit of a sequence of Poincaré discs $P_k(\lambda)$, as noted in (3.39). We now extend this construction to bounded symmetric domains.

Naturally, one needs to replace the Poincaré metric by another invariant distance on a bounded symmetric domain D. We choose the Kobayashi distance k_D. Analogous to (3.38), we define a *Kobayashi ball* $D_z(r)$, centred at $z \in D$ with radius $\tanh^{-1} r > 0$, to be the set

$$D_z(r) := \{x \in D : k_D(x, z) < \tanh^{-1} r\} = g_{-z}(D(0, r))$$

where g_{-z} is the Möbius transformation and, to avoid confusion with the notation for Bergman operators, $D(0, r)$ denotes the (norm) open ball centred at 0 of radius $r > 0$ throughout this section. We note that the Kobayashi ball $D_z(r)$ is convex by Corollary 3.5.10.

We define the *limit* of a sequence $(D_{z_k}(r_k))$ of Kobayashi balls to be the set

$$\lim_k D_{z_k}(r_k) := \{x \in \overline{D} : x = \lim_k x_k, \ x_k \in D_{z_k}(r_k)\}. \qquad (3.41)$$

Now, given a fixed-point free *compact* holomorphic map $f : D \longrightarrow D$, we can find a sequence (z_k) in D converging to a boundary point $\xi \in \partial D$, as before. For each $\lambda > 0$, one wishes to find a suitable sequence of Kobayashi balls *'converging'* to the horoball $H(\xi, \lambda)$, that is, in view of (3.39), $H(\xi, \lambda)$ is the interior of the limit of these balls.

Mimicking the one-dimensional construction again, given $\lambda > 0$, we can choose $r_k \in (0, 1)$ satisfying

$$\lambda(1 - \|z_k\|^2) = 1 - r_k^2$$

from some k onwards. To simplify notation, we write

$$D_k(\lambda) := D_{z_k}(r_k)$$

for the Kobayashi ball centred at z_k with radius $\tanh^{-1} r_k$. We denote the limit of the sequence $(D_k(\lambda))$ by

$$S(\xi, \lambda) := \lim_k D_k(\lambda) \qquad (3.42)$$

and the interior by $S_0(\xi, \lambda)$. We note that $S(\xi, \lambda) \neq \emptyset$ since $\xi = \lim_k z_k \in S(\xi, \lambda)$.

For $D = \mathbb{D}$, both $H(\xi, \lambda)$ and $S_0(\xi, \lambda)$ just defined are identical to the ones in (3.39). What is an appropriate generalisation of the disc

$$\frac{\lambda}{1 + \lambda}\xi + \frac{1}{1 + \lambda}\mathbb{D}$$

in (3.39) to higher dimensions? A simple computation reveals that this disc can be written as

$$\frac{\lambda}{1 + \lambda}\xi + B\left(\sqrt{\frac{\lambda}{1 + \lambda}}\xi, \sqrt{\frac{\lambda}{1 + \lambda}}\xi\right)^{1/2}(\mathbb{D}) \qquad (3.43)$$

where $B\left(\sqrt{\frac{\lambda}{1+\lambda}}\xi, \sqrt{\frac{\lambda}{1+\lambda}}\xi\right)$ is the Bergman operator on $\mathbb{D}$. The formulation in (3.43) can be carried over to all bounded symmetric domains.

Question: with the notions in (3.42) and (3.43), do we have, as in (3.39), that for all bounded symmetric domains D,

$$H(\xi, \lambda) = S_0(\xi, \lambda) = \frac{\lambda}{1 + \lambda}\xi + B\left(\sqrt{\frac{\lambda}{1 + \lambda}}\xi, \sqrt{\frac{\lambda}{1 + \lambda}}\xi\right)^{1/2}(D)$$

with perhaps some appropriate modification of the last set? The answer is affirmative for finite-rank domains.

First, we always have the inclusion $H(\xi, \lambda) \subset S_0(\xi, \lambda)$.

Proposition 3.5.26. *For a fixed-point free compact holomorphic self-map f on any bounded symmetric domain D and for $\lambda > 0$, we have*

$$H(\xi, \lambda) \subset S_0(\xi, \lambda) \quad \text{and} \quad S(\xi, \lambda) \cap D \subset \{x \in D : F(x) \leq 1/\lambda\}$$

where F is the function in Lemma 3.5.20 and $S(\xi, \lambda) \cap D$ is also an f-invariant domain.

Proof. Let $x \in H(\xi, \lambda)$. Then we have, from some k onwards,

$$\frac{1 - \|z_k\|^2}{1 - \|g_{-z_k}(x)\|^2} < \frac{1}{\lambda} = \frac{1 - \|z_k\|^2}{1 - r_k^2}$$

which implies $\|g_{-z_k}(x)\| < r_k$, that is, $x \in D_k(\lambda)$. Hence $x \in S(\xi, \lambda)$. We have shown $H(\xi, \lambda) \subset S(\xi, \lambda)$ and therefore $H(\xi, \lambda) \subset S_0(\xi, \lambda)$ since $H(\xi, \lambda)$ is open.

For the second assertion, let $x \in S(\xi, \lambda) \cap D$ with $x = \lim_k x_k$ and $x_k \in D_k(\lambda)$. Since $\|g_{-z_k}(x_k)\| < r_k$ for each k, we have from Lemma 3.5.21 that

$$\limsup_{k \to \infty} \frac{1 - \|z_k\|^2}{1 - \|g_{-z_k}(x)\|^2} = \limsup_{k \to \infty} \frac{1 - \|z_k\|^2}{1 - \|g_{-z_k}(x_k)\|^2}$$

$$\leq \limsup_{k \to \infty} \frac{1 - \|z_k\|^2}{1 - r_k^2} = \frac{1}{\lambda}.$$

This proves $S(\xi, \lambda) \cap D \subset \{x : F(x) \leq 1/\lambda\}$.

To see that $S(\xi, \lambda) \cap D$ is f-invariant, write $f_k = \alpha_k f$ where as before, $\alpha_k \in (0, 1)$ is chosen so that $f_k(z_k) = z_k$. Then for each $x \in S(\xi, y) \cap D$ with $x = \lim_k x_k$ and $x_k \in D_k(\lambda)$, the Schwarz-Pick lemma yields

$$\|g_{-z_k}(f_k(x_k))\| = \|g_{-f_k(z_k)}(f_k(x_k))\| \leq \|g_{-z_k}(x_k)\| \leq \|g_{-z_k}(y)\|$$

and hence $f_k(x_k) \in D_k(\lambda)$. It follows that

$$f(x) = \lim_k f(x_k) = \lim_k f_k(x_k) \in S(\xi, y) \cap D.$$

$\square$

Although the question raised before Proposition 3.5.26 remains open for *all* bounded symmetric domains, a complete answer *can* be given for all *finite-rank* bounded symmetric domains, and in particular, for finite dimensional ones. This has been proved in [50, Theorem 5.13, Theorem 5.14]. We state the result below, but suppress the lengthy proof which relies substantially on Jordan theory.

Theorem 3.5.27. *Let f be a fixed-point free compact holomorphic self-map on a bounded symmetric domain D of finite rank p. Then there is a sequence (z_k) in D converging to a boundary point*

$$\xi = \sum_{j=1}^{m} \alpha_j e_j \qquad (\alpha_j > 0,\ m \in \{1, \ldots, p\})$$

where $e_1, \ldots, e_m$ are orthogonal minimal tripotents in ∂D, such that for each $\lambda > 0$, the f-invariant horoball $H(\xi, \lambda)$ is given by

$$H(\xi, \lambda) = S_0(\xi, \lambda)$$

$$= \sum_{j=1}^{m} \frac{\sigma_j \lambda}{1 + \sigma_j \lambda} e_j + B\left(\sum_{j=1}^{m} \sqrt{\frac{\sigma_j \lambda}{1 + \sigma_j \lambda}} e_j, \sum_{j=1}^{m} \sqrt{\frac{\sigma_j \lambda}{1 + \sigma_j \lambda}} e_j \right)^{1/2} \quad (D)$$

which is affinely homeomorphic to D, where $\sigma_j \geq 0$ and $\max\{\sigma_j : j = 1, \ldots, m\} = 1$.

Remark 3.5.28. For $p = 1$, that is, D is a Hilbert ball, in the above theorem, we have

$$S_0(\xi, \lambda) = \frac{\lambda}{1 + \lambda} \xi + B\left(\sqrt{\frac{\lambda}{1 + \lambda}} \xi, \sqrt{\frac{\lambda}{1 + \lambda}} \xi \right)^{1/2} \quad (D)$$

and in one dimension, $D = \mathbb{D}$, it reduces to Wolff's horodisc in (3.39) as noted earlier:

$$S_0(\xi, \lambda) = \frac{\lambda}{1 + \lambda} \xi + \frac{1}{1 + \lambda} \mathbb{D}.$$

We note that both Theorems 3.5.22 and 3.5.27 assume the compactness property of the holomorphic self-map f in infinite dimension. The question remains open whether this condition can be removed. For Hilbert balls, however, the answer is affirmative for the following reason.

Let f be a fixed-point free holomorphic self-map on a bounded symmetric domain D. We observe that, in the previous construction of the horoballs $H(\xi, \lambda)$, the existence of the sequence (z_k) in D does not depend on the compactness of f, but the existence of the limit $\xi \in \partial D$ does require the compactness assumption. In the case of a Hilbert ball D, one can circumvent this requirement. To see this, let (z_k) be chosen, via the Earle-Hamilton fixed-point theorem as before, satisfying $\alpha_k f(z_k) = z_k$ with $0 < \alpha_k \uparrow 1$.

By relatively weak compactness of the Hilbert ball D, we may assume, by choosing a subsequence if necessary, that (z_k) converges weakly to some point $\xi \in \overline{D}$. We show $\xi \in \partial D$, otherwise we have

$$\|g_{-f(\xi)}(f(z_k))\| \leq \|g_{-\xi}(z_k)\|.$$

Using (3.28) and substituting $f(z_k) = z_k/\alpha_k$, one deduces

$$\frac{1 - \alpha_k^{-2}\|z_k\|^2}{1 - \|z_k\|^2} \frac{|1 - \langle z_k, \xi \rangle|^2}{1 - \|\xi\|^2} \geq \frac{|1 - \langle \alpha_k^{-1} z_k, f(\xi) \rangle|^2}{1 - \|f(\xi)\|^2}.$$

Since $\frac{1 - \alpha_k^{-2}\|z_k\|^2}{1 - \|z_k\|^2} \leq 1$, we have

$$1 \geq \frac{1}{1 - \|g_{-f(\xi)}(\xi)\|^2}$$

by letting $k \to \infty$. This gives $\|g_{-f(\xi)}(\xi)\| = 0$ and $f(\xi) = \xi$, contradicting the non-existence of a fixed point in D. It follows that (z_k) actually norm converges to ξ since $\limsup_{k\to\infty} \|z_k\| \leq 1 = \|\xi\| \leq \liminf_{k\to\infty} \|z_k\|$.

Now one can follow the same proofs for Theorems 3.5.22 and 3.5.27, but without the compactness assumption of f in which case, the horoball $H(\xi, \lambda)$ is indeed 'internally tangent to' D.

Theorem 3.5.29. *Let $f : D \longrightarrow D$ be a fixed-point free holomorphic map on a Hilbert ball D in Theorem 3.5.27. Then for each $\lambda > 0$, we have*

$$\overline{H(\xi, \lambda)} \cap \partial D = \{\xi\}.$$

Proof. We have $\overline{H(\xi,\lambda)} = \overline{S_0(\xi,\lambda)} = S(\xi,\lambda)$. Let $p \in \overline{H(\xi,\lambda)} \cap \partial D$.

Retaining the previous notation, we have from (3.42), $p = \lim_k p_k$ for some $p_k \in D_k(\lambda)$, where $\|g_{-z_k}(p_k)\| < r_k$ and $\lambda(1 - \|z_k\|^2) = 1 - r_k^2$. Using (3.28), we deduce

$$\begin{aligned}
|1 - \langle p_k, z_k \rangle|^2 &\leq (1 - \|p_k\|^2) \left(\frac{1 - \|z_k\|^2}{1 - \|g_{-z_k}(p_k)\|^2} \right) \\
&< (1 - \|p_k\|^2) \left(\frac{1 - \|z_k\|^2}{1 - r_k^2} \right) = \frac{1 - \|p_k\|^2}{\lambda} \to 0
\end{aligned}$$

as $k \to \infty$, which gives $\langle p, \xi \rangle = 1$ and $p = \xi$. $\qquad\square$

To conclude this section, we discuss briefly the dynamics of a holomorphic self-map f on a bounded symmetric domain D, which concerns the asymptotic behaviour of the iterates (f^n) of f, where

$$f^n = \overbrace{f \circ \cdots \circ f}^{n\text{-times}}.$$

In other words, one considers (f, D) as a discrete-time dynamical system and studies its limit sets, which can be described as the images of subsequential limits of the iterates (f^n). We will call a subsequential limit

$$h = \lim_{k \to \infty} f^{n_k}$$

in the topology of locally uniform convergence, a *limit function* of (f^n). The case of the disc $\mathbb{D}$ is well understood.

Let $f : D \longrightarrow D$ be a holomorphic map. Then either f has a fixed-point in D or has none. In the former case, let $g_a : D \longrightarrow D$ be the Möbius transformation induced by the fixed-point $a \in D$. Then $g_{-a} \circ f \circ g_a(0) = 0$. In particular, if f is biholomorphic, then it is conjugate to a linear isometry $\varphi : D \longrightarrow D$ (cf. Lemma 1.1.8) and the dynamics of f is essentially the same as φ. For $D = \mathbb{D}$, the isometry φ is a rotation: $\varphi(z) = e^{i\theta} z$ for some $\theta \in [0, 2\pi)$. It follows that the orbit $\{f^n(z)\}$ is finite if θ is a rational multiple of π, in which case f^p is the identity map for some $p \in \mathbb{N}$. If θ is an irrational multiple of π, then $\{e^{in\theta} : n = 1, 2, \ldots\}$ is dense in $\partial\mathbb{D}$ and from this, one can infer the dynamics of f.

Remark 3.5.30. For $D = \mathbb{D}$ and a biholomorphic map $f = \alpha g_c$ with a fixed-point in D that is not the identity map, where $c \in D$ and $\alpha \in \partial\mathbb{D}$, it has been proved by Rigby in [148] that f^p is the identity map exactly when

$$\sum_{r=0}^{p-1} \alpha^r \sum_{s=0}^{\min\{r,\,p-1-r\}} |c|^{2s} \binom{r}{s} \binom{p-1-r}{s} = 0.$$

For a Hilbert ball D and $c \in D\backslash\{0\}$, it has also been shown in [148] that αg_c has a fixed-point in D if and only if $|1 - \alpha| > 2\|c\|$.

Hence there are two remaining cases to be discussed, namely,

(I) f has a fixed-point in D and is not biholomorphic,

(II) f has no fixed-point in D.

Case (I). We begin with the unit disc $\mathbb{D}$. Again, Schwarz lemma provides a solution. The following result is well-known.

Theorem 3.5.31. *Let $f : \mathbb{D} \longrightarrow \mathbb{D}$ be holomorphic with a fixed-point $a \in \mathbb{D}$. If f is not biholomorphic, then the sequence (f^n) of iterates converge locally uniformly to a constant function with value a.*

Proof. Conjugating with a Möbius transformation as before, we may assume $a = 0$. Since f is not biholomorphic, Schwarz lemma says that we must have $|f(z)| < |z|$ for all $z \neq 0$. Let $0 < r < 1$ and let

$$M(r) = \sup\{|f(z)| : |z| \leq r\}.$$

Then we have $M(r) < r$ and we may assume $M(r) > 0$. Define a holomorphic map $h : \mathbb{D} \longrightarrow \mathbb{D}$ by

$$h(z) = \frac{f(rz)}{M(r)}.$$

Then apply the Schwarz lemma once more, we obtain

$$|f(z)| \leq \frac{M(r)}{r}|z| \qquad (|z| \leq r).$$

Write $c := M(r)/r \in (0,1)$. Iterating the above inequality gives

$$|f^n(z)| \leq c^n|z| \leq c^n r$$

for all z in the closed disc $\overline{\mathbb{D}}(0,r) \subset \mathbb{D}$. This implies (f^n) converges uniformly on $\overline{\mathbb{D}}(0,r)$ to 0, and hence locally uniformly on $\mathbb{D}$. □

Remark 3.5.32. An interesting consequence of the preceding theorem is that a holomorphic map $f : \mathbb{D} \longrightarrow \mathbb{D}$, which is not biholomorphic, can have at most one fixed-point in $\mathbb{D}$.

It is natural to ask if the preceding result and proof can be extended to bounded symmetric domains D. For this, one would need the crucial inequality $\|f(z)\| < \|z\|$ in the case of a holomorphic self-map f on D, which has a fixed-point and is not biholomorphic. Unfortunately, it is unavailable even for the 2-dimensional Euclidean ball $B = \{(z,w) \in \mathbb{C}^2 : |z|^2 + |w|^2 < 1\}$. The holomorphic map $f : (z,w) \in B \mapsto (z,0) \in B$ is an example, where $f(z,0) = (z,0)$.

One may try an alternative approach to Theorem 3.5.31 for $\mathbb{D}$ using instead the derivative $f'(a)$ at the fixed-point a and the Schwarz lemma. Indeed, under the assumption of the theorem with $a = 0$, we would have $|f'(0)| < 1$ by the Schwarz lemma, since f is not biholomorphic. It follows from continuity that $|f'(z)| < 1$ on a closed disc $\overline{\mathbb{D}}(0,r)$ with $0 < r < 1$. Hence we have

$$|f(z) - f(w)| \leq c|z - w|$$

for some $c \in (0,1)$ and all $z, w \in \overline{\mathbb{D}}(0,r)$. From this, one concludes that (f^n) converges locally uniformly to 0, using the well-known contraction mapping theorem.

Although the above example of $f(z, w) = (z, 0)$ also shows that the inequality $\|f'(0)\| < 1$ is unavailable for a holomorphic self-map f on the ball B, which fixes 0 and is not biholomorphic, nevertheless, Cauchy inequality implies that, if (f^n) converges to 0 locally uniformly, then we have the weaker inequality

$$\rho(f'(0)) = \lim_{n \to \infty} \|(f^n)'(0)\|^{1/n} = \lim_{n \to \infty} \|(f')^n(0)\|^{1/n} < 1$$

where $\rho(f'(0))$ denotes the spectral radius of $f'(0)$. This inequality turns out to be also sufficient for the convergence of the iterates (f^n). The proof of sufficiency and detailed references for the following theorem, due to Vesentini, Khatskevich and Shoikhet, can be found in [146, Proposition 5.3].

Theorem 3.5.33. *Let D be a bounded domain in a complex Banach space and let $f : D \longrightarrow D$ be a holomorphic map with a fixed-point $a \in D$. The following conditions are equivalent.*

(i) *(f^n) converges locally uniformly to a constant function with value a.*

(ii) *The spectral radius $\rho(f'(a))$ is strictly less than 1.*

Case (II). We now consider a *fixed-point free* holomorphic self-map f on a bounded symmetric domain D. In the case of $D = \mathbb{D}$, we have the following celebrated Denjoy-Wolff theorem [55, 178] which, together with Theorem 3.5.31, determines completely the dynamics of a holomorphic self-map on $\mathbb{D}$ that is not biholomorphic (cf. Remark 3.5.32).

Theorem 3.5.34. *Let $f : \mathbb{D} \longrightarrow \mathbb{D}$ be a fixed-point free holomorphic map. Then there is a unique boundary point $\xi \in \partial\mathbb{D}$ such that the iterates (f^n) converge locally uniformly to the constant function $f(\cdot) = \xi$.*

Can this result be generalised to bounded symmetric domains? This question has been investigated by many authors, and also for other domains. It is impossible to discuss details of all these works. One can find, for example, substantial literature on this topic in [2, 145, 146]. Some recent literature, by no means complete, is included in [39].

For infinite dimensional bounded symmetric domains, we consider as before compact holomorphic maps. Let us begin the discussion with a useful criterion for the existence of a fixed-point for a compact holomorphic self-map on a domain, stated below, which has been proved in [114].

Lemma 3.5.35. *Let B be the open unit ball of a complex Banach space and let $f : B \longrightarrow B$ be a compact holomorphic map. The following conditions are equivalent.*

(i) *f is fixed-point free.*

(ii) *There exists $a \in B$ such that $\sup_k \|f^{n_k}(a)\| = 1$ for every subsequence (f^{n_k}) of the iterates of f.*

An important consequence of this criterion and Lemma 3.4.11 is that, for a fixed-point free compact holomorphic self-map f on a bounded symmetric domain D, realised as the open unit ball of a JB*-triple, the image $h(D)$ of a limit function $h = \lim_k f^{n_k}$ is entirely contained in a single boundary component of the boundary ∂D.

Hérver has shown in [82] that the Denjoy-Wolff theorem can be extended to finite dimensional Euclidean balls (see also [126]), but also shown in [83] that the theorem fails in the case of the bidisc $\mathbb{D} \times \mathbb{D}$. The iterates of a fixed-point free holomorphic map $f : \mathbb{D} \times \mathbb{D} \longrightarrow \mathbb{D} \times \mathbb{D}$ need not converge to a constant function. Instead, we have a phenomenon which can be formulated as follows.

Theorem 3.5.36. *Given a fixed-point free holomorphic map* $f : \mathbb{D} \times \mathbb{D} \longrightarrow \mathbb{D} \times \mathbb{D}$, *there is one single boundary component* $\Gamma \subset \partial(\mathbb{D} \times \mathbb{D})$ *such that the image* $h(\mathbb{D} \times \mathbb{D})$ *of all subsequential limits* $h = \lim_k f^{n_k}$ *of* (f^n) *is contained in the closure* $\overline{\Gamma}$.

The closed boundary components of $\partial(\mathbb{D} \times \mathbb{D})$ are the singletons $\{(\xi, \eta)\}$ and sets of the form $\{\xi\} \times \overline{\mathbb{D}}$ and $\overline{\mathbb{D}} \times \{\eta\}$, where $|\xi| = |\eta| = 1$. Hence there are three distinct possibilities for the iterates (f^n) on the bidisc. On the other hand, the boundary components of a Hilbert ball are exactly the boundary points. View in these perspectives, a formulation of a possible generalisation of the *Denjoy-Wolff theorem* emerges. We state it as a *conjecture* since it has not been completely proved at present.

> **Conjecture.** Let D be a bounded symmetric domain in a complex Banach space V and let $f : D \longrightarrow D$ be a fixed-point free *compact* holomorphic map. Then there is one single boundary component Γ in the boundary ∂D such that $h(D) \subset \overline{\Gamma}$ for *all* limit functions h of (f^n).

This conjecture is true for finite dimensional Euclidean balls as well as the bidisc, as already discussed. In infinite dimension, the conjecture is false without the assumption of *compactness* of the self-map f. Stachura [160] has given an example of a fixed-point free biholomorphic map f on an infinite dimensional Hilbert ball, with a subsequence (f^{n_k}) satisfying $\limsup_n \|f^n(0)\| = 1$ and $\lim_k f^{n_k}(0) = 0$.

Nevertheless, the conjecture is true for all Hilbert balls. This has been proved by Chu and Mellon [47]. In fact, compactness of the self-map f is not a necessary condition for a Denjoy-Wolff theorem for Hilbert balls. We will now prove a more general result without the compactness assumption, which includes the Chu-Mellon result as a special case. Actually, we will give necessary and sufficient conditions for the Denjoy-Wolff theorem to hold on a Hilbert ball. We prove two lemmas first.

Lemma 3.5.37. *Let D be the open unit ball of a Hilbert space with inner product $\langle \cdot, \cdot \rangle$. Given a sequence (a_k) in D norm converging to $a \in \partial D$ and a sequence (v_k) in D weakly convergent to some $v \in D$, we have*

$$\lim_{k \to \infty} \|g_{-a_k}(v_k)\| = 1. \tag{3.44}$$

If there is a sequence (b_k) in D norm converging to $b \in \partial D$ and $a \neq b$, then

$$\lim_{k \to \infty} \|g_{-a_k}(b_k)\| = 1. \tag{3.45}$$

Proof. By Example 3.2.29, the Bergman operators $B(a_k, a_k)$ norm converges to 0. For $c \in D$, the Möbius transformation $g_c(v) = c + B(c, c)^{1/2}(I + v \,\square\, c)^{-1}(v)$ can be written as

$$g_c(v) = c + \frac{B(c, c)^{1/2}(v)}{1 + \langle v, c \rangle} \qquad (v \in D).$$

It follows that

$$
1 > \|g_{-a_k}(v_k)\| = \left\| -a_k + \frac{1}{1 - \langle v_k, a_k \rangle} B(a_k, a_k)^{1/2}(v_k) \right\|
$$
$$
\geq \|a_k\| - \frac{1}{|1 - \langle v_k, a_k \rangle|} \|B(a_k, a_k)^{1/2}(v_k)\| \longrightarrow 1, \quad \text{as} \quad k \to \infty
$$

where $\langle v, a \rangle \neq 1$. This proves (3.44).

To show (3.45), observe from Lemma 3.2.28 that

$$
1 - \|g_{-a_k}(b_k)\|^2 = \|B(b_k, b_k)^{-1/2} B(b_k, a_k) B(a_k, a_k)^{-1/2}\|^{-1}
$$
$$
\leq \frac{\|B(b_k, b_k)^{1/2}\| \|B(a_k, a_k)^{1/2}\|}{\|B(b_k, a_k)\|} \longrightarrow 0 \text{ as } k \to \infty
$$

where, by Example 3.2.29, $\lim_k B(b_k, a_k) = B(b, a) \neq 0$ while

$$\lim_k B(b_k, b_k) = B(b, b) = 0$$

and $B(a, a) = 0$.   □

Lemma 3.5.38. *Let D be a Hilbert ball and $f : D \longrightarrow D$ a holomorphic map such that the iterates f^n converge pointwise to a constant map $h : D \longrightarrow \partial D$. Then (f^n) converges locally uniformly to h.*

Proof. Let $h(D) = \{\xi\}$. Let $B \subset D$ be an open ball strictly contained in D so that $\mathrm{dist}(B, \partial D) > 0$. Let $D(\xi, \varepsilon)$ be an open ball of radius $\varepsilon > 0$ such that $\overline{D(\xi, \varepsilon)} \cap B = \emptyset$. We show that $f^n(B) \subset D(\xi, \varepsilon) \cap D$ from some n onwards. This would complete the proof.

Suppose what we claim to show is false. Then there is a subsequence (f^{n_k}) such that $f^{n_k}(b_k) \notin D(\xi, \varepsilon)$, where $b_k \in B$ and $f^{n_k}(b_k)$ converges weakly to some $v \in \overline{D}$, by weak compactness of $\overline{D}$.

Fix a point $y \in D$. We first note that $v \in \partial D$, for otherwise, (3.44) implies

$$k_D(y, b_k) \geq k_D(f^{n_k}(y), f^{n_k}(b_k)) = \tanh^{-1} \|g_{-f^{n_k}(y)}(f^{n_k}(b_k))\| \longrightarrow \infty$$

which contradicts the fact that

$$\sup_k \{k_D(y, b_k)\} = \sup_k \{\tanh^{-1} \|g_{-y}(b_k)\|\} < \infty$$

since $b_k \in B$ for all k. Hence $\|v\| = 1$ and the sequence $(f^{n_k}(b_k))$ norm converges to v. Therefore $v \notin D(\xi, \varepsilon)$.

To complete the proof, compare the two sequences $(f^{n_k}(b_k))$ and $(f^{n_k}(y))$, having limits in the boundary ∂D. Since

$$k_D(f^{n_k}(b_k), f^{n_k}(y)) \leq k_D(b_k, y) \leq \sup_k \{k_D(y, b_k)\} < \infty$$

we must have $v = \xi$ by (3.45), which contradicts $v \notin D(\xi, \varepsilon)$.   □

By means of the preceding two lemmas, we arrive at the following dichotomy for a fixed-point free holomorphic self-map on a Hilbert ball.

Proposition 3.5.39. *Let* $f : D \longrightarrow D$ *be a fixed-point free holomorphic map on a Hilbert ball* D *and let* $a \in D$. *Then either* $\liminf_{n \to \infty} \| f^{2n}(a) \| < 1$ *or* (f^n) *converges locally uniformly to a constant map taking value at the boundary* ∂D.

Proof. Given $\liminf_{n} \| f^{2n}(a) \| \not< 1$, we must have $\lim_{n \to \infty} \| f^{2n}(a) \| = 1$. Let $a \in H(\xi, \lambda)$ for some $\lambda > 0$, where

$$\overline{H(\xi, \lambda)} \cap \partial D = \{\xi\}$$

by Theorem 3.5.29.

We first show that $(f^{2n}(a))$ converges to ξ. Indeed, $f^{2n}(a) \in H(\xi, \lambda)$ and (3.28) implies

$$
\begin{aligned}
\limsup_{n} |1 - \langle f^{2n}(a), \xi \rangle|^2 &= \limsup_{n} (1 - \| f^{2n}(a) \|^2) \left(\frac{1 - \| z_k \|^2}{1 - \| g_{-z_k}(a) \|^2} \right) \\
&\leq \frac{1}{\lambda} (1 - \| f^{2n}(a) \|^2) \longrightarrow 0 \quad \text{as } n \to \infty
\end{aligned}
$$

which gives $\lim_{n} \langle f^{2n}(a), \xi \rangle = 1$ and $\lim_{n} f^{2n}(a) = \xi$.

We next show $\lim_{n} f^{2n+1}(a) = \xi$. For this, it suffices to show that ξ is the only weak limit point of the sequence $(f^{2n+1}(a))$. Let $(f^{2n_k+1}(a))$ be a subsequence of $(f^{2n+1}(a))$ weakly convergent to $\zeta \in \overline{D}$. If $\zeta \in D$, (3.44) implies

$$
\begin{aligned}
k_D(a, f(a)) &\geq k_D(f^{2n_k}(a), f^{2n_k+1}(a)) \\
&= \tanh^{-1} \| g_{-f^{2n_k}(a)}(f^{2n_k+1}(a)) \| \longrightarrow \infty
\end{aligned}
$$

which is impossible. Hence we have $\zeta \in \overline{H(\xi, \lambda)} \cap \partial D = \{\xi\}$ and $\lim_{n} f^{n}(a) = \xi$.

Using Lemma 3.5.38, we complete the proof by showing that (f^n) converges pointwise to the constant map $h(\cdot) = \xi$. Let $y \in D$ and $v \in \overline{D}$ be any weak limit point of the sequence $(f^n(y))$. Then $(f^n(y))$ admits a subsequence $(f^{n_k}(y))$ weakly converging to v and $v \in \overline{H(\xi, \mu)}$ for some $\mu > 0$. We show that $\|v\| = 1$. Otherwise, $v \in D$ implies

$$
\begin{aligned}
k_D(a, y) &\geq k_D(f^{n_k}(a), f^{n_k}(y)) \\
&= \tanh^{-1} \|g_{-f^{n_k}(a)}(f^{n_k}(y))\| \to \infty
\end{aligned}
$$

by (3.44), which is a contradiction. Hence we have $v \in \overline{(\xi, \mu)} \cap \partial D = \{\xi\}$, by Theorem 3.5.29. This shows that $\xi \in \partial D$ is the only weak limit point of the sequence $(f^n(y))$. Therefore $\lim_n f^n(y) = \xi$. As $y \in D$ was arbitrary, we have shown that (f^n) converges pointwise to the constant map $h(\cdot) = \xi$. □

If a fixed-point free self-map f on a Hilbert ball D has a convergent orbit, then its limit must lie in the boundary and the theorem below now follows immediately from the preceding proposition.

Theorem 3.5.40. *Let $f : D \longrightarrow D$ be a fixed-point free holomorphic map on a Hilbert ball D. The following conditions are equivalent:*

(i) $\displaystyle \lim_{n \to \infty} \|f^{2n}(a)\| = 1$ *for some $a \in D$;*

(ii) *an orbit $(f^n(a))$ converges for some $a \in D$;*

(iii) (f^n) *converges locally uniformly to a constant map $h(\cdot) = \xi \in \partial D$.*

Remark 3.5.41. The Denjoy-Wolff theorem proved in [47] for *compact* holomorphic maps on Hilbert balls D is a special case of the above result. Indeed, given a fixed-point free compact holomorphic self-map f on D, we have, for some $a \in D$, that $\sup_k \|f^{n_k}(a)\| = 1$ for all subsequences (f^{n_k}) of (f^n) by Lemma 3.5.35. The example in [160] reveals that condition (i) above cannot be weakened to $\lim_k \|f^{n_k}(a)\| = 1$ for some subsequence (f^{n_k}) since there is a biholomorphic map f [160] on an infinite dimensional Hilbert ball such that $\lim_k \|f^{n_k}(0)\| = 1$ for some subsequence (f^{n_k}) of (f^n), but failing the Denjoy-Wolff theorem.

We note that having one convergent orbit is a strong condition, as shown by the following result.

Proposition 3.5.42. *Let f be a fixed-point free holomorphic self-map on a bounded symmetric domain D such that one orbit $(f^n(a))$ converges for some $a \in D$. Then all limit functions of (f^n) take values in one single boundary component Γ in $\overline{D}$.*

Proof. Let h and h_1 be limit functions of (f^n). By Lemma 3.4.11, $h(D)$ is contained in some boundary component $\Gamma \subset \overline{D}$ and $h_1(D)$ is contained in another boundary component, say $\Gamma_1 \subset \overline{D}$. By assumption, we have $h(a) = h_1(a) \in \Gamma \cap \Gamma_1$ and hence $\Gamma = \Gamma_1$. $\qquad\square$

We also note that the existence of a *unique* limit function of the iterates (f^n) of a compact holomorphic self-map f, with or without fixed-point, would imply locally uniform convergence of (f^n). It may be of interest to compare this fact with Lemma 3.5.38, where compactness of the self-map f is not assumed.

Proposition 3.5.43. *Let f be a compact holomorphic self-map on a bounded symmetric domain D. If the iterates (f^n) has a unique limit function h, then (f^n) converges locally uniformly to h.*

Proof. If (f^n) does not converge to h locally uniformly, then there exists $\varepsilon > 0$, and a subsequence (f^{n_k}) of (f^n) such that

$$\|f^{n_k} - h\|_B \geq \varepsilon \qquad (k = 1, 2, \ldots)$$

on some closed ball B strictly contained in D, where $\|\cdot\|_B$ denotes the supremum norm on B. By [47, Lemma 1], (f^{n_k}) has a subsequence (f^{m_k}) converging locally uniformly to a holomorphic map on D, which must be h by the uniqueness assumption. This contradicts $\|f^{m_k} - h\|_B \geq \varepsilon$ for all m_k. $\qquad\square$

Example 3.5.44. Let B be the open unit ball of the Hilbert space ℓ_2 of square summable complex sequences. Define $f : B \longrightarrow B$ by

$$f(x_1, x_2, \ldots) = \left(\frac{1+x_1}{2}, \left(\frac{1-x_1}{2} \right) \frac{x_1}{2}, \left(\frac{1-x_1}{2} \right) \frac{x_2}{3}, \ldots \right)$$

$$= \left(\frac{1+x_1}{2}, 0, 0, \ldots \right) + \frac{1-x_1}{2} \left(0, \frac{x_1}{2}, \frac{x_2}{3}, \ldots \right).$$

Then f is holomorphic and fixed-point free. Moreover, f is compact since it is the sum of two compact maps. The iterates (f^n) converge to $h(\cdot) = (1, 0, 0, \ldots)$.

Finally, returning to the previous Conjecture for bounded symmetric domains, we prove the following result for domains of finite rank.

Theorem 3.5.45. *Let D be a bounded symmetric domain of finite rank in a complex Banach space V and let $f : D \longrightarrow D$ be a fixed-point free compact holomorphic map. Then there is one single boundary component Γ in the boundary ∂D such that for each limit function h of (f^n), we have $h(D) \subset \overline{\Gamma}$ whenever $h(D)$ is weakly closed.*

More precisely, there is a boundary point $\xi \in \partial D$ of the form

$$\xi = \sum_{j=1}^{m} \alpha_j e_j \quad (\alpha_j > 0, \ m \le p = \operatorname{rank} D)$$

for some orthogonal minimal tripotents $e_1, \ldots, e_m \in \partial D$, such that for each limit function h of (f^n) with weakly closed range, we have $h(D) \subset \overline{\Gamma}_e$, where Γ_e is the boundary component of $e = e_1 + \cdots + e_m$.

Remark 3.5.46. The conjecture would be completely proved if the conditions of D and $h(D)$ being finite-rank and weakly closed respectively are removed from the above theorem.

Proof. Let $p = \operatorname{rank} D$ and let $\xi = \sum_{j=1}^{m} \alpha_j e_j$ be the boundary point obtained in Theorem 3.5.27, where $\alpha_j > 0$, $m \le p$ and $e_1, \ldots, e_m$ are orthogonal minimal tripotents. Let

$$e = e_1 + \cdots + e_m$$

which is a tripotent in ∂D.

Let h be a limit function such that $h(D)$ is weakly closed. Since f is a compact map, the remark following Lemma 3.5.35 implies $h(D) \subset \partial D$. By Lemma 3.4.11, $h(D)$ is contained in a boundary component Γ_u of $\overline{D}$ for some tripotent $u \in \partial D$.

For $n = 1, 2, \ldots$, pick y_n in the horoball $H(\xi, n) = S_0(\xi, n)$. By f-invariance, we have $h(y_n) \in S(\xi, n)$, which, by Theorem 3.5.27, is of the form

$$h(y_n) = \sum_{j=1}^{m} \frac{\sigma_j n}{1 + \sigma_j n} e_j + B \left(\sum_{j=1}^{m} \sqrt{\frac{\sigma_j n}{1 + \sigma_j n}} e_j, \sum_{j=1}^{m} \sqrt{\frac{\sigma_j n}{1 + \sigma_j n}} e_j \right)^{1/2} (w_n)$$

for some $w_n \in \overline{D}$. Let (w_{n_k}) be a subsequence of (w_n) weakly converging to $w \in \overline{D}$, say. Then the sequence $(h(y_{n_k}))$ weakly converges to

$$\sum_{j=1}^{m} e_j + B \left(\sum_{j=1}^{m} e_j, \sum_{j=1}^{m} e_j \right)^{1/2} (w) = \sum_{j=1}^{m} e_j + P_0 \left(\sum_{j=1}^{m} e_j \right) (w)$$

$$\subset e + P_0(\overline{D}) = \overline{\Gamma}_e$$

where Γ_e is the boundary component in ∂D containing the tripotent e (cf. Example 3.4.15). Since $h(D)$ is weakly closed, we have $\emptyset \neq h(D) \cap \overline{\Gamma}_e \subset \Gamma_u \cap \overline{\Gamma}_e$ and Γ_u meets either Γ_e or a boundary component of $\partial \Gamma_e$. By Example 3.4.15, the latter is also a boundary component of $\overline{D}$. It follows that either $\Gamma_u = \Gamma_e$ or Γ_u is a boundary component of $\partial \Gamma_e$, that is, $\Gamma_u \subset \overline{\Gamma}_e$ which gives $h(D) \subset \overline{\Gamma}_e$. □

In the special case of a finite product of Hilbert balls, a careful examination reveals that the preceding theorem has been proved in [49, Theorem 3.2], which generalises Hervé's result for the bidisc.

We close the section with an example of limit functions of which the image can be a singleton or a whole boundary component, which is not closed.

Example 3.5.47. Let D be a finite-rank bounded symmetric domain of rank p. Pick any non-zero $a \in D$, with spectral decomposition $a = \alpha_1 e_1 + \cdots + \alpha_p e_p$, where $\|a\| = \alpha_1 \geq \cdots \geq \alpha_p \geq 0$.

Let $g_a : D \to D$ be the Möbius transformation induced by a, which is not a compact map if D is infinite dimensional. Let $x = \beta_1 e_1 + \beta_2 e_2 + \cdots + \beta_p e_p$, where $\beta_1, \beta_2, \ldots, \beta_p \in \mathbb{D}$ so that $x \in D$. By orthogonality, we have

$$
\begin{aligned}
x \,\square\, a \;&=\; (\beta_1 e_1 + \beta_2 e_2 + \cdots + \beta_p e_p) \,\square\, (\alpha_1 e_1 + \cdots + \alpha_p e_p) \\
&=\; \beta_1 \alpha_1 e_1 \,\square\, e_1 + \cdots + \beta_p \alpha_p e_p \,\square\, e_p
\end{aligned}
$$

and $(x \,\square\, a)^n (x) = \beta_1^{n+1} \alpha_1^n e_1 + \cdots + \beta_p^{n+1} \alpha_p^n e_p$ for $n = 1, 2, \ldots$. It follows that

$$
\begin{aligned}
g_a(x) &= a + B(a,a)^{1/2}(1 + x \,\square\, a)^{-1}(x) \\
&= a + B(a,a)^{1/2}\big(1 - x \,\square\, a + (x \,\square\, a)^2 - (x \,\square\, a)^3 + \cdots\big)(x) \\
&= a + B(a,a)^{1/2}\big(\beta_1 e_1 + \beta_2 e_2 + \cdots + \beta_p e_p \\
&\qquad\qquad - (\beta_1^2 \alpha_1 e_1 + \cdots + \beta_p^2 \alpha_p e_p) + \cdots\big) \\
&= a + B(a,a)^{1/2}\big[(1 - \beta_1 \alpha_1 + \beta_1^2 \alpha_1^2 + \cdots)\beta_1 e_1 \\
&\qquad\qquad + \cdots + (1 - \beta_p \alpha_p + \beta_p^2 \alpha_p^2 + \cdots)\beta_p e_p\big] \\
&= a + B(a,a)^{1/2}\left(\frac{\beta_1 e_1}{1 + \beta_1 \alpha_1} + \cdots + \frac{\beta_p e_p}{1 + \beta_p \alpha_p}\right) \\
&= \alpha_1 e_1 + \cdots + \alpha_p e_p + \frac{(1 - \alpha_1^2)\beta_1 e_1}{1 + \beta_1 \alpha_1} + \cdots + \frac{(1 - \alpha_p^2)\beta_p e_p}{1 + \beta_p \alpha_p} \\
&= \frac{\alpha_1 + \beta_1}{1 + \alpha_1 \beta_1} e_1 + \cdots + \frac{\alpha_p + \beta_p}{1 + \alpha_p \beta_p} e_p \\
&= g_{\alpha_1}(\beta_1) e_1 + \cdots + g_{\alpha_p}(\beta_p) e_p
\end{aligned}
$$

where $\boldsymbol{g}_{\alpha_j}$ is the Möbius transformation on the complex disc $\mathbb{D}$, induced by α_j for $j = 1, \ldots, p$. If $\alpha_j = 0$, then $\boldsymbol{g}_{\alpha_j}$ is the identity map. If $\alpha_j > 0$, then the iterates $(\boldsymbol{g}_{\alpha_j}^n)$ converge locally uniformly to the constant map with value $\alpha_j/|\alpha_j| = 1$. Hence the iterates

$$g_a^n(x) = \boldsymbol{g}_{\alpha_1}^n(\beta_1)e_1 + \cdots + \boldsymbol{g}_{\alpha_p}^n(\beta_p)e_p \qquad (n = 2, 3, \ldots)$$

converge to

$$e_1 + \gamma_2 e_2 + \cdots + \gamma_p e_p, \quad \gamma_j = \left\{ \begin{array}{ll} 1 & (\alpha_j > 0) \\ \beta_j & (\alpha_j = 0) \end{array} \right. \qquad (j = 2, \ldots, p).$$

In particular, if $\alpha_j > 0$ for all j, then the iterates (g_a^n) converge pointwise to the constant map $g(\cdot) = \xi = e_1 + \cdots + e_p$, in which case $h(D) = \{\xi\}$ for every limit function h.

On the other hand, if $J = \{j : \alpha_j > 0\}$ is a proper subset of $\{1, \ldots, p\}$, then

$$\lim_n g_a^n(x) = \sum_{j \in J} e_j + \sum_{j \notin J} \beta_j e_j \in e + D_e$$

where $e = \sum_{j \in J} e_j$ is a tripotent in ∂D and $D_e = V_0(e) \cap D$. It follows that, in this case, the image of every limit function h of (g_a^n) is the whole boundary component $e + D_e$ since for any $e + z \in e + D_e$ with $z \in D_e$ and spectral decomposition $z = \sum_{j \notin J} \beta_j u_j$, we have

$$h\Big(\sum_{j \in J} \alpha_j e_j + \sum_{j \notin J} \beta_j u_j\Big) = e + \sum_{j \notin J} \beta_j u_j.$$

Notes. The Carathéodory distance was introduced by Carathéodory [33], and the Kobayashi distance was first introduced in [107] for finite dimensional complex manifolds. In Proposition 3.5.17, the completeness of the Carathéodory distance on a bounded homogeneous domain has been proved in [172, p. 279]. The domination of the Bergman metric over the Carathéodory metric stated in Example 3.5.13 has been proved essentially in the paper [122]. For more details of invariant metrics on infinite dimensional domains, we refer to the book [63].

A succinct exposition of iteration of a holomorphic self-map on the disc $\mathbb{D}$ and historical remarks has been given in [29]. For Euclidean balls in $\mathbb{C}^n$, this topic has been treated thoroughly in [2, Chapter 2.2]. Various forms of generalisation of Wolff's theorem and the Denjoy-Wolff theorem to other domains in higher dimensions have been shown by many authors (see, for example, references listed in [2, 38, 39, 50, 145, 146]). The invariant domains for finite dimensional bounded symmetric domains obtained in [129] have a similar form to the one in Theorem 3.5.27. Ellipsoids as invariant domains in Hilbert balls has been shown in [67]. The unified treatment for all bounded symmetric domains in Theorem 3.5.22 was given in [50]. Horospheres have also been used to extend the Denjoy-Wolff theorem to other domains. For instance, a notion of horospheres has been used in [26] and [4] to prove a Denjoy-Wolff theorem for bounded strictly convex domains and weakly convex domains in $\mathbb{C}^n$, respectively. The only strictly convex bounded symmetric domains are the Hilbert balls. Nevertheless, the result in [4] makes use of the concept of a *sequence horosphere*. For polydiscs, the sequence horospheres with pole at the origin [4, p. 1516] are the same as our horoballs (cf. [49, Proposition 2.4]).

Equivalent conditions in Theorem 3.5.40 for the Denjoy-Wolff theorem have been established in [49], where one only assumes that the self-maps contract the Kobayashi distance, but need not be holomorphic. Example 3.5.44 is taken from [47] and, Example 3.5.47 as well as Theorem 3.5.45 have been shown in [50]. Needless to say, the study of holomorphic dynamics on bounded symmetric domains is incomplete and the closely related topic of angular derivatives in infinite dimension has yet to be explored further.

3.6 Siegel domains

The holomorphic equivalence of the open unit disc and the upper half-plane in $\mathbb{C}$, via the Cayley transform, is fundamental in classical complex analysis. Siegel domains generalise the notion of the upper half-plane and play an important role in several complex variables. Indeed, a seminal result of Vinberg, Gindikin and Piatetski-Shapiro [171] asserts that every bounded homogeneous domain in $\mathbb{C}^d$ is biholomorphically equivalent to a homogeneous Siegel domain. In this section, we discuss bounded symmetric domains which are biholomorphic to a Siegel domain, from a Jordan perspective.

Given a real Banach space V, one can equip its complexification $V_c = V \oplus iV$ with a norm $\| \cdot \|_c$ so that $(V_c, \| \cdot \|_c)$ is a complex Banach space and the isometric embedding $v \in V \mapsto (v, 0) \in V \oplus iV$ identifies V as a real closed subspace of V_c. Although there are many choices of the norm $\| \cdot \|_c$, they are all equivalent to the ℓ_∞-norm $\| \cdot \|_{\max}$ if we require

$$\|u + iv\|_c \geq \|u + iv\|_{\max} := \max(\|u\|, \|v\|) \qquad (u, v \in V)$$

by the open mapping theorem. We will always assume V_c is equipped with such a norm and by a slight abuse of language, call $(V_c, \| \cdot \|_c)$ *the complexification* of V. We denote the *conjugation* in V_c by

$$\overline{u + iv} := u - iv$$

so that the *imaginary part* $\mathrm{Im}\, z$ of an element $z \in V_c$ is given by $\mathrm{Im}\, z = \frac{1}{2}(z - \overline{z})$.

Definition 3.6.1. Let V be a *real* Banach space with complexification V_c and let $\Omega \subset V$ be a (non-empty) open cone. Let W be a complex Banach space and $F : W \times W \longrightarrow V_c$ a continuous mapping, which is conjugate linear in the first variable, linear in the second variable and satisfies

$$\overline{F(w_1, w_2)} = F(w_2, w_1).$$

The set

$$D(\Omega, F) := \{(z, w) \in V_c \oplus W : \operatorname{Im} z - F(w, w) \in \Omega\}$$

is called a *Siegel domain* (*of the second kind*). If $W = \{0\}$, then $D(\Omega, F)$ reduces to

$$D(\Omega) := \{z \in V_c : \operatorname{Im} z \in \Omega\} = V \oplus i\Omega$$

which is called a *tube domain* over the cone Ω (or a Siegel domain of the first kind).

Example 3.6.2. The motivating example of Siegel domains is of course the complex upper half-plane which is a tube domain over $(0, \infty)$. More generally, let V be the space of $n \times n$ real symmetric matrices and $\Omega \subset V$ the cone of positive definite matrices. The tube domain $D(\Omega)$ over Ω is also called the *Siegel upper half-plane* (of degree n).

To discuss Siegel domains, we begin with cones and partial ordering. Let Ω be an open cone in a real Banach space V with norm $\| \cdot \|$. Then we have $\operatorname{int} \overline{\Omega} = \Omega$ by the following lemma.

Lemma 3.6.3. *Let C be an open convex set in a real topological vector space V. Then $\operatorname{int} \overline{C} = C$.*

Proof. There is nothing to prove if C is empty. Pick any $q \in C$. Let $p \in \operatorname{int} \overline{C}$. Then p is an internal point of $\overline{C}$, that is, every line through p meets $\overline{C}$ in a set containing an interval around p (cf. [58, p. 410, 413]). In particular, for the line joining p and q, there exists $\delta \in (0, 1)$ such that $p \pm \delta(q - p) \in \overline{C}$. Since C is open and q is an interior point of C, we have $\lambda(p - \delta(q - p)) + (1 - \lambda)q \in C$ for $0 < \lambda < 1$ (cf. [58, p. 413]). Hence

$$p = \frac{1}{1 + \delta}(p - \delta(q - p)) + \frac{\delta}{1 + \delta}q \in C. \qquad \square$$

Trivially, V is a cone in itself. In the sequel, we shall exclude this case since Siegel domains of interest to us are the ones biholomorphic to *bounded* domains, but $D(V) = V \oplus iV$ is not biholomorphic to a bounded domain!

If Ω is an open cone *properly* contained in V, then we must have $0 \notin \Omega$ although the closure $\overline{\Omega}$ contains 0. The closure $\overline{\Omega}$ of Ω is also a cone, which induces a partial ordering $\leq$ in V so that

$$x \leq y \Leftrightarrow y - x \in \overline{\Omega}.$$

We also write $y \geq x$ for $x \leq y$. As usual, a continuous linear functional $f \in V^*$ is called *positive* if $f(\overline{\Omega}) \subset [0, \infty)$. By the Hahn-Banach separation theorem, we have

$$\overline{\Omega} = \{v \in V : f(v) \geq 0 \text{ for each } f \in V^* \text{ with } f(\overline{\Omega}) \subset [0, \infty)\}.$$

We note that each element $e \in \Omega$ is an *order unit*, that is, for each $v \in V$, we have

$$-\lambda v \leq v \leq \lambda e$$

for some $\lambda > 0$. Indeed, since Ω is open, $e - \Omega$ is a neighbourhood of $0 \in V$ and therefore one can find $r > 0$ such that

$$B(0, r) = \{x \in V : \|x\| < r\} \subset e - \Omega.$$

For $v \neq 0$, we have $\pm(r/2\|v\|)v \in B(0, r)$ which implies

$$-\frac{2\|v\|}{r}e \leq v \leq \frac{2\|v\|}{r}e. \qquad (3.46)$$

The preceding argument also implies

$$V = \Omega - \Omega. \qquad (3.47)$$

An order unit $e \in \Omega$ induces a semi-norm $\| \cdot \|_e$ on V, defined by

$$\|x\|_e = \inf\{\lambda > 0 : -\lambda e \leq x \leq \lambda e\} \qquad (x \in V)$$

which satisfies

$$-\|x\|_e e \leq x \leq \|x\|_e e \tag{3.48}$$

and

$$\{x \in V : \|x\|_e \leq 1\} = \{x \in V : -e \leq x \leq e\}. \tag{3.49}$$

Since $\{x \in V : \|x\|_e = 0\} = \overline{\Omega} \cap -\overline{\Omega}$, the semi-norm $\| \cdot \|_e$ is a norm if and only if $\overline{\Omega} \cap -\overline{\Omega} = \{0\}$.

Definition 3.6.4. An open cone Ω in a real Banach space V is called *regular* if $\overline{\Omega} \cap -\overline{\Omega} = \{0\}$ (in which case, Ω is properly contained in V).

If Ω is a regular cone, then as noted above, $\| \cdot \|_e$ is a norm, called the *order-unit norm* induced by e. All order-unit norms induced by elements in Ω are mutually equivalent.

Henceforth, let Ω be a regular open cone in V. It follows from (3.49) that every linear map $\psi : V \longrightarrow V$ which is *positive*, meaning $\psi(\overline{\Omega}) \subset \overline{\Omega}$, is continuous with respect to the order-unit norm $\| \cdot \|_e$ and moreover, $\|\psi\|_e = \|\psi(e)\|_e$, where the former denotes the norm of ψ with respect to $\| \cdot \|_e$. In particular, if $\psi : V \longrightarrow \mathbb{R}$ is a positive linear functional, then $\|\psi\|_e = \psi(e)$. A positive linear map $\psi : V \longrightarrow V$ is an isometry if and only if $\psi(e) = e$ [40, Proposition 2.3].

Lemma 3.6.5. *Let Ω be a regular open cone in a real Banach space* $(V, \| \cdot \|)$, *partially ordered by* $\overline{\Omega}$. *Then for each $e \in \Omega$, the order-unit norm $\| \cdot \|_e$ satisfies*

$$\| \cdot \|_e \leq c \| \cdot \|. \tag{3.50}$$

for some $c > 0$.

Proof. Let $v \in V \backslash \{0\}$. By (3.46), we have $\|v\|_e \leq (2/r)\|v\|$ for some $r > 0$. $\qquad\qquad\qquad\qquad\qquad\qquad\qquad\qquad\qquad\qquad\qquad\qquad$ $\square$

Let $(V, \|\cdot\|_e)$ denote the vector space V equipped with the order-unit norm $\|\cdot\|_e$, and $(V, \|\cdot\|_e)^*$ its dual space. It follows from (3.50) that every $\|\cdot\|_e$-continuous linear functional on V is also $\|\cdot\|$-continuous. On the other hand, given $f \in V^*$ satisfying $f(e) = 1 = \|f\|_e$, then f is positive and hence continuous with respect to the norm $\|\cdot\|_e$.

Denote the *state space* (with respect to the order unit e) by

$$S \;=\; \{f \in (V, \|\cdot\|_e)^* : f(e) = 1 = \|f\|_e\}$$
$$\;=\; \{f \in V^* : f(e) = 1, f \text{ is positive}\} \qquad (3.51)$$

which is a weak* compact convex set in the dual V^* and we have

$$\|x\|_e = \sup\{|f(v)| : f \in S\} \qquad (x \in V)$$

(cf. [77, Lemma 1.2.5]).

Lemma 3.6.6. *Let Ω be a regular open cone in a real Banach space V and let $e \in \Omega$, which induces an order-unit norm $\|\cdot\|_e$ on V. Then we have*

$$\Omega = \bigcap_{f \in S} f^{-1}(0, \infty).$$

Proof. Given that V is partially ordered by the closure $\overline{\Omega}$, we have

$$\overline{\Omega} = \bigcap_{f \in S} f^{-1}[0, \infty) \qquad (3.52)$$

since $f/f(e) \in S$ for each non-zero positive linear functional $f \in V^*$.

Let $a \in \Omega$. Then for each $f \in S$, we have $f(a) > 0$ since a is an order unit, which implies $e \leq \lambda a$ for some constant $\lambda > 0$ and hence $1 \leq \lambda f(a)$. This proves

$$\Omega \subset \bigcap_{f \in S} f^{-1}(0, \infty).$$

Conversely, let $a \in V$ and $f(a) > 0$ for all $f \in S$. Then $a \in \overline{\Omega}$ and by weak* compactness of S, one can find some $\delta > 0$ such that $f(a) \geq \delta$ for all $f \in S$. Let

$$N = \left\{ x \in V : \|x - a\| < \frac{\delta}{2c} \right\} \subset \left\{ x \in V : \|x - a\|_e < \frac{\delta}{2} \right\}$$

where $c > 0$ is given in (3.50). Then N is an open neighbourhood of a and, $N \subset \overline{\Omega}$ since

$$x \in N \Rightarrow -\frac{\delta}{2}e \leq x - a \Rightarrow a - \frac{\delta}{2}e \leq x \Rightarrow \frac{\delta}{2} \leq f(x)$$

for all $f \in S$. Hence a belongs to the interior $\overline{\Omega}^0$ of $\overline{\Omega}$ and, as Ω is open and convex, we have $a \in \Omega = \overline{\Omega}^0$. $\qquad\square$

We see from Lemma 3.6.5 that if Ω is a regular open cone in a finite dimensional Banach space V, then the order-unit norm $\| \cdot \|_e$ induced by $e \in \Omega$ is equivalent to the norm of V, by the open mapping theorem. In fact, the equivalence of the two norms is related to the basic concept of a normal cone in the theory of partially ordered topological vector spaces.

Lemma 3.6.7. *Let Ω be a regular open cone in a real Banach space V with norm $\| \cdot \|$. Then the order-unit norm $\| \cdot \|_e$ induced by $e \in \Omega$ is equivalent to $\| \cdot \|$ if and only if Ω is a normal cone in V, that is, there is a constant $\gamma > 0$ such that $0 \leq x \leq y$ implies $\|x\| \leq \gamma \|y\|$ for all $x, y \in V$. In particular, $(V, \| \cdot \|_e)$ is a Banach space if Ω is a normal cone.*

Proof. By the definition of the order-unit norm $\| \cdot \|_e$, we have $0 \leq x \leq y$ in V implies $\|x\|_e \leq \|y\|_e$. Hence Ω is normal in $(V, \| \cdot \|_e)$. If $\| \cdot \|$ is equivalent to $\| \cdot \|_e$, then evidently Ω is also normal in $(V, \| \cdot \|)$.

Conversely, let Ω be normal in $(V, \| \cdot \|)$. We have already noted in (3.50) that $\| \cdot \|_e \leq c \| \cdot \|$ for some constant $c > 0$. By (3.49) and normality of Ω, there is a constant $\gamma > 0$ such that

$$\|x\|_e \leq 1 \Leftrightarrow -e \leq x \leq e \Rightarrow 0 \leq x + e \leq 2e$$
$$\Rightarrow \|x + e\| \leq 2\gamma\|e\| \Rightarrow \|x\| < 2(\gamma + 1)\|e\|$$

which implies $\| \cdot \| \leq 2(\gamma + 1)\|e\|\| \cdot \|_e$ and the equivalence of $\| \cdot \|$ and $\| \cdot \|_e$. $\qquad\qquad\square$

We note that a self-dual cone Ω in a Hilbert space H is regular, and also normal since it has been shown in [40, Lemma 2.6] that the order-unit norms induced by elements in Ω are all equivalent to the norm of H.

The partially ordered Banach spaces related to symmetric Siegel domains are the JB-algebras. The partial ordering $\leq$ in a JB-algebra A is defined by the closed cone

$$\{a^2 : a \in A\}$$

and if A is unital, then the identity e is an order-unit in the interior of the cone, which is regular, and the induced order-unit norm coincides with the original norm of A [77, Proposition 3.3.10].

We now discuss Siegel domains which are biholomorphic to bounded symmetric domains in Banach spaces.

Theorem 3.6.8. *Let Ω be a regular open cone in a real Banach space V. The following conditions are equivalent.*

(i) *The tube domain $V \oplus i\Omega$ is biholomorphic to a bounded symmetric domain.*

(ii) *V_c is a unital JB*-algebra with an equivalent norm and*

$$\overline{\Omega} = \{x^2 : x \in V\}.$$

(iii) *V is a unital JB-algebra with an order-unit norm and*

$$\overline{\Omega} = \{x^2 : x \in V\}.$$

In this case, the tube domain $V \oplus i\Omega$ is biholomorphic to the open unit ball D of the JB*-algebra V_c via the Cayley transform

$$z \in V \oplus i\Omega \mapsto (z - ie)(z + ie)^{-1} \in D$$

where e is the identity of V_c, and the symmetry at $ie \in V \oplus i\Omega$ is given by the map $z \mapsto -z^{-1}$.

Proof. (i) $\Rightarrow$ (ii). We make use of the results in [103]. By Theorem 3.2.18, there is a Jordan triple product on V_c, induced by a symmetry of $V \oplus i\Omega$, such that V_c becomes a JB*-triple in an equivalent norm $\| \cdot \|_{sp}$.

Let $e \in \Omega$. Using the symmetry s_{ie} at $ie \in V \oplus i\Omega$, it has been shown in [103, Theorem 2.5] that V_c is the Jordan triple *associated* to the symmetric domain $V \oplus i\Omega$, where the associated Jordan triple product $\{\cdot, \cdot, \cdot\}$ in V_c satisfies

$$\{e, a, e\} = a \quad \text{for all} \quad a \in V. \tag{3.53}$$

By Remark 3.2.6, this Jordan triple product coincides with the triple product of the JB*-triple $(V_c, \| \cdot \|_{sp})$.

It follows from (3.53) that e is a tripotent in V_c and $P_2(e)(z) = z$ for all $z \in V_c$. Hence, by Example 2.4.18, $V_c = P_2(e)(V_c)$ is a JB*-algebra with identity e, Jordan product $z \circ w = \{z, e, w\}$ and involution $z^* = \{e, z, e\}$. Moreover, it has been shown in [103, (4.1), (4.6)] that $\overline{\Omega} = \{x^2 : x \in V\}$ and the symmetry at ie is the inverse map $z \in V \oplus i\Omega \mapsto -z^{-1} \in V \oplus i\Omega$.

(ii) $\Leftrightarrow$ (iii). This has been proved in [179].

(iii) $\Rightarrow$ (i). By assumption, the identity e of the JB-algebra V is an order-unit in the interior $\text{int}\{x^2 : x \in V\} = \text{int}\,\overline{\Omega} = \Omega$. By Lemma

3.6.5, the complete order-unit norm $\| \cdot \|_e$ is equivalent to the original norm of V. Hence the complexification V_c of $(V, \| \cdot \|_e)$ is a JB*-algebra in a norm $\| \cdot \|$ equivalent to the norm $\| \cdot \|_c$ of V_c.

By Theorem 3.2.20, the open unit ball $D = \{z \in V_c : \|z\| < 1\}$ is a bounded symmetric domain. To complete the proof, we show that the tube domain $V \oplus i\Omega$ is biholomorphic to D via the Cayley transform

$$z \in V \oplus i\Omega \mapsto (z - ie)(z + ie)^{-1} \in D. \tag{3.54}$$

By Corollary 2.4.20, $z + ie \in V \oplus i\Omega$ is invertible. We need to show $\|(z - ie)(z + ie)^{-1}\| < 1$.

Let $z = v + i\omega \in V \oplus i\Omega$, where ω is a positive invertible element in the JB-algebra V. Let $\mathcal{B}$ be the closed *-subalgebra of V_c generated by the self-adjoint elements v, ω, e. Then $\mathcal{B}$ is isometrically *-isomorphic to a closed *-subalgebra of the JB*-algebra $L(H)$ of bounded linear operators on some Hilbert space H, with inner product $\langle \cdot, \cdot \rangle$, (see, for example, [179, Corollary 2.2]). Hence we can identify v and ω as self-adjoint operators on H. Let $\xi \in H$ and $\eta = (z + ie)^{-1}\xi \in H$. Then we have

$$\|(z + ie)\eta\|^2 - \|(z - ie)\eta\|^2$$
$$= \langle (v - i(e + \omega))(v - i(e + \omega))\eta, \eta \rangle - \langle (v + i(e - \omega))(v - i(e - \omega))\eta, \eta \rangle$$
$$= 4\langle \omega\eta, \eta \rangle$$

which gives

$$\|\xi\|^2 - \|(z - ie)(z + ie)^{-1}\xi\|^2 = 4\langle \omega\eta, \eta \rangle = 4\langle (z^* - ie)^{-1}\omega(z + ie)^{-1}\xi, \xi \rangle$$

where $\langle \omega\eta, \eta \rangle > 0$ by positivity and invertibility of ω. Therefore we have proved

$$I_H - ((z - ie)(z + ie)^{-1})^*(z - ie)(z + ie)^{-1} = 4(z^* - ie)^{-1}\omega(z + ie)^{-1},$$

where $I_H \in L(H)$ is the identity, and it follows that $\|(z-ie)(z+ie)^{-1}\| < 1$.

Hence the Cayley transform in (3.54) is well-defined and holomorphic. In fact, it is biholomorphic since the holomorphic map

$$\gamma : z \in D \mapsto i(e+z)(e-z)^{-1} = -ie + 2i(e-z)^{-1} \in V \oplus i\Omega$$

is its inverse. Indeed, the inverse $(e-z)^{-1}$ exists since $\|z\| < 1$. Also, $0 \le z^*z \le e$ implies

$$i(e+z)(e-z)^{-1} - (i(e+z)(e-z)^{-1})^* = 2i(e-z^*)^{-1}(e-z^*z)(e-z)^{-1}$$

and $i(e+z)(e-z)^{-1} \in V \oplus i\Omega$. So γ is well-defined and

$$\begin{aligned}
\gamma((z-ie)(z+ie)^{-1}) &= -ie + 2i(e - (z-ie)(z+ie)^{-1})^{-1} \\
&= -ie + 2i(2i(z+ie)^{-1})^{-1} = z
\end{aligned}$$

for $z \in V \oplus i\Omega$. $\qquad\qquad\qquad\qquad\qquad\qquad\qquad\qquad\qquad\square$

Remark 3.6.9. In [144], Siegel domains are defined over *regular* open cones and are finite dimensional. They are biholomorphic to bounded domains (see also [108, Chapter II, Sec. 5]) but this need not be true for infinite dimensional Siegel domains (cf. [72]). In fact, condition (i) in the preceding theorem is equivalent to saying that $V \oplus i\Omega$ is biholomorphic to a bounded domain and there is a symmetry at one point in $V \oplus i\Omega$, for the latter condition already implies that the domain is homogeneous (and hence symmetric), which has been shown in [103, Theorem 2.5]. In this case, Ω is actually *linearly homogeneous* (cf. [24, (2.1)]).

Remark 3.6.10. Considering V_c as a JB*-triple in the proof of Theorem 3.6.8, the Cayley transform can be derived from the vector field $(e - \{z, e, z\})\frac{\partial}{\partial z}$ on the tube domain, as shown in [103]. In fact, the exponential $\exp(-\frac{\pi i}{4}(e - \{z, e, z\})\frac{\partial}{\partial z})$ is the biholomorphic map $z \mapsto i(z - ie)(z + ie)^{-1}$ on $V \oplus i\Omega$.

We have seen in Theorem 3.6.8 that the open unit ball of a JB*-algebra is biholomorphic to a tube domain if and only if the algebra has an identity. A natural question is: can the open unit ball of a JB*-triple be realised as a Siegel domain? This question has been answered in [103] to which we refer the interested reader for a proof of the following theorem.

Theorem 3.6.11. *Let Z be a JB*-triple. Then the open unit ball of Z is biholomorphic to a Siegel domain (of the second kind) if and only if Z has a maximal tripotent e, in which case the Siegel domain can be constructed from the Peirce decomposition $Z = Z_2(e) \oplus Z_1(e)$:*

$$D = \{(z, w) \in Z_2(e)_{sa} \oplus Z_1(e) : \operatorname{Im} z - F(w, w) \in \{\{x, e, x\} : x \in Z_2(e)\}\}$$

where $Z_2(e)_{sa} = \{z \in Z_2(e) : z = \{e, z, e\}\}$ and $F(w_1, w_2) = 2\{w_1, w_2, e\}$.

If we call a Siegel domain *symmetric* whenever it is biholomorphic to a bounded symmetric domain, then one can summarise the preceding two theorems by saying that symmetric Siegel domains of the first kind are exactly (via biholomorphism) the open unit balls of unital JB*-algebras, those of the second kind are the open unit balls of JB*-triples containing a maximal tripotent.

Tube domains biholomorphic to bounded symmetric domains in Hilbert spaces are exactly the ones over linearly homogeneous self-dual cones. By Theorem 2.4.14, these cones are of the form $\Omega = \operatorname{int}\{x^2 : x \in \mathcal{H}\}$ for some unital JH-algebra $\mathcal{H}$. In fact, unital JH-algebras also carry the structure of a JB-algebra, and can be classified.

Lemma 3.6.12. *Let $\mathcal{A}$ be a unital JH-algebra. Then $\mathcal{A}$ decomposes into a finite direct sum*

$$\mathcal{A} = \mathcal{A}_1 \oplus \cdots \oplus \mathcal{A}_d$$

of unital JB-algebras $\mathcal{A}_1, \ldots, \mathcal{A}_d$, *where each* $\mathcal{A}_j$ *is either a finite dimensional JB-algebra or a real spin factor, for* $j = 1, \ldots, d$.

Proof. Let $e \in \mathcal{A}$ be the identity. Then e is an order-unit in the linearly homogeneous self-dual cone int $\{a^2 : a \in \mathcal{A}\}$ and the order-unit norm $\| \cdot \|_e$ is equivalent to the Hilbert space norm $\| \cdot \|_h$.

We show that $\mathcal{A}$, when equipped with the order-unit norm, is a unital JB-algebra. By [77, Proposition 3.1.6], it suffices to show that $-e \leq a \leq e$ implies $0 \leq a^2 \leq e$ for each $a \in \mathcal{A}$.

We first observe that, for each projection $p \in \mathcal{A}$, we have $p, e - p \leq e$. Indeed, we have $p, e - p \in \mathcal{C}$ since $p = p^2$ and $e - p = (e-p)^2$. Let $-e \leq a \leq e$. Then by [37, p.108], there are mutually orthogonal (nonzero) projections $\{p_k\}_k$ in $\mathcal{A}$ such that $a = \sum_{k=1}^\infty \lambda_k p_k$. By self-duality, $0 \leq e - a$ implies $0 \leq \langle e - a, p_k \rangle$ and

$$\lambda_k \langle p_k, p_k \rangle = \langle a, p_k \rangle \leq \langle e, p_k \rangle = \langle p_k, p_k \rangle$$

for all k. Likewise $0 \leq e + a$ implies $-\langle p_k, p_k \rangle \leq \lambda_k \langle p_k, p_k \rangle$ for all k. Hence we have $-1 \leq \lambda_k \leq 1$ and $\lambda_k^2 \leq 1$. It follows that

$$a^2 = \sum_{k=1}^\infty \lambda_k^2 p_k \leq \sum_{k=1}^\infty p_k \leq e.$$

This proves that $(\mathcal{A}, \| \cdot \|_e)$ is a unital JB-algebra. Moreover, it is a reflexive Banach space since it is isomorphic to the Hilbert space $(\mathcal{A}, \| \cdot \|_h)$. By Corollary 3.3.6, $(\mathcal{A}, \| \cdot \|_e)$ must be a finite ℓ_∞-sum of unital JB-algebras, each of which is either a finite dimensional algebra or a real spin factor. $\square$

The classification of finite dimensional unital JB-algebras has been listed in (2.9).

Corollary 3.6.13. *Let Ω be a regular open cone in a real Banach space V. The following conditions are equivalent.*

(i) $V \oplus i\Omega$ *is biholomorphic to a bounded symmetric domain in a Hilbert space.*

(ii) V *is a reflexive JB-algebra with an equivalent norm and*
$$\overline{\Omega} = \{x^2 : x \in V\}.$$

(iii) V *is a unital JH-algebra with an equivalent norm and*
$$\overline{\Omega} = \{x^2 : x \in V\}.$$

(iv) V *is a Hilbert space in an equivalent norm, where Ω is a linearly homogeneous self-dual cone.*

Proof. (i) $\Rightarrow$ (ii). By Theorem 3.6.8 and condition (i), V carries the structure of a reflexive (unital) JB-algebra in an equivalent norm and $\overline{\Omega} = \{x^2 : x \in V\}$.

(ii) $\Rightarrow$ (iii). By the proof of the preceding lemma, V is a finite ℓ_∞-sum $\bigoplus_j^{\ell_\infty} A_j$, where A_j is either a real spin factor or a finite dimensional JB-algebra. Each A_j ia a unital JH-algebra (in an equivalent norm) and hence the ℓ_2-sum $V = \bigoplus_j^{\ell_2} A_j$ is a unital JH-algebra.

(iii) $\Rightarrow$ (i). This follows from the fact that a unital JH-algebra is also a JB-algebra in an equivalent norm, shown in Lemma 3.6.12.

(iii) $\Leftrightarrow$ (iv). See Theorem 2.4.14. $\square$

We have noted after Theorem 2.4.14 that a linearly homogeneous self-dual cone in a real Hilbert space carries the structure of a Riemannian symmetric space. In fact, if a tube domain $V \oplus i\Omega$ is biholomorphic to a bounded symmetric domain, we have seen that V carries the structure of a unital JB-algebra and Ω is the regular open cone of squares in V. By Remark 3.6.9, Ω is linearly homogeneous. It is also a normal cone since the norm of V is the order-unit norm $\| \cdot \|_e$ induced by the identity $e \in \Omega$. Moreover, we can equip Ω with a compatible tangent norm $b : T\Omega \longrightarrow [0, \infty)$, induced by $\| \cdot \|_e$. With respect to this tangent norm, Ω is a real symmetric Banach manifold, where the symmetry at e is the inverse map $x \in \Omega \mapsto x^{-1} \in \Omega$. To see this, we first define the tangent norm b, following [168, 12.31].

Let $L(V)$ be the real Banach Lie algebra of bounded linear operators on V, in the usual Lie brackets

$$[S, T] = ST - TS \qquad (S, T \in L(V)).$$

The open subgroup $GL(V)$ of $L(V)$, consisting of invertible elements, is a real Banach Lie group with Lie algebra $L(V)$. The linear maps $g \in GL(V)$ satisfying $g(\Omega) = \Omega$ form a subgroup of $GL(V)$, denoted by

$$G(\Omega) = \{g \in GL(V) : g(\Omega) = \Omega\}. \tag{3.55}$$

We shall call $G(\Omega)$ the *linear automorphism group* of Ω so that linear homogeneity of Ω is saying that $G(\Omega)$ acts transitively on Ω. An element $g \in GL(V)$ belongs to $G(\Omega)$ if and only if $g(\overline{\Omega}) = \overline{\Omega}$. Hence $G(\Omega)$ is a closed subgroup of $GL(V)$ and can be topologised to a real Banach Lie group with Lie algebra

$$\mathfrak{g}(\Omega) = \{X \in L(V) : \exp tX \in G(\Omega), \forall t \in \mathbb{R}\} \tag{3.56}$$

(cf. [168, p. 387]).

By a remark before Lemma 3.6.5, a linear isomorphism $g \in G(\Omega)$ satisfying $g(e) = e$ is an isometry with respect to the order unit norm $\| \cdot \|_e$ and hence one can define

$$b(p, v) = \|h(v)\|_e \qquad ((p, v) \in T\Omega) \tag{3.57}$$

for any $h \in G(\Omega)$ satisfying $h(p) = e$. The tangent norm b is $G(\Omega)$-*invariant*, in other words, each $g \in G(\Omega)$ is a b-isometry. Indeed, given

$g \in G(\Omega)$, and $(p, v) \in T\Omega$ with $b(p, v) = \|h(v)\|_e$ for some $h(p) = e$, pick $h_1 \in G(\Omega)$ such that $h_1(g(p)) = e$. Then $h_1 g h^{-1}(e) = e$ and hence $h_1 g h$ is an isometry with respect to $\| \cdot \|_e$, which gives

$$b(g(p), g'(p)(v)) = b(g(p), g(v)) = \|h_1(g(v))\|_e$$
$$= \|h_1 g h^{-1}(h(v))\|_e = \|h(v)\|_e.$$

Now, in the case where $V \oplus i\Omega$ is biholomorphic to a bounded symmetric domain, the symmetry $z \in V \oplus i\Omega \mapsto -z^{-1} \in V \oplus i\Omega$ at ie restricts to the inverse map $z \in i\Omega \mapsto -z^{-1} \in i\Omega$ since $z^{-1} = Q_z^{-1}(z) \in iV \cap (V \oplus i\Omega) = i\Omega$, where $Q_z = \{z, \cdot, z\}$ is the quadratic operator on the Jordan algebra $V \oplus iV$ and $Q_z(iV) = iV$ for $z \in i\Omega$, by invertibility. It follows that the inverse map

$$s_e : x \in \Omega \mapsto x^{-1} = Q_x^{-1}(x) \in \Omega$$

is a symmetry at e for the $G(\Omega)$-invariant tangent norm b since $s'_e(x) = -Q_x^{-1}$ and $Q_x^{-1} \in G(\Omega)$ (cf. [77, 3.2.11, 3.3.6]). We have therefore proved, in view of Remark 3.6.9, the following corollary.

Corollary 3.6.14. *Let Ω be a regular open cone in a real Banach space V. For the two conditions below, we have* (i) $\Rightarrow$ (ii).

 (i) *$V \oplus i\Omega$ is biholomorphic to a bounded symmetric domain.*

 (ii) *Ω is a normal linearly homogeneous cone and a symmetric Banach manifold in a $G(\Omega)$-invariant tangent norm.*

Problem 3.6.15. The formulation of the preceding corollary hints at the question of the converse (ii) $\Rightarrow$ (i). Although there is tangible evidence to suggest the validity of this implication, it remains unresolved. For a regular open cone Ω in a finite dimensional Euclidean space V, it has been shown in [158] and [166] that if Ω is linearly homogeneous and also a symmetric space in the canonical Riemannian metric defined in (2.52), then it is self-dual and hence $V \oplus i\Omega$ is indeed biholomorphic to a bounded symmetric domain by Corollary 3.6.13. However, it seems that the implication (ii) $\Rightarrow$ (i) could be false without the homogeneity condition on Ω.

We note that, besides the Riemannian metric, one can ask the question of symmetry with respect to other compatible tangent norms on a linearly homogeneous regular open cone Ω in a real Banach space V.

By Lemma 3.6.7, the order-unit norms induced by the order units in Ω are all equivalent to the norm of V and we can define a tangent norm $\tau : T\Omega \longrightarrow [0, \infty)$ by

$$\tau(p, v) = \|v\|_p \qquad ((p, v) \in \Omega \times V) \tag{3.58}$$

where $\| \cdot \|_p$ denotes the order-unit norm induced by the order unit $p \in \Omega$. To see that τ is continuous, let (p_n) converge to p in Ω and (v_n) converge to v in V. Given $1 > \varepsilon > 0$, $\|p_n - p\|_p \to 0$ implies $-\varepsilon p \le p_n - p \le \varepsilon p$ and $(1 - \varepsilon)p \le p_n \le (1 + \varepsilon)p$ from some n onwards, which gives

$$-(1 + \varepsilon)\|v_n\|_{p_n}p \le -\|v_n\|_{p_n}p_n \le v_n \le \|v_n\|_{p_n}p_n \le (1 + \varepsilon)\|v_n\|_{p_n}p$$

and hence $\|v_n\|_p \le (1 + \varepsilon)\|v_n\|_{p_n}$. Likewise $p \le \frac{p_n}{1-\varepsilon}$ implies $\|v_n\|_{p_n} \le \frac{\|v_n\|_p}{1-\varepsilon}$ and therefore

$$1 - \varepsilon \le \frac{\|v_n\|_p}{\|v_n\|_{p_n}} \le 1 + \varepsilon.$$

Since $\|v_n\|_p \to \|v\|_p$, we conclude $\|v_n\|_{p_n} \to \|v\|_p$.

In fact, the tangent norm τ coincides with $b : T\Omega \longrightarrow [0, \infty)$ defined in (3.57) (in the setting of unital JB-algebras). This follows from the fact that τ is $G(\Omega)$-invariant, which implies $\tau = b$. For if $h \in G(\Omega)$, then we have, for $v \in T_p\Omega = V$,

$$\tau(h(p), h'(p)(v)) = \tau(h(p), h(v)) = \|h(v)\|_{h(p)} = \|v\|_p = \tau(v, p)$$

where the third identity follows from the equivalent conditions

$$-\lambda h(p) \leq h(v) \leq \lambda h(p) \Leftrightarrow \lambda p \leq v \leq \lambda p \qquad (\lambda > 0).$$

By [138, Lemma 1.3, Theorem 1.1], the integrated distance d_τ of τ on Ω coincides with Thompson's metric

$$d_\tau(x, y) = \max\{\log M(x/y), \log M(y/x)\} \qquad (x, y \in \Omega)$$

where

$$M(a/b) := \inf\{\beta > 0 : \beta a \geq b\} \qquad (a, b \in \Omega).$$

It is interesting that the distance d_τ is related to the Carathéodory distance on the corresponding tube domain $V \oplus i\Omega$. Let $V \oplus i\Omega$ be biholomorphic to a bounded domain and let

$$V_+^* = \{f \in V^* : f(\Omega) \subset (0, \infty)\}$$

be the cone of 'strictly positive' continuous linear functionals on V. Following Vessentini [169], we define a *Carathéodory-type distance* $\delta :$ $\Omega \times \Omega \to [0, \infty)$ by

$$\delta(x, y) = \sup\left\{\left|\log \frac{f(x)}{f(y)}\right| : f \in V_+^*\right\} \qquad (x, y \in \Omega).$$

By regularity of Ω, it can be shown that δ is a distance on Ω, which is invariant under all affine automorphisms of Ω [169, Proposition 4.3]. Further, δ actually coincides with the restriction of the Carathéodory distance c on $V \oplus i\Omega$ to $i\Omega$:

$$\delta(x, y) = c(ix, iy) \qquad (x, y \in \Omega)$$

(cf. [169, Theorems I, II]), as well as the distance d_τ which is shown below.

Lemma 3.6.16. *Let Ω be a linearly homogeneous regular open cone in a real Banach space V. Then the Carathéodory-type distance δ on Ω coincides with d_τ.*

Proof. For each $u \in \Omega$, we denote the *state space* with respect to the order-unit u, with order-unit norm $\|\cdot\|_u$, by

$$S_u = \{f \in V^* : f(u) = \|f\|_u = 1\} \subset V_+^*$$

where $\|f\|_u = \sup\{|f(x)| : \|x\|_u \leq 1\}$. For each $f \in V_+^*$, we have $\|f\|_u = f(u)$ and hence $f/\|f\|_u \in S_u$.

We make use of the identity

$$\delta(x, y) = \sup\left\{\left|\log \frac{f(x)}{f(y)}\right| : f \in V_+^*\right\} \qquad (x, y \in \Omega)$$

where

$$\frac{f(x)}{f(y)} = \frac{f(x)/\|f\|_y}{f(y)/\|f\|_y} = f(x)/\|f\|_y \quad \text{and} \quad f/\|f\|_y \in S_y.$$

Hence δ can be written as

$$
\begin{aligned}
\delta(x, y) &= \sup\{|\log f(x)| : f \in S_y\} \\
&= \sup\left\{\max\left(\log f(x), \log \frac{1}{f(x)}\right) : f \in S_y\right\} \\
&= \log\left\{\max\left(\sup\{f(x) : f \in S_y\}, \sup\left\{\frac{1}{f(x)} : f \in S_y\right\}\right)\right\}.
\end{aligned}
$$

We have $\sup\{f(x) : f \in S_y\} = \|x\|_y = \inf\{\beta > 0 : x \leq \beta y\} = M(y/x)$. We complete the proof by showing

$$M(x/y) = \sup\left\{\frac{1}{f(x)} : f \in S_y\right\}.$$

Indeed, $y \leq \beta x$ implies $1 \leq \beta f(x)$ for each $\beta > 0$ and $f \in S_y$, which gives

$$\sup\{1/f(x) : f \in S_y\} \leq M(x/y).$$

To reverse the inequality, let $\mu = \inf\{f(x) : f \in S_y\}$. Since $f(x) = \|f\|_x > 0$ for all $f \in S_y$, we have $\mu > 0$ by weak* compactness of S_y. Observe that $1/\mu = \sup\{1/f(x) : f \in S_y\}$ and $\mu y \leq x$ since $f(x) \geq \mu = \mu f(y)$ for all $f \in S_y$. It follows that $1/\mu \geq M(x/y)$ as desired.   □

We conclude this section with a final remark relating to Problem 3.6.15, by the following proposition.

Proposition 3.6.17. *Let Ω be a regular open cone in a real Banach space V such that $V \oplus i\Omega$ is biholomorphic to a bounded domain. Then the following conditions are equivalent.*

(i) *$V \oplus i\Omega$ is biholomorphic to a bounded symmetric domain.*

(ii) *Ω admits a symmetry, with respect to a compatible tangent norm, which extends to a symmetry of the domain $\Omega \oplus iV$.*

If V is a Hilbert space, then these conditions are equivalent to Ω being a linearly homogeneous self-dual cone.

Proof. We note that $\Omega \oplus iV$ is biholomorphic to the tube domain $D(\Omega) = V \oplus i\Omega$. If a symmetry $s : \Omega \to \Omega$ at $e \in \Omega$ extends to a symmetry $\tilde{s} : \Omega \oplus iV \to \Omega \oplus iV$, then $D(\Omega)$ admits a symmetry and by Remark 3.6.9, $D(\Omega)$ is already biholomorphic to a bounded symmetric domain.

Conversely, if $V \oplus i\Omega$ is biholomorphic to a bounded symmetric domain, then V is a JB-algebra with identity $e \in \Omega$, and the inverse map $z \in \Omega \oplus iV \mapsto z^{-1} \in \Omega \oplus iV$ restricts to the symmetry $x \in \Omega \mapsto x^{-1} \in \Omega$, with respect to the tangent norm b defined in (3.57).   □

Notes. Theorem 3.6.8 was proved in [24] prior to the appearance of [99], without using the underlying JB*-triple structure of a bounded symmetric domain. This theorem can be viewed as an infinite dimensional extension of Koecher's seminal work. Indeed, in his 'elementary approach to bounded symmetric domains' (in finite dimensions) [111], Koecher showed that the tube domain over the open cone of squares of a finite dimensional JB-algebra, that is, a formally real Jordan algebra, is biholomorphic to a bounded symmetric domain obtained from the TKK construction. Finite dimensional linearly homogeneous self-dual cones are also discussed in [5] in connection with compactification of locally symmetric varieties. (*Note added in proofs. The author has recently given an affirmative answer to Problem 3.6.15 in a preprint (arXiv:2006.06449) entitled 'Siegel domains over Finsler symmetric cones'.*)

3.7 Holomorphic homogeneous regular domains

The notion of a *holomorphic homogeneous regular* (*HHR*) complex manifold M (of finite dimension) has been introduced by Liu, Sun and Yau [119, Definition 7.2] in connection with the estimation of several canonical metrics on the moduli and Teichmüller spaces of Riemann surfaces. It can be described by saying that a particular function

$$\sigma : M \to (0, 1]$$

called the *squeezing function*, has a strictly positive lower bound (cf. [53]). These manifolds possess many important geometric properties (e.g. all classical metrics on them are equivalent) [119, 120] and have also been studied by several authors (see, for example, [53, 54, 62, 104, 182]) in the case of complex domains. In particular, it has been shown in [182] that a holomorphic homogeneous regular bounded domain D in $\mathbb{C}^n$ must be pseudoconvex and all bounded strongly convex domains in $\mathbb{C}^n$ are holomorphic homogeneous regular. It has been shown recently in [104] that actually all bounded convex domains in $\mathbb{C}^n$ are holomorphic homogeneous regular.

The squeezing function on a bounded homogeneous domain in $\mathbb{C}^n$ is constant, by its holomorphic invariance, and has been computed explicitly for the four series of classical Cartan domains in [115]. In view of these interesting works, it is natural to ask if they can be extended to the setting of infinite dimensional domains. In this section, we do just that.

We extend the concept of a *holomorphic homogeneous regular domain* and generalise the aforementioned results to the infinite dimensional setting. In particular, we determine completely which bounded symmetric domains are HHR and compute the squeezing functions for all these domains, including the two exceptional domains, which were left untreated in [115].

The concept of the squeezing function for finite dimensional complex manifolds involves comparing a given manifold with various Euclidean balls via holomorphic embeddings. For infinite dimensional domains, we consider their holomorphic embeddings in Hilbert balls, as a natural infinite dimensional generalisation.

Definition 3.7.1. A map $f : D_1 \to D_2$ between two domains is called a *holomorphic embedding* of D_1 into D_2 if $f(D_1)$ is a domain in D_2 and f is biholomorphic onto $f(D_1)$. The set of all holomorphic embeddings of D_1 into D_2 is denoted by $H_{emb}(D_1, D_2)$.

Let D be a domain in a complex Banach space V. In this section, we denote by

$$B_H = \{x \in H : \|x\| < 1\}$$

the open unit ball of a Hilbert space H. The set $H_{emb}(D, B_H)$ of holomorphic embeddings may be empty. For instance, if D is the open unit ball of the Banach space ℓ_∞ of bounded complex sequences, then $H_{emb}(D, B_H) = \emptyset$ for any Hilbert ball B_H.

In fact, $H_{emb}(D, B_H) \neq \emptyset$ if and only if the ambient Banach space V of D is linearly homeomorphic to H. Indeed, if there is a holomorphic embedding $f : D \to B_H$, then V, as the tangent space at a point $p \in D$, must be linearly homeomorphic to H, which is the tangent space of $f(D)$ at $f(p)$. Conversely, if $\varphi : V \to H$ is a linear homeomorphism, then we have $\varphi(D) \subset RB_H$ for some $R > 0$, and for each $p \in D$, the map $f : z \in D \mapsto \varphi(z - p)/2R \in B_H$ is a biholomorphic map onto the domain $f(D)$ in B_H, with $f(p) = 0$ and $rB_H \subset f(D) \subset B_H$ for some $r > 0$.

Given $H_{emb}(D, B_H) \neq \emptyset$, then for each $p \in D$, the set

$$\mathcal{F}(p, D) = \{f \in H_{emb}(D, B_H) : f(p) = 0\}$$

is non-empty, as just noted. Hence we can define the *squeezing function*

$$\sigma_D : D \to (0, 1]$$

by

$$\sigma_D(p) = \sup_{f \in \mathcal{F}(p,D)} \{r > 0 : rB_H \subset f(D)\}.$$

The *squeezing constant* $\hat{\sigma}_D$ for D is defined to be

$$\hat{\sigma}_D = \inf_{p \in D} \sigma_D(p).$$

Both the squeezing function and squeezing constant are biholomorphic invariants.

Remark 3.7.2. We note that, if $H_{emb}(D, B_H) \neq \emptyset$, then the definition of the squeezing function for a domain $D \subset V$ does not depend on the chosen Hilbert ball B_H. Indeed, if there is a holomorphic embedding of D into another Hilbert ball B_K of a Hilbert space K, then the previous remarks imply that there is a continuous linear isomorphism $T : H \to K$. Denote the inner products of H and K by $\langle \cdot, \cdot \rangle_H$ and $\langle \cdot, \cdot \rangle_K$ respectively. Let $\alpha : H^* \to H$ and $\beta : K \to K^*$ be the canonical isometries. Then the linear isomorphism $\alpha T^* \beta T : H \to H$ satisfies

$$\langle \alpha T^* \beta T x, y \rangle_H = \langle Tx, Ty \rangle_K \qquad (x, y \in H)$$

and the linear isomorphism $T(\alpha T^* \beta T)^{-1/2} : H \to K$ is an isometry. It follows that the squeezing functions σ_D defined in terms B_H and B_K respectively are identical.

We now extend the concept of a finite dimensional HHR manifold introduced in [119, 120] to infinite dimensional domains. A finite dimensional HHR domain is also called a domain with *uniform squeezing property* in [182].

Definition 3.7.3. A domain D in a complex Banach space V is called *holomorphic homogeneous regular* (HHR) if D admits a holomorphic embedding into some Hilbert ball B_H and its squeezing function $\sigma_D :$ $D \to (0, 1]$ has a strictly positive lower bound, that is, $\hat{\sigma}_D > 0$.

Remark 3.7.4. If D is an HHR domain in a Banach space V, then as noted previously, V must be linearly homeomorphic to a Hilbert space. We call V an *isomorph of a Hilbert space*. The class of these Banach spaces has been characterised by many authors, for instance, it has been shown in [116] that a Banach space is an isomorph of a Hilbert space if and only if it is of type 2 and cotype 2. We refer to [141, Chapter IV] for more details.

We begin our discussion of infinite dimensional HHR domains by showing some properties of the squeezing function. We will make use of the Carathéodory pseudo-distance c_D on a domain D. We first show that the squeezing function is continuous.

Proposition 3.7.5. *Let D be a domain in a Banach space V linearly homeomorphic to a Hilbert space H. Then the squeezing function $\sigma_D :$ $D \to (0, 1]$ is continuous.*

Proof. Let (z_k) be a sequence converging to $a \in D$. We show

$$\liminf_{k \to \infty} \sigma_D(z_k) \geq \sigma_D(a) \geq \limsup_{k \to \infty} \sigma_D(z_k).$$

Let $0 < 2\varepsilon < \sigma_D(a)$ and pick $\sigma_D(a) \geq \rho > \sigma_D(a) - \varepsilon$ such that there is a holomorphic embedding $f : D \to B_H$ satisfying $f(a) = 0$ and $\rho B_H \subset f(D)$. By continuity, we have

$$\|f(z_k)\| < \varepsilon$$

for $k > K$, for some $K > 0$. Consider the holomorphic embedding $f_k : D \to B_H$ given by

$$f_k(\omega) = \frac{f(\omega) - f(z_k)}{1 + \varepsilon} \qquad (\omega \in D)$$

which satisfies $f_k(z_k) = 0$ and

$$\frac{\rho - \varepsilon}{1 + \varepsilon} B_H \subset f_k(D).$$

This gives

$$\sigma_D(z_k) \geq \frac{\rho - \varepsilon}{1 + \varepsilon} > \frac{\sigma_D(a) - 2\varepsilon}{1 + \varepsilon}$$

for $k > K$ and hence $\lim_{k\to\infty} \inf \sigma_D(z_k) \geq \sigma_D(a)$ since $\varepsilon > 0$ was arbitrary.

For the upper limit, let $0 < 2\varepsilon < \lim_k \inf \sigma_D(z_k)$ and let $f_k : D \to B_H$ be a holomorphic embedding satisfying $f_k(z_k) = 0$ and $\rho_k B_H \subset f_k(D)$ for some $\sigma_D(z_k) \geq \rho_k > \sigma_D(z_k) - \varepsilon$. Since the Carathéodory pseudo-distance $c_D(z_k, a) \to 0$ as $k \to \infty$, we have

$$\tanh^{-1} \|f_k(a)\| = c_{B_H}(0, f_k(a)) \leq c_D(z_k, a) \to 0$$

and hence there exists some $M > 0$ such that $\|f_k(a)\| < \varepsilon$ for $k > M$. By analogous arguments as before, one obtains

$$\sigma_D(a) \geq \frac{\rho_k - \varepsilon}{1 + \varepsilon} > \frac{\sigma_D(z_k) - 2\varepsilon}{1 + \varepsilon}$$

for $k > M$, which gives $\sigma_D(a) \geq \lim_{k\to\infty} \sup \sigma_D(z_k)$. $\qquad\square$

Although the continuity of the squeezing function implies readily that if there is a sequence (p_k) in a *finite dimensional* bounded domain D with $\lim_k \sigma_D(p_k) = 0$, then the sequence admits a subsequence (p_j) converging to a boundary point $p \in \partial D$, this is not immediately clear for infinite dimensional domains. Nevertheless, one can still show, in infinite dimension, (p_k) has a subsequence (p_j) for which the distance $d(p_j, \partial D)$ to the boundary tends to 0. We prove a lemma first.

Lemma 3.7.6. *Let Ω be a bounded domain in an isomorph V of a Hilbert space H and $\varphi : V \to H$ a linear homeomorphism. Then there is a constant $m > 0$ such that for each $q \in \Omega$ satisfying $B_V(q, s) \subset \Omega$ for some $s > 0$, we have*

$$\sigma_\Omega(q) \geq \frac{s}{m^2 \|\varphi\| \|\varphi^{-1}\|}.$$

Proof. By a translation, we may assume $q = 0$. Since Ω is bounded, we have $\Omega \subset B_V(0, m)$ for some $m > 0$ and

$$\frac{1}{\|\varphi^{-1}\|} B_H(0, m) \subset \varphi(B_V(0, m)) \subset B_H(0, m\|\varphi\|) = m\|\varphi\| B_H. \quad (3.59)$$

The restriction of φ to Ω, still denoted by φ, is a holomorphic embedding of Ω into $m\|\varphi\| B_H$ satisfying $\varphi(q) = 0$. It follows from (3.59) that

$$\frac{s}{m\|\varphi^{-1}\|} B_H(0, m) \subset \varphi(B_V(0, s)) \subset \varphi(\Omega) \subset \varphi(B_V(0, m)) \subset m\|\varphi\| B_H.$$

Hence we have

$$\sigma_\Omega(q) \geq \frac{s}{m^2 \|\varphi\| \|\varphi^{-1}\|}.$$

$\square$

Lemma 3.7.7. *Let (p_k) be a sequence in a bounded domain Ω in an isomorph V of a Hilbert space such that $\lim_{k \to \infty} \sigma_\Omega(p_k) = 0$. Then there is a subsequence (p_j) of (p_k) such that*

$$\lim_{j \to \infty} d(p_j, \partial \Omega) = 0$$

and for each j, there exists a boundary point $q_j \in \partial\Omega$ with $\|p_j - q_j\| = d(p_j, \partial\Omega)$.

Proof. Let (p_k) be the given sequence satisfying

$$\lim_{k \to \infty} \sigma_\Omega(p_k) = 0. \qquad (3.60)$$

Since the bounded domain Ω is relatively weakly compact in V, there is a subsequence (p_j) in Ω converging weakly to some point $p \in \overline{\Omega}$. (We do not know if the squeezing function σ_Ω is weakly continuous on Ω.)

Let $r_j = d(p_j, \partial\Omega)$ denote the distance from p_j to the boundary $\partial\Omega$. We first show that $\lim_{j \to \infty} r_j = 0$. Otherwise, we may assume (by choosing a subsequence)

$$r_j \geq s, \quad \text{for some} \quad s > 0$$

for all j. For all $z \in \partial\Omega$, we have $\|z - p_j\| \geq r_j$. Observe that $B_V(p_j, r_j) \subset \Omega$, for if there exists some $\omega \in B_V(p_j, r_j) \backslash \Omega$, then we must have $\omega \notin \overline{\Omega}$. Therefore the (real) line joining p_j and ω must intersect $\partial\Omega$ at a point z_0 say, which gives a contradiction that

$$r_j \leq \|z_0 - p_j\| \leq \|\omega - p_j\| < r_j.$$

By Lemma 3.7.6, there exists $m > 0$ such that

$$\sigma_\Omega(p_j) \geq \frac{r_j}{m^2 \|\varphi\| \|\varphi^{-1}\|} \geq \frac{s}{m^2 \|\varphi\| \|\varphi^{-1}\|} > 0,$$

contradicting $\lim_j \sigma_\Omega(p_j) = 0$. Therefore we have established

$$r_j = d(p_j, \partial\Omega) \to 0 \quad \text{as} \quad j \to \infty.$$

We next show that there exists $q_j \in \partial\Omega$ such that $\|p_j - q_j\| = r_j$. Indeed, there is a sequence (q_n) in $\partial\Omega$ satisfying $\lim_n \|q_n - p_j\| = r_j$. By weak compactness of the boundary $\partial\Omega$, we may assume, by choosing a subsequence, that the sequence (q_n) weakly converges to some $q_j \in \partial\Omega$. Then we have

$$r_j = \liminf_n \|q_n - p_j\| \geq \|q_j - p_j\| \geq r_j.$$

$\square$

It has been shown in [119, Theorem 7.2] that the Bergman, Cara-théodory and Kobayashi metrics are equivalent on HHR manifolds, where two differential metrics ω_1 and ω_2 are said to be *equivalent* if they are *quasi-isometric* in the sense that

$$C^{-1}\omega_1 \leq \omega_2 \leq C\omega_2$$

for some constant $C > 0$. Likewise, the Carathéodory and Kobayashi pseudo-metrics are equivalent on HHR domains in Banach spaces.

Proposition 3.7.8. *Let D be an HHR domain in an isomorph V of a Hilbert space H, with the squeezing constant $\hat{\sigma}_D$. Then we have*

$$\hat{\sigma}_D \mathcal{K}_D(x, v) \leq \mathcal{C}_D(x, v) \leq \mathcal{K}_D(x, v)$$

for all $x \in D$ and $v \in T_x D$.

Proof. We need only show the first inequality. Let $x \in D$ and $v \in V$, where we identify the tangent space $T_x D$ with V. Let

$$f_x : D \to B_H$$

be a holomorphic embedding such that $f_x(x) = 0$ and $rB_H \subset f_x(D) \subset B_H$ for some $r > 0$.

For each $\varphi \in H^*$ with $\|\varphi\| \leq 1$, the composite map $\varphi f_x : D \to \mathbb{D}$ satisfies $\varphi f_x(x) = 0$ and hence $|\varphi f_x'(x)(v)| = |(\varphi f_x)'(x)(v)| \leq \mathcal{C}_D(x, v)$. It follows that $\|f_x'(x)(v)\| \leq \mathcal{C}_D(x, v)$.

Define a holomorphic map $\gamma : \mathbb{D} \to rB_H$ by

$$\gamma(\alpha) = \frac{\alpha r f_x'(x)(v)}{\|f_x'(x)(v)\|} \qquad (\alpha \in \mathbb{D}).$$

The restriction of the inverse $f_x^{-1} : f_x(D) \to D$ to $rB_H \subset f_x(D)$ is well-defined and its composite with γ gives a holomorphic map $f_x^{-1}\gamma : \mathbb{D} \to D$ satisfying $f_x^{-1}\gamma(0) = x$ and

$$
\begin{aligned}
(f_x^{-1}\gamma)'(0) \left(\frac{\|f_x'(x)(v)\|}{r} \right) &= (f_x^{-1})'(0)\gamma'(0) \left(\frac{\|f_x'(x)(v)\|}{r} \right) \\
&= f_x'(x)^{-1}(f_x'(x)(v)) = v.
\end{aligned}
$$

This gives $\mathcal{K}(x,v) \le \frac{\|f_x'(x)(v)\|}{r} \le \mathcal{C}_D(x,v)/r$ and hence

$$\hat{\sigma}_D \, \mathcal{K}_D(x,v) \le \mathcal{C}_D(x,v)$$

by the definition of the squeezing constant. $\qquad \square$

In finite dimensions, bounded HHR domains are pseudoconvex, which has been proved in [182, Lemma 2]. To extend this result to infinite dimension, let us recall the definition of pseudoconvexity. For this, we first introduce the concept of a *plurisubharmonic function*. Loosely speaking, these functions are 'subharmonic on complex lines'.

Let D be a domain in a complex Banach space V. A function $f : D \longrightarrow \mathbb{R} \cup \{-\infty\}$ is called *plurisubharmonic* if it satisfies the following conditions:

(i) f is upper semi-continuous,

(ii) f restricts to a subharmonic function on every complex line in D, that is, on the open set $U = \{\alpha \in \mathbb{C} : a + \alpha b \in D\}$, where $a, b \in V$, the function $\alpha \in U \mapsto f(a + \alpha b)$ is subharmonic.

Definition 3.7.9. A domain D in a complex Banach space V is called *pseudoconvex* if the function $z \in D \mapsto -\log d(z, \partial D) \in \mathbb{R} \cup \{-\infty\}$ is plurisubharmonic, where $d(z, \partial D)$ is the distance from z to ∂D.

In finite dimensions, a celebrated result of Oka [139] reveals that pseudoconvex domains are exactly the domains of holomorphy, which solves the classical Levi problem (see also [25, 136]). In all Banach spaces, domains of holomorphy are pseudoconvex [133, 11.4, 37.7]. The converse is true in separable Banach spaces with the bounded approximation property (cf. [137] and [133, 5.8]), but false in general [95].

We show that bounded HHR domains are domains of holomorphy, which extends the result of [182, Lemma 2].

Theorem 3.7.10. *Let D be a bounded HHR domain in a complex Banach space V. Then D is a domain of holomorphy.*

Proof. In view of Proposition 3.5.16, we need only show that the Carathéodory distance in D is complete.

By the hypothesis, the squeezing constant $\hat{\sigma}_D$ takes the value, say, $r \in (0, 1]$. Let (x_n) be a c_D-Cauchy sequence in D. We show that (x_n) c_D-converges. Let $\varepsilon = \tanh^{-1} \frac{r}{2}$. Then there is a number $N > 0$ such that $c_D(x_n, x_N) < \varepsilon$ for $n > N$.

Let $f \colon D \to B_H$ be a holomorphic embedding into a Hilbert ball B_H with $f(x_N) = 0$ and $B_H(0, \frac{3r}{4}) \subset f(D)$. Then the inverse holomorphic map $g := f^{-1} \colon f(D) \to D$ is well-defined on the ball $B_H(0, \frac{3r}{4})$.

We have, for $n > N$,

$$c_{B_H}(0, f(x_n)) = c_{B_H}(f(x_N), f(x_n)) \leq c_D(x_N, x_n) < \varepsilon = \tanh^{-1} \frac{r}{2}$$

as well as

$$\lim_{n,m \to \infty} c_{B_H}(f(x_m), f(x_n)) \leq \lim_{n,m \to \infty} c_D(x_m, x_n) = 0.$$

Since B_H is complete in the Carathéodory distance, there is a subsequence (x_{n_k}) of (x_n) such that $f(x_{n_k})$ converges to some $y_0 \in B_H$ with respect to c_{B_H}, and $c_{B_H}(0, y_0) \leq \varepsilon$. Hence, as noted previously, we have

$y_0 \in \overline{B}_H(0, \frac{r}{2}) \subset B_H(0, \frac{3r}{4}) \subset f(D)$ and also,

$$\lim_{k \to \infty} c_D(x_{n_k}, g(y_0)) \leq \lim_{k \to \infty} c_D(g(y_{n_k}), g(y_0))$$

$$\leq \lim_{k \to \infty} \frac{4}{3r} c_{B_H}(f(g(y_{n_k})), f(g(y_0)))$$

$$= \lim_{k \to \infty} \frac{4}{3r} c_{B_H}(y_{n_k}, y_0) = 0.$$

It follows that the sequence (x_n) converges to $g(y_0)$ in D with respect to c_D and the proof is complete. $\qquad \square$

We now consider bounded symmetric domains. In finite dimensions, it is well-known that a bounded symmetric domain of rank ℓ contains a polydisc of dimension ℓ as a totally geodesic submanifold [108, p. 41]. To see that this is also the case for infinite dimensional bounded symmetric domains of finite rank, we only need to consider the *irreducible* ones. There are only two classes of such domains, namely, the Type IV domains realisable as the Lie balls, which are of rank 2, and the Type I domains of rank ℓ, which can be realised as the open unit ball of the JB*-triple $L(\mathbb{C}^\ell, K)$ of bounded linear operators between Hilbert spaces $\mathbb{C}^\ell$ and K, with $\ell \leq \dim K \leq \infty$ and $\ell < \infty$.

Given $\ell < \infty$, every operator $T \in L(\mathbb{C}^\ell, K)$ is a Hilbert-Schmidt operator in the Hilbert-Schmidt norm

$$\|T\|_2 = \left(\sum_{k=1}^{\ell} \|Te_k\|^2 \right)^{1/2}$$

satisfying $\|T\| \leq \|T\|_2 \leq \sqrt{\ell} \|T\|$, where $\{e_1, \ldots, e_\ell\}$ is the standard orthonormal basis in $\mathbb{C}^\ell$.

Let $\overline{\mathbb{D}}$ be the closure of $\mathbb{D} = \{z \in \mathbb{C} : |z| < 1\}$ and $\overline{D}$ the closure of the open unit ball

$$D = \{T \in L(\mathbb{C}^\ell, K) : \|T\| < 1\}.$$

Fix orthonormal basis vectors $u_{\alpha_1}, \ldots, u_{\alpha_\ell}$ from an orthonormal basis $\{u_\alpha\}$ in K. Then the continuous map $\varphi : \overline{\mathbb{D}} \times \cdots \times \overline{\mathbb{D}} \to \overline{D}$, defined by

$$\varphi(z_1, \ldots, z_\ell) = \sum_{k=1}^{\ell} z_k (e_k \otimes u_{\alpha_k}) \qquad (z_1, \ldots, z_\ell) \in \overline{\mathbb{D}}^\ell, \qquad (3.61)$$

restricts to an injective holomorphic map

$$\varphi : \mathbb{D} \times \cdots \times \mathbb{D} \to D$$

with $\varphi(0, \ldots, 0) = 0$, where $e_k \otimes u_{\alpha_k} : \mathbb{C}^\ell \to K$ is the rank-one operator

$$e_k \otimes u_{\alpha_k}(h) = \langle h, e_k \rangle u_{\alpha_k} \qquad (h \in \mathbb{C}^\ell)$$

with $\|e_k \otimes u_{\alpha_k}\| = \|e_k \otimes u_{\alpha_k}\|_2 = 1$. This also implies that φ maps the boundary $\partial \mathbb{D}^\ell$ of $\mathbb{D}^\ell$ into the boundary $\partial D = \{T \in L(\mathbb{C}^\ell, K) : \|T\| = 1\}$.

Let D be the open unit ball of a spin factor V, which is of rank 2. Let e_1 and e_2 be two mutually (triple) orthogonal minimal tripotents in V. Then we have $\|\lambda e_1 + \mu e_2\| = \max\{|\lambda|, |\mu|\}$ for $\lambda, \mu \in \mathbb{C}$, by Corollary 3.2.24. Hence one can define a continuous map

$$\varphi : (z_1, z_2) \in \overline{\mathbb{D}}^2 \mapsto z_1 e_1 + z_2 e_2 \in \overline{D} \qquad (3.62)$$

which restricts to an injective holomorphic map from $\mathbb{D}^2$ to D satisfying $\varphi(0) = 0$ and $\varphi(\partial \mathbb{D}^2) \subset \partial D$.

Given a Hilbert space H, a holomorphic map $f : \mathbb{D}^n \to H$ admits a power series representation in terms of homogeneous polynomials from $\mathbb{C}^n$ to H. We recall that a homogeneous polynomial p of degree d from $\mathbb{C}^n$ to H is given by

$$p(z_1, \ldots, z_n) = P((z_1, \ldots, z_n), \ldots, (z_1, \ldots, z_n)) \in H, \quad (z_1, \ldots, z_n) \in \mathbb{C}^n$$

where $P : \underbrace{\mathbb{C}^n \times \cdots \times \mathbb{C}^n}_{d\text{-times}} \to H$ is a d-linear map. Let $\{e_\alpha\}$ be an orthonormal basis in H. We can write

$$p(z_1, \ldots, z_n) = \sum_\alpha \mathfrak{q}_\alpha(z_1, \ldots, z_n) e_\alpha$$

where $\mathfrak{q}_\alpha(z_1, \ldots, z_n)$ is a homogeneous polynomial of degree d in n complex variables $z_1, \ldots, z_n$ and has the form

$$\mathfrak{q}_\alpha(z_1, \ldots, z_n) = \sum_{j_1 + \cdots + j_n = d} c_{\alpha; j_1, \ldots, j_n} z_1^{j_1} \cdots z_n^{j_n} \qquad (c_{\alpha; j_1, \ldots, j_n} \in \mathbb{C}).$$

Hence a holomorphic map $f : \mathbb{D}^n \to H$ has a representation

$$f(z_1, \ldots, z_n) = f(0) + \sum_{d=1}^\infty p^d(z_1, \ldots, z_n), \qquad (z_1, \ldots, z_n) \in \mathbb{D}^n$$

where p^d is a homogeneous polynomial of degree d from $\mathbb{C}^n$ to H and has the from

$$p^d(z_1, \ldots, z_n) = \sum_\alpha \sum_{j_1 + \cdots + j_n = d} c_{\alpha; j_1, \ldots, j_n}^d z_1^{j_1} \cdots z_n^{j_n} e_\alpha \quad (c_{\alpha; j_1, \ldots, j_n}^d \in \mathbb{C}).$$

$$(3.63)$$

Let $h : D \to D'$ be a biholomorphic map between two *open unit balls* D, D' of Banach spaces V and V' respectively. If $h(0) = 0$, then by Cartan's uniqueness theorem, h is the restriction of the derivative $h'(0) : V \to V'$, which is a linear isometry. In particular, h extends to a continuous map $\bar{h} : \bar{D} \to \bar{D}'$ between the closures $\bar{D}$ and $\bar{D}'$, where $\bar{h} = h'(0)|_{\bar{D}}$. Moreover, $\bar{h}(\partial D) = \partial D'$.

Let D be a bounded symmetric domain, realised as the open unit ball of a JB*-triple V. Given a holomorphic embedding $f : D \to B_H$ of D into a Hilbert ball B_H, the image $f(D)$ is a bounded symmetric domain and hence there is an equivalent norm $\| \cdot \|_{sp}$ on H such that $(H, \| \cdot \|_{sp})$ is a JB*-triple and $f(D)$ identifies (via a biholomorphic map) as the open unit ball of $(H, \| \cdot \|_{sp})$. If $f(0) = 0$, then the previous remark implies that f extends to a continuous map $\bar{f}$, which maps ∂D onto the boundary $\partial f(D)$ of the domain $f(D)$.

The following lemma is a simple infinite dimensional extension of Alexander's result in [10, Proposition 1] (see also [115, Lemma 1]).

Lemma 3.7.11. *Let D be a bounded domain with boundary ∂D and B a Hilbert ball such that the following two continuous maps*

$$\overline{\mathbb{D}}^\ell \overset{\varphi}{\to} \overline{D} \overset{f}{\to} \overline{B}$$

on the closures restrict to holomorphic maps

$$\mathbb{D}^\ell \overset{\varphi}{\to} D \overset{f}{\to} B$$

with open image $f(D)$, satisfying $\varphi(\partial \mathbb{D}^\ell) \subset \partial D$ and $f(\partial D) \subset \partial f(D)$. If $\rho B \subset f(D)$ for some $\rho > 0$, then $\ell \rho^2 \leq 1$.

Proof. Let $\{e_\alpha\}$ be an orthonormal basis in the Hilbert space containing the ball B. By (3.63), the holomorphic map $f \circ \varphi$ on $\mathbb{D}^\ell$ has a power series representation

$$f \circ \varphi(z_1, \ldots, z_\ell) = \sum_{d=1}^{\infty} p^d(z_1, \ldots, z_\ell)$$

where $p^d(z_1, \ldots, z_\ell)$ is a d-homogeneous polynomial of the form

$$p^d(z_1, \ldots, z_\ell) = \sum_{\alpha} \sum_{j_1 + \cdots + j_\ell = d} c^d_{\alpha; j_1, \ldots, j_\ell} z_1^{j_1} \cdots z_\ell^{j_\ell} e_\alpha \qquad (c^d_{\alpha; j_1, \ldots, j_\ell} \in \mathbb{C}).$$

Since $\rho B \subset f(D)$, we have $\|f(w)\| \geq \rho$ for each $w \in \partial D$. Noting that $f \circ \varphi(\partial \mathbb{D}^\ell) \subset \partial f(D)$, we deduce

$$
\begin{aligned}
\rho^2 &\leq \frac{1}{2\pi} \int_0^{2\pi} \|f \circ \varphi(0, \ldots, e^{i\theta_j}, 0, \ldots, 0)\|^2 d\theta_j \\
&= \frac{1}{2\pi} \lim_{r \to 1} \int_0^{2\pi} \|f \circ \varphi(0, \ldots, re^{i\theta_j}, 0, \ldots, 0)\|^2 d\theta_j \\
&= \frac{1}{2\pi} \lim_{r \to 1} \int_0^{2\pi} \sum_\alpha \left| \sum_d c_{\alpha;\, 0,\ldots,0,d,0,\ldots,0}^d \, r^d e^{id\theta_j} \right|^2 d\theta_j \\
&= \frac{1}{2\pi} \lim_{r \to 1} \sum_\alpha \int_0^{2\pi} \left| \sum_d c_{\alpha;\, 0,\ldots,0,d,0,\ldots,0}^d \, r^d e^{id\theta_j} \right|^2 d\theta_j \\
&= \lim_{r \to 1} \sum_\alpha \sum_d \left| c_{\alpha;\, 0,\ldots,0,d,0,\ldots,0}^d \right|^2 r^{2d} \\
&= \sum_\alpha \sum_d \left| c_{\alpha;\, 0,\ldots,0,d,0,\ldots,0}^d \right|^2.
\end{aligned}
$$

It follows that

$$
\begin{aligned}
1 &\geq \lim_{r \to 1} \left(\frac{1}{2\pi} \right)^\ell \int_0^{2\pi} \cdots \int_0^{2\pi} \|f \circ \varphi(re^{i\theta_1}, \ldots, re^{i\theta_\ell})\|^2 d\theta_1 \cdots d\theta_\ell \\
&= \lim_{r \to 1} \sum_\alpha \sum_d \sum_{\nu_1 + \cdots + \nu_\ell = d} \left| c_{\alpha;\, \nu_1, \ldots, \nu_\ell}^d \right|^2 r^{2d} \\
&= \sum_\alpha \sum_d \sum_{\nu_1 + \cdots + \nu_\ell = d} \left| c_{\alpha;\, \nu_1, \ldots, \nu_\ell}^d \right|^2 \\
&\geq \sum_\alpha \sum_d \left| c_{\alpha;\, d,0,\ldots,0}^d \right|^2 + \cdots + \sum_\alpha \sum_d \left| c_{\alpha;\, 0,\ldots,0,d}^d \right|^2 \geq \ell \rho^2.
\end{aligned}
$$

$\square$

In finite dimensions, the squeezing constant of the classical Cartan domains has been computed by Kubota in [115]. We will now compute the squeezing constants of the remaining finite rank bounded symmetric domains of all dimensions.

We begin with the two exceptional domains which are realised as the open unit balls of the JB*-triples $M_{1,2}(\mathcal{O})$ and $H_3(\mathcal{O})$ respectively, where $\dim M_{1,2}(\mathcal{O}) = 16$ and $\dim H_3(\mathcal{O}) = 27$. Both JB*-triples carry a Hilbert space structure with the trace form and one can define an inner product by

$$\langle x, y \rangle = \frac{1}{9} \operatorname{Trace}(x \square y) \qquad (x, y \in H_3(\mathcal{O})). \tag{3.64}$$

so that $\langle e, e \rangle = 1$ for each minimal tripotent $e \in H_3(\mathcal{O})$ (cf. [150, Proposition 2.8, Corollary 2.14]). If e and u are two mutually triple orthogonal tripotents in $H_3(\mathcal{O})$, then $\langle e, u \rangle = 0$.

The 27-dimensional domain $D_{27} \subset H_3(\mathcal{O})$ has rank 3 whereas the 16-dimensional domain $D_{16} \subset M_{1,2}(\mathcal{O})$ has rank 2.

The following two propositions, together with Kubota's results in [115], give a complete list of squeezing constants of all finite dimensional irreducible bounded symmetric domains.

Proposition 3.7.12. *The squeezing constant of the exceptional domain D_{27} is given by $\hat{\sigma}_{D_{27}} = 1/\sqrt{3}$.*

Proof. We compute $\sigma_{D_{27}}(0) = \hat{\sigma}_{D_{27}}$. We have $D_{27} = \{z \in H_3(\mathcal{O}) : \|z\| < 1\}$, where $\|\cdot\|$ is the norm of the JB*-triple $H_3(\mathcal{O})$. Given $z \in H_3(\mathcal{O})$, we have the spectral decomposition

$$z = \alpha_1 e_1 + \alpha_2 e_2 + \alpha_3 e_3 \qquad (\alpha_1 \geq \alpha_2 \geq \alpha_3 \geq 0),$$

and $\|z\| = \alpha_1$, where the minimal tripotents e_1, e_2, e_3 are mutually orthogonal with respect to the inner product in (3.64). The Hilbert space norm $\|z\|_2$ of z is given by

$$\|z\|_2^2 = \langle z, z \rangle = \alpha_1^2 + \alpha_2^2 + \alpha_3^2.$$

It follows that

$$\|z\| \leq \|z\|_2 \leq \sqrt{3}\|z\|$$

for all $z \in H_3(\mathcal{O})$. This implies

$$B_{27} \subset D_{27} \subset \sqrt{3} B_{27}$$

where $B_{27} = \{z \in H_3(\mathcal{O}) : \|z\|_2 < 1\}$ is the Hilbert ball in $H_3(\mathcal{O})$. Hence we have $\hat{\sigma}_{D_{27}} \geq 1/\sqrt{3}$. To show the reverse inequality, we define a continuous map $\varphi : \overline{\mathbb{D}}^3 \to \overline{D}_{27}$ by

$$\varphi(z_1, z_2, z_3) = \begin{pmatrix} z_1 & 0 & 0 \\ 0 & z_2 & 0 \\ 0 & 0 & z_3 \end{pmatrix} = z_1 e_{11} + z_2 e_{22} + z_3 e_{33}$$

where e_{jj} is the diagonal matrix in $H_3(\mathcal{O})$ with 1 in the jj-entry and 0 elsewhere. Since e_{11}, e_{22}, e_{33} are mutually triple orthogonal minimal tripotents in $H_3(\mathcal{O})$, we see that φ restricts to an injective holomorphic map from $\mathbb{D}^3$ into D_{27} with $\varphi(0) = 0$ and $\varphi(\partial \mathbb{D}^3) \subset \partial D_{27}$. By Lemma 3.7.11 and the remarks before it, for each holomorphic embedding $f : D_{27} \to B_{27}$ with $f(0) = 0$ and $\rho B_{27} \subset f(D_{27})$, we must have $3\rho^2 \leq 1$. This proves the reverse inequality. $\qquad\square$

Proposition 3.7.13. *The squeezing constant of the exceptional domain* D_{16} *is given by* $\hat{\sigma}_{D_{16}} = 1/\sqrt{2}$.

Proof. The arguments are similar to those in the proof of Proposition 3.7.12, we recapitulate for completeness. We consider $M_{1,2}(\mathcal{O})$ as a subtriple of $H_3(\mathcal{O})$. It suffices to show $\sigma_{D_{16}}(0) = 1/\sqrt{2}$. We have $D_{16} = \{z \in M_{1,2}(\mathcal{O}) : \|z\| < 1\}$, where $\| \cdot \|$ is the norm of the JB*-triple $M_{1,2}(\mathcal{O})$. Each $z \in M_{1,2}(\mathcal{O})$ has a spectral decomposition

$$z = \alpha_1 e_1 + \alpha_2 e_2 \qquad (\alpha_1 \geq \alpha_2 \geq 0),$$

and $\|z\| = \alpha_1$, where the minimal tripotents e_1, e_2 are mutually orthogonal with respect to the inner product in (3.64). The Hilbert space norm $\|z\|_2$ of z is given by

$$\|z\|_2^2 = \langle z, z \rangle = \alpha_1^2 + \alpha_2^2$$

and

$$\|z\| \leq \|z\|_2 \leq \sqrt{2}\|z\|$$

for all $z \in M_{1,2}(\mathcal{O})$. This implies

$$B_{16} \subset D_{16} \subset \sqrt{2}B_{16},$$

where $B_{16} = \{z \in M_{1,2}(\mathcal{O}) : \|z\|_2 < 1\}$ is the Hilbert ball in $M_{1,2}(\mathcal{O})$. Hence $\hat{\sigma}_{D_{16}} \geq 1/\sqrt{2}$. For the reverse inequality, one defines a continuous map $\varphi : \overline{\mathbb{D}}^2 \to \overline{D}_{16}$ by

$$\varphi(z_1, z_2) = z_1 e_{11} + z_2 e_{22}$$

where $e_{11} = (1, 0)$ and $e_{22} = (0, 1)$ are mutually triple orthogonal minimal tripotents in $M_{1,2}(\mathcal{O})$, and φ restricts to an injective holomorphic map from $\mathbb{D}^2$ into D_{16} with $\varphi(0) = 0$ and $\varphi(\partial\mathbb{D}^2) \subset \partial D_{16}$. As before, for each holomorphic embedding $f : D_{16} \to B_{16}$ satisfying $f(0) = 0$ and $\rho B_{16} \subset f(D_{16})$, we must have $2\rho^2 \leq 1$. This proves the reverse inequality. $\square$

The proof of the two preceding propositions is analogous to Kobota's in [115] for the classical Cartan domains. We will not repeat Kubota's computation for the classical domains in the following corollary.

Corollary 3.7.14. *Let D be a finite dimensional irreducible bounded symmetric domain of rank p. Then its squeezing constant is given by $\hat{\sigma}_D = 1/\sqrt{p}$.*

The following result, which characterises HHR bounded symmetric domains, reveals the connection between the rank of a symmetric domain and the extent to which a Hilbert ball can be squeezed inside it.

Theorem 3.7.15. *Let D be a bounded symmetric domain in a complex Banach space. Then D is HHR if and only if it is of finite rank. In this* *case, D is biholomorphic to a finite product*

$$D_1 \times \cdots \times D_k$$

of irreducible bounded symmetric domains and we have

$$\hat{\sigma}_D = \left(\frac{1}{\hat{\sigma}_{D_1}^2} + \cdots + \frac{1}{\hat{\sigma}_{D_k}^2} \right)^{-1/2}. \tag{3.65}$$

If $\dim D_j < \infty$, *then* D_j *is a classical Cartan domain or an exceptional domain, and* $\hat{\sigma}_{D_j} = 1/\sqrt{p_j}$ *where* p_j *is the rank of* D_j.

If $\dim D_j = \infty$, *then* D_j *is either a Lie ball or a Type I domain of finite rank* p_j. *For a Lie ball* D_j, *we have* $\hat{\sigma}_{D_j} = 1/\sqrt{2}$. *For a rank* p_j *Type I domain* D_j, *we have* $\hat{\sigma}_{D_j} = 1/\sqrt{p_j}$.

Proof. Let D be HHR, realised as the open unit ball of a JB*-triple V. Then V is linearly homeomorphic to some Hilbert space H. In particular, V is reflexive and hence D is of finite rank. Conversely, by Theorem 3.3.5, a finite-rank bounded symmetric domain D decomposes into a finite Cartesian product $D = D_1 \times \cdots \times D_k$ of irreducible bounded symmetric domains, where each D_j is of finite rank p_j and realised as the open unit ball of a Cartan factor V_j for $j = 1, \ldots, k$.

To complete the proof, we show that each domain D_j of rank p_j has squeezing constant $\hat{\sigma}_{D_j} = 1/\sqrt{p_j}$ and $\hat{\sigma}_D = (p_1 + \cdots + p_k)^{-1/2}$.

By Corollary 3.7.14, we have $\hat{\sigma}_{D_j} = 1/\sqrt{p_j}$ if $\dim V_j < \infty$. In fact, this is also the case even if V_j is infinite dimensional, in which case V_j is either a spin factor or the Type I Cartan factor $L(\mathbb{C}^\ell, K)$ with $\dim K = \infty > \ell$. We now compute the squeezing constant in these two cases.

First, let D_j be a Lie ball, that is, the open unit ball of a spin factor $(V, \|\cdot\|)$, which has rank 2. In this case, V is a Hilbert space with norm $\|\cdot\|_h$ satisfying

$$\|\cdot\| \leq \|\cdot\|_h \leq \sqrt{2}\|\cdot\|$$

(cf. (2.55) and (2.58)). This gives $\hat{\sigma}_{D_j} \geq 1/\sqrt{2}$. Making use of the map φ in (3.62) and analogous arguments in the proof of Proposition 3.7.13, one concludes that $\hat{\sigma}_{D_j} = \hat{\sigma}_{D_j}(0) = 1/\sqrt{2}$.

Next, let D_j be a Type I domain of rank p_j, realised as the open unit ball

$$D_j = \{T \in L(\mathbb{C}^{p_j}, K) : \|T\| < 1\}$$

of $L(\mathbb{C}^{p_j}, K)$ with $\dim K = \infty$. Equipped with the Hilbert-Schmidt norm $\| \cdot \|_2$, the vector space $L(\mathbb{C}^{p_j}, K)$ is a Hilbert space. Let $B = \{T \in L(\mathbb{C}^{p_j}, K) : \|T\|_2 < 1\}$ be its open unit ball. Since $\| \cdot \| \leq \| \cdot \|_2 \leq \sqrt{p_j} \| \cdot \|$, we have $B \subset D_j \subset \sqrt{p_j} B$ and therefore $\hat{\sigma}_{D_j}(0) \geq 1/\sqrt{p_j}$. As before, using the map φ in (3.61) and similar arguments, we deduce $\hat{\sigma}_{D_j} = \hat{\sigma}_{D_j}(0) = 1/\sqrt{p_j}$.

It remains to establish (3.65). The domain $D = D_1 \times \cdots \times D_k$ is the open unit ball of the ℓ_∞-sum

$$V_1 \oplus \cdots \oplus V_k$$

of Cartan factors, where D_j is the open unit ball of V_j of rank p_j for $j = 1, \ldots, k$. We observe from the previous arguments that for each domain D_j, one can construct a continuous map $\varphi_j : \overline{\mathbb{D}}^{p_j} \to \overline{D}_j$ which restricts to a holomorphic map from $\mathbb{D}^{p_j}$ to D_j satisfying $\varphi_j(0) = 0$ and $\varphi_j(\partial \mathbb{D}^{p_j}) \subset \partial D_j$. Hence the product map

$$\varphi := \varphi_1 \times \cdots \times \varphi_k : \overline{\mathbb{D}}^{p_1} \times \cdots \times \overline{\mathbb{D}}^{p_k} \to \overline{D}_1 \times \cdots \times \overline{D}_k = \overline{D}$$

is continuous, which restricts to a holomorphic map from $\mathbb{D}^{p_1} \times \cdots \times \mathbb{D}^{p_k}$ to $D_1 \times \cdots \times D_k$ satisfying $\varphi(0, \ldots, 0) = (0, \ldots, 0)$ and maps the boundary of $\mathbb{D}^{p_1} \times \cdots \times \mathbb{D}^{p_k}$ into the boundary of $D_1 \times \cdots \times D_k = D$. Applying Lemma 3.7.11 again, we deduce that

$$\hat{\sigma}_D \leq \frac{1}{\sqrt{p_1 + \cdots + p_k}} = \frac{1}{\sqrt{\hat{\sigma}_{D_1}^{-2} + \cdots + \hat{\sigma}_{D_k}^{-2}}}.$$

For each $j = 1, \ldots, k$, the previous arguments reveal that there is a Hilbert space H_j with open unit ball B_j such that

$$B_j \subset D_j \subset \sqrt{p_j} B_j.$$

Let B be the open unit ball of the Hilbert space direct sum $H_1 \oplus_2 \cdots \oplus_2 H_k$. Then we have

$$B \subset D_1 \times \cdots \times D_k \subset \sqrt{p_1 + \cdots + p_k}\, B.$$

This implies that

$$\hat{\sigma}_D \geq \frac{1}{\sqrt{p_1 + \cdots + p_k}}$$

which completes the proof. $\square$

We have noted before that finite dimensional bounded convex domains are HHR. This is of course false in infinite dimension, for instance, the open unit ball of ℓ_∞ is not HHR. However, an appropriate extension of this result would be the assertion that bounded convex domains in *isomorphs* of Hilbert spaces are HHR. It is unclear at present if this is true. We conclude this section by introducing a large class of bounded convex domains including the strongly convex domains, which we call *uniformly elliptic domains* and show that these domains are HHR in Hilbert spaces.

Recall that a finite dimensional bounded domain $D \subset \mathbb{C}^n$ with a C^2 boundary ∂D is called *strongly convex* if all normal curvatures of ∂D are positive (cf. [2, p.108]). Such a domain is a manifold with curvature pinched which entails the existence of two positive constants $R > r > 0$ such that for each $q \in \partial D$, there are two points q', q'' in D with the property that q is a common boundary point of the Euclidean balls $B_{\mathbb{C}^n}(q', r)$ and $B_{\mathbb{C}^n}(q'', R)$ satisfying $B_{\mathbb{C}^n}(q', r) \subset D \subset B_{\mathbb{C}^n}(q'', R)$. For fixed r and R, it can be seen that q' and q'' are unique and colinear with q. For instance, an ellipsoid is strongly convex and has this property.

In view of the fact that Hilbert balls are the only bounded symmetric domains with a C^2 boundary, we generalise the concept of strong convexity to infinite dimension without the assumption of a smooth boundary, to cover a wider class of domains, as follows.

Definition 3.7.16. A bounded convex domain Ω in a complex Banach space V is called *uniformly elliptic* if there exist universal constants r, R with $0 < r < R$ such that to each $q \in \partial\Omega$, there correspond two unique points $q', q'' \in \Omega$, colinear to q, satisfying

(3.7.16.1) $B_V(q', r) \subset \Omega \subset B_V(q'', R)$;

(3.7.16.2) $q \in \partial B_V(q', r) \cap \partial B_V(q'', r)$, that is, q is a common
boundary point of $B_V(q', r), B_V(q'', R)$ and Ω.

Evidently, the definition of uniform ellipticity depends on the norm of the ambient Banach space. By the previous remarks, strongly convex domains are uniformly elliptic, but the converse is false. In fact, all open balls in Banach spaces are uniformly elliptic. Indeed, if say, $\Omega = B_V$ is the open unit ball of a Banach space V, then for each boundary point $q \in \partial\Omega$, we have $\|q\| = 1$ and

$$B_V(q/2, 1/2) \subset \Omega = B_V(0, 1)$$

and $q \in \partial B_V(\frac{q}{2}, \frac{1}{2}) \cap \partial\Omega \cap \partial B_V(0, 1)$. For $R = 1$ and $r = 1/2$, the points $q' = q/2$ and $q'' = 0$ are unique and colinear to q.

By definition, each point p in a uniformly elliptic domain Ω in a Banach space V lies in the ball $B_V(q'', R)$ for all $q \in \partial\Omega$, as in (3.7.16.1) above, although p need not be colinear with q and q''. We consider the question of colinearity below.

Lemma 3.7.17. *Let Ω be a uniformly elliptic domain in a Banach space V and for each $q \in \partial\Omega$, let*

$$B_V(q', r) \subset \Omega \subset B_V(q'', R), \quad q \in \partial B_V(q', r) \cap \partial B_V(q'', r)$$

be as in the definition of uniform ellipticity. Then for each $p \in \Omega$ and $q \in \partial\Omega$ with $\|p - q\| = d(p, \partial\Omega)$, either p is colinear with q and q'' or, there exists $q_1 \in \partial\Omega$ such that p is colinear with q_1 and $q_1'' = q''$ satisfying $\|p - q_1\| = \|p - q\|$.

Proof. Let $q \in \partial\Omega \cap \partial B_V(q'', R)$ satisfy $\|p-q\| = d(p, \partial\Omega)$. Suppose p is not colinear with q and q''. We show the existence of q_1 in the lemma.

Consider $p \in \Omega \subset B_V(q'', R)$. Extend the (real) line through q'' and p to a point $q_1 \in \partial B_V(q'', R)$. Then we have $\|p - q_1\| = d(p, \partial B_V(q'', R)) \leq \|p - q\|$. We show that $q_1 \in \partial\Omega$, which would imply $\|p - q_1\| \geq \|p - q\|$ and complete the proof by uniqueness of q_1' and q_1''.

If $q_1 \notin \partial\Omega$, we deduce a contradiction. Since $q_1 \notin \overline{\Omega}$ and $p \in \Omega$, the line joining p and q_1 must intersect $\partial\Omega$ at some point ω, say. Now we have the contradiction

$$\|p - q\| \geq \|p - q_1\| > \|p - \omega\| \geq d(p, \partial\Omega) = \|p - q\|.$$

$\square$

Theorem 3.7.18. *Let Ω be a uniformly elliptic domain in a Hilbert space H. Then Ω is HHR.*

Proof. We need to show that the squeezing function σ_Ω of Ω has a strictly positive lower bound. Suppose, to the contrary, that there is a sequence (p_ν) in Ω such that

$$\lim_{\nu \to \infty} \sigma_\Omega(p_\nu) = 0. \tag{3.66}$$

We deduce a contradiction. By Lemma 3.7.7, we may assume, by choosing a subsequence, that $d(p_\nu, \partial\Omega)$ converges to 0 as $\nu \to \infty$ and one can find a boundary point $q_\nu \in \partial\Omega$ such that

$$\|q_\nu - p_\nu\| = d(p_\nu, \partial\Omega) > 0.$$

Write $\lambda_\nu = d(p_\nu, \partial\Omega)$ and let

$$B_V(q'_\nu, r) \subset \Omega \subset B_V(q''_\nu, R), \quad q_\nu \in \partial B_V(q'_\nu, r) \cap \partial B_V(q''_\nu, R)$$

be as in the definition of uniformly ellipticity of Ω where, by Lemma 3.7.17, q_ν can be chosen so that p_ν lies on the line through q_ν and q''_ν.

We complete the proof by a contradiction that there is a subsequence $(p_{\nu'})$ of (p_ν) and a constant $\delta > 0$ satisfying

$$\sigma_\Omega(p_{\nu'}) > \delta \quad \text{for all } \nu'.$$

In fact, δ depends only on r and R.

For each ν, we define a holomorphic embedding

$$\Phi \circ L_\nu : \Omega \to H$$

as follows. Let $\mathbf{e}^1$ be the unit vector

$$\mathbf{e}^1 := \frac{q''_\nu - q_\nu}{\|q''_\nu - q_\nu\|}.$$

We have

(S1) $q''_\nu = R\mathbf{e}^1 + q_\nu$, $q'_\nu = r\mathbf{e}^1 + q_\nu$,

(S2) $p_\nu = \lambda_\nu \mathbf{e}^1 + q_\nu$ $(\lambda_\nu \to 0$ as $\nu \to \infty)$.

Since $\sigma_\Omega(p_\nu) = \sigma_{\Omega - q_\nu}(p_\nu - q_\nu)$, taking a translation, we may assume $q_\nu = 0$. Then we have

(S1^0) $q''_\nu = R\mathbf{e}^1$, $\varphi q'_\nu = r\mathbf{e}^1$,

(S2^0) $p_\nu = \lambda_\nu \mathbf{e}^1$.

We now have

$$p_\nu = \lambda_\nu \mathbf{e}^1 \in B_H(r\mathbf{e}^1, r) \subset \Omega \subset B_H(R\mathbf{e}^1, R) \tag{3.67}$$

where

$$q'_\nu = r\mathbf{e}^1, \quad q''_\nu = R\mathbf{e}^1.$$

Extend $\{\mathbf{e}^1\}$ to an orthonormal basis $\{\mathbf{e}^\gamma\}_{\gamma \in \Gamma}$ in H. For each $z \in H$, we will write

$$z = \sum_{\gamma \in \Gamma} z_\gamma \mathbf{e}^\gamma = z_1 \mathbf{e}^1 + \sum_{\gamma \neq 1} z_\gamma \mathbf{e}^\gamma$$

with $z_\gamma \in \mathbb{C}$. We have

$$z \in B_H(r\mathbf{e}^1, r) \Leftrightarrow \|z - r\mathbf{e}^1\| < r$$

where

$$\|z - r\mathbf{e}^1\|^2 = |z_1 - r|^2 + \sum_{\gamma \neq 1} |z_\gamma|^2 = |z_1|^2 - 2r\,\mathrm{Re}\,z_1 + r^2 + \sum_{\gamma \neq 1} |z_\gamma|^2. \quad (3.68)$$

We definite a dilation $L_\nu : H \to H$ by

$$L_\nu(z) = \frac{z_1}{\lambda_\nu} \mathbf{e}^1 + \frac{1}{\sqrt{\lambda_\nu}} \sum_{\gamma \neq 1} z_\gamma \mathbf{e}^\gamma, \quad z = \sum_{\gamma \in \Gamma} z_\gamma \mathbf{e}^\gamma$$

which satisfies $L_\nu(p_\nu) = \mathbf{e}^1$. The map L_ν is a linear homeomorphism of H, with inverse

$$L_\nu^{-1}(z) = \lambda_\nu z_1 \mathbf{e}^1 + \sqrt{\lambda_\nu} \sum_{\gamma \neq 1} z_\gamma \mathbf{e}^\gamma.$$

Define a Cayley transform $\Phi : \{z \in H : \mathrm{Re}\,z_1 > 0\} \to H$ by

$$\Phi(z) := \frac{z_1 - 1}{z_1 + 1} \mathbf{e}^1 + \sum_{\gamma \neq 1} \frac{\sqrt{2} z_\gamma}{z_1 + 1} \mathbf{e}^\gamma, \quad z = \sum_\gamma z_\gamma \mathbf{e}^\gamma$$

and the holomorphic embedding

$$\Phi \circ L_\nu : \Omega \to H$$

where $\Phi(L_\nu(p_\nu)) = 0$. Although Φ depends on ν, we omit the subscript ν indicating this, to simplify notation, since confusion is unlikely.

We will show that

$$B_H\left(0, \sqrt{\frac{r}{2+2r}}\,\right) \subset \Phi(L_\nu(\Omega)) \subset B_H(0, \sqrt{1+R})$$

for sufficiently large ν.

Substituting R for r in (3.68), we see that

$$\begin{aligned} B_H(R\mathbf{e}^1, R) &= \{z \in H : \|z - R\mathbf{e}^1\|^2 < R^2\} \\ &= \{z \in H : \sum_{\gamma \in \Gamma} |z_\gamma|^2 < 2R\,\mathrm{Re}\,z_1\}. \end{aligned}$$

Given $\zeta = \sum_\gamma \zeta_\gamma \mathbf{e}^\gamma \in B_H$, we have

$$\Phi^{-1}(\zeta) = \frac{1+\zeta_1}{1-\zeta_1}\mathbf{e}^1 + \sum_{\gamma \neq 1} \frac{\sqrt{2}\zeta_\gamma}{1-\zeta_1}\mathbf{e}^\gamma.$$

Hence

$$\begin{aligned} &\zeta \in \Phi L_\nu(B_H(R\mathbf{e}^1, R)) \Leftrightarrow L_\nu^{-1}\Phi^{-1}\zeta \in B_H(R\mathbf{e}^1, R) \\ &\Leftrightarrow \lambda_\nu\left(\frac{1+\zeta_1}{1-\zeta_1}\right)\mathbf{e}^1 + \sum_{\gamma \neq 1} \frac{\sqrt{2}\lambda_\nu}{1-\zeta_1}\mathbf{e}^\gamma \in B_H(R\mathbf{e}^1, R) \\ &\Leftrightarrow \alpha\lambda_\nu^2|1+\zeta_1|^2 + 2\lambda_\nu \sum_{\gamma \neq 1}|\zeta_\gamma|^2 < 2R\lambda_\nu(1-|\zeta_1|^2) \\ &\Rightarrow \sum_{\gamma \neq 1}|\zeta_\gamma|^2 < R(1-|\zeta_1|^2) \\ &\Rightarrow \|\zeta\|^2 = |\zeta_1|^2 + \sum_{\gamma \neq 1}|\zeta_\gamma|^2 < 1 + R \end{aligned}$$

and therefore we have, by (3.67),

$$\Phi L_\nu(\Omega) \subset \Phi L_\nu(B_H(R\mathbf{e}^1, R)) \subset B_H(0, \sqrt{1+R}). \tag{3.69}$$

We now show that $B_H(0, \sqrt{\frac{r}{2+2r}}) \subset \Phi L_\nu(\Omega)$ for sufficiently large ν. For this, we will make use of the inclusion $B_H(r\mathbf{u}, r) \subset \Omega$.

We have $L_\nu^{-1}\Phi^{-1}(\zeta) \in B_H(r\mathbf{u}, r)$ if and only if $\|L_\nu^{-1}\Phi^{-1}(\zeta) - r\mathbf{u}\| < r$, where

$$\|L_\nu^{-1}\Phi^{-1}(\zeta) - r\mathbf{u}\|^2 < r^2$$

$$\Leftrightarrow \left| \lambda_\nu \left(\frac{1+\zeta_1}{1-\zeta_1} \right) - r \right|^2 + \frac{2\lambda_\nu}{|1-\zeta_1|^2} \sum_{\gamma \neq 1} |\zeta_\gamma|^2 < r^2$$

$$\Leftrightarrow \lambda_\nu (\lambda_\nu |1+\zeta_1|^2 - 2r(1-|\zeta_1|^2) + 2\sum_{\gamma \neq 1} |\zeta_\gamma|^2) < 0. \quad (3.70)$$

For $\zeta \in B_H(0, \sqrt{\frac{r}{2+2r}})$, we have $2r - (2r|\zeta_1|^2 + 2\|\zeta\|^2) > r$ and $|1+\zeta_1|^2 \leq \left(1 + \sqrt{\frac{r}{2+2r}}\right)^2$. Since $\lambda_\nu \to 0$ as $\nu \to \infty$, there exists ν_0 such that $\nu \geq \nu_0$ implies

$$\lambda_\nu < \frac{r}{2\left(1 + \sqrt{\frac{r}{2+2r}}\right)^2}$$

and hence

$$\lambda_\nu |1+\zeta_1|^2 - 2r(1-|\zeta_1|^2) + 2\sum_{\gamma \neq 1} |\zeta_\gamma|^2$$

$$\leq \lambda_\nu |1+\zeta_1|^2 - 2r + 2r|\zeta_1|^2 + 2\|\zeta\|^2$$

$$< \frac{r}{2\left(1 + \sqrt{\frac{r}{2+2r}}\right)^2} |1+\zeta_1|^2 - r < -r/2$$

which gives $\lambda_\nu (\lambda_\nu |1+\zeta_1|^2 - 2r(1-|\zeta_1|^2) + 2\sum_{\gamma \neq 1} |\zeta_\gamma|^2) < -r\lambda_\nu/2 < 0$ and by (3.70),

$$\|L_\nu^{-1}\Phi^{-1}(\zeta) - r\mathbf{u}\|^2 < r^2.$$

We have therefore shown that, for $\nu \geq \nu_0$, the inclusions

$$B_H(0, \sqrt{r/(2+2r)}) \subset \Phi L_\nu(B_H(r\mathbf{e}^1, r)) \subset \Phi L_\nu(\Omega)$$

are satisfied.

Now it follows from this and (3.69) that

$$\sigma_\Omega(p_\nu) \geq \sqrt{\frac{r}{2(1+r)(1+R)}} > 0$$

for all $\nu \geq \nu_0$, which contradicts $\lim_{\nu \to \infty} \sigma_\Omega(p_\nu) = 0$ and completes the proof.   □

Notes. HHR manifolds can be regarded as a generalisation of Teichmüller spaces $\mathcal{T}$ of Riemann surfaces. By Bers embedding theorem, given $p \in \mathcal{T}$, there is an embedding $f_p : \mathcal{T} \longrightarrow \mathbb{C}^d$ so that $f_p(p) = 0$ and $f_p(\mathcal{T})$ is sandwiched between two Euclidean balls of positive radii, which provides the uniform squeezing property (see the proof of [119, Theorem 7.1]). We refer to Siu's lecture [159] for a useful survey of the classical Levi problem and its generalisations. Proposition 3.7.5 and its proof are direct generalisation of the finite dimensional result and arguments in [53, Theorem 3.1]. Theorem 3.7.18 generalises the finite dimensional result for strongly convex domains in [182, Proposition 1]. Apart from Proposition 3.7.8, all results in this section have been proven in [46].

3.8   Classification

We discuss a Jordan approach to the classification of bounded symmetric domains, up to biholomorphisms. We have already formulated in Section 3.3 É. Cartan's classification of finite dimensional bounded symmetric domains in terms of Cartan factors, which has been extended in Corollary 3.3.7 to the classification of finite-rank bounded symmetric domains. However, the formulation in Section 3.3 depends on some results to be proved in this section. We will now see in detail that Cartan's classification can be viewed as a special case of the *Jordan* classification developed in what follows.

Our approach is based on the realisation of bounded symmetric domains as open unit balls of JB*-triples and the following classification theorem, already hinted by the remarks before Theorem 3.2.22.

Theorem 3.8.1. *Let V and W be JB*-triples with open unit balls B_V and B_W respectively. Then B_V is biholomorphic to B_W if and only if V is isometrically triple isomorphic to W.*

Proof. This follows from Theorem 3.2.22 and the remarks preceding it. □

Classifying bounded symmetric domains, therefore, amounts to classifying JB*-triples. This is a formidable task in infinite dimension. Indeed, classifying the subclass of C*-algebras is already a vast enterprise. Nevertheless, analogous to Murray and von Neumann's classification of von Neumann algebras, one can still classify (to some extent) a large class of JB*-triples, namely, the JBW*-triples. This enables us to classify bounded symmetric domains in dual Banach spaces.

Lemma 3.8.2. *A bounded symmetric domain in a dual Banach space is biholomorphic to the open unit ball of a JBW*-triple.*

Proof. We recall that a JBW*-triple is a JB*-triple which is a dual Banach space. Let D be a bounded symmetric domain in a dual Banach space V. Then V can be equipped with an equivalent norm $\|\cdot\|$ so that $(V, \|\cdot\|)$ is a JB*-triple and D is biholomorphic to the open unit ball of $(V, \|\cdot\|)$. Since V is a dual Banach space, it is a JBW*-triple. □

Definition 3.8.3. A JBW*-triple W is called a *JBW*-factor* or simply, a *factor,* if it does not contain any weak* closed triple ideal other than $\{0\}$ and W.

Cartan factors are indeed JBW*-factors. We note that, however, a factor *can* contain non-trivial *norm* closed triple ideal. For example, the factor $L(H)$ of bounded operators on a Hilbert space H contains the norm closed ideal $K(H)$ of compact operators on H.

A JB*-triple V is called *simple* if does not contain any *norm closed* triple ideal other than $\{0\}$ and V. The preceding remark says that the JBW*-factor $L(H)$ is not a simple JBW*-triple.

To verify if a *norm closed* subspace J of a JB*-triple V is a triple ideal, it suffices to verify the inclusion $\{V, J, J\} \subset J$ or $\{V, J, V\} \subset J$, by [28, Proposition 1.3]. It follows that the weak* closure of a triple ideal in a JBW*-triple is also a triple ideal since the triple product is separately weak* continuous in JBW*-triples (cf. [37, Theorem 3.3.9]).

Definition 3.8.4. Given an element u in a JBW*-triple V, we denote by $J(u)$ the smallest weak* closed triple ideal of V containing u. A tripotent $u \in V$ is called *abelian* if the Peirce 2-space $V_2(u)$ is an abelian Jordan triple system as defined in Definition 2.2.13.

Remark 3.8.5. A minimal tripotent u in a JBW*-triple is clearly abelian. We take this opportunity of correcting a misprint in the definition of abelian tripotents in [37, p. 211] where the triple ideal $J(u)$ should be replaced by the Peirce 2-space $V_2(u)$.

Lemma 3.8.6. *Let $\mathcal{A}$ be a von Neumann algebra. If $\mathcal{A}$ is an abelian JBW*-triple, then it is an abelian, that is, a commutative, algebra.*

Proof. It suffices to show $pq = qp$ for two projections p and q in $\mathcal{A}$, by Example 2.4.7. Let $e \in \mathcal{A}$ be the identity. By assumption, we have

$$\{p, q, \{p, e, e\}\} = \{p.\{q, p, e\}, e\} \tag{3.71}$$

which simplifies to $pqp = \frac{1}{2}(pq + qp)$ and therefore $pqp = qp$. Changing the role of p and q in (3.71), we get $qpq = pq$ and hence $pq = qp$. □

Definition 3.8.7. A JBW*-triple W is called *discrete* if it admits an abelian tripotent u such that $W = J(u)$.

Remark 3.8.8. A discrete JBW*-triple is also known as a *type I JBW*-triple* [86]. To avoid confusion with the notion of a Type I Cartan factor, we will not use this terminology. Our nomenclature, however, is consistent with that in operator algebras where type I von Neumann algebras are also called discrete von Neumann algebras (cf. [154, 2.2.9]).

Following a common approach in classification theory, we decompose a JBW*-triple into *simpler* ones to classify. We first decompose a JBW*-triple into two parts, one atomic and the other non-atomic. For this, we need two lemmas.

Lemma 3.8.9. *Let J be a weak* closed triple ideal in a JBW*-triple V. Then there is a weak* closed triple ideal $J^\square$ in V such that $J^\square \,\square\, J = \{0\}$ and $V = J \oplus J^\square$.*

Proof. We have noted after Theorem 3.4.8 that JBW*-triples contain a maximal tripotent. Let u be a maximal tripotent in J. Then u is a tripotent in V. Let $P_j(u) : V \longrightarrow V$ be the Peirce projections, $j = 0, 1, 2$. For $x = x_1 + x_2 \in P_1(u)V + P_2(u)V$, we have $jx_j = 2\{u, u, x\} \in J$ and hence

$$J = P_1(u)V + P_2(u)V.$$

Let $J^\square = P_0(u)V$ which is a weak* closed subtriple of V. We have $J^\square \,\square\, P_2(u)V = \{0\}$. For $y \in J^\square$ and $z \in P_1(u)V$, we have $\{z, z, y\} \in P_0(u)V \cap J = \{0\}$. Hence $J^\square \,\square\, J = \{0\}$ and it follows that

$$\{V, J^\square, J^\square\} = \{J \oplus J^\square, J^\square, J^\square\} \subset J^\square$$

that is, $J^\square$ is a triple ideal in V. $\qquad\square$

Remark 3.8.10. In the above lemma, it is easy to see that $J^\square = \{v \in V : v \,\square\, J = \{0\}\}$ since the latter set contains $J^\square$ and has intersection $\{0\}$ with J.

Lemma 3.8.11. *Let u be a minimal tripotent in a JBW*-triple V. Then the smallest weak* closed triple ideal $J(u)$ in V containing u is a discrete factor. If $v \in V$ is another minimal tripotent, then either $J(u) = J(v)$ or $J(u) \square J(v) = \{0\}$.*

Proof. If J is a non-zero proper weak* closed triple ideal in $J(u)$, then Lemma 3.8.9 implies $J(u) = J \oplus J^{\square}$ for some non-zero proper weak* closed triple ideal $J^{\square}$ with $J^{\square} \square J = \{0\}$. Hence $u = u_1 + u_2 \in J \oplus J^{\square}$ and u_1, u_2 must be tripotents, contradicting minimality of u. Therefore $J(u)$ is a discrete factor. The second assertion is an immediate consequence. $\square$

Lemma 3.8.12. *Let W_1 and W_2 be discrete JBW*-triples. Then their ℓ_∞-sum $W_1 \oplus W_2$ is a discrete JBW*-triple.*

Proof. Let $W_j = J(u_j)$ for $j = 1, 2$, where u_j is an abelian tripotent in W_j. Then $u_1 \oplus u_2$ is an abelian tripotent in $W_1 \oplus W_2$. Let $J(u_1 \oplus u_2)$ be the smallest weak* closed triple ideal in $W_1 \oplus W_2$ containing $u_1 \oplus u_2$. We show $W_1 \oplus W_2 = J(u_1 \oplus u_2)$, which would complete the proof.

First observe that $u_1 \oplus 0 = \{u_1 \oplus 0, u_1 \oplus u_2, u_1 \oplus u_2\} \in J(u_1 \oplus u_2)$. Let $P_1 : W_1 \oplus W_2 \longrightarrow W_1$ be the natural projection, which is a triple homomorphism. Then the weak* closure $\overline{P_1(J(u_1 \oplus u_2))}$ is a weak* closed triple ideal in W_1 containing u_1 and hence $\overline{P_1(J(u_1 \oplus u_2))} \supset J(u_1)$.

Let $x_1 \oplus x_2 \in J(u_1 \oplus u_2)^{\square}$. Then for each $y \in J(u_1)$, there is a net $(a_\alpha \oplus b_\alpha) \in J(u_1 \oplus u_2)$ such that (a_α) weak* converges to y in W_1. Since x_1 is triple orthogonal to each a_α, it is also triple orthogonal to y. As $y \in J(u_1)$ was arbitrary, we have shown that $x_1 \in J(u_1)^{\square} = \{0\}$. Likewise, $x_2 = 0$. It follows that $J(u_1 \oplus u_2)^{\square} = \{0\}$ and hence $J(u_1 \oplus u_2) = W_1 \oplus W_2$. $\square$

Let V be a JBW*-triple. It follows from Lemma 3.8.9, Lemma

3.8.11 and Corollary 3.2.24 that V can be decomposed into an ℓ_∞-sum

$$V = \left(\bigoplus_u J(u)\right) \oplus \left(\bigoplus_u J(u)\right)^\square \tag{3.72}$$

of two weak* closed triple ideals, where the first summand sums over all minimal tripotents u in V and is called the *atomic part* of V. It is an ℓ_∞-sum of discrete factors. The second summand does not contain any minimal tripotent, called the *non-atomic part* of V. A JBW*-triple is called *atomic* if its non-atomic part vanishes and by (3.72), it is an ℓ_∞-sum of discrete JBW*-factors. On the other hand, a JBW*-triple may not contain any minimal tripotent in which case, its atomic part vanishes and it is called *non-atomic* when this happens.

Corollary 3.8.13. *A JB*-triple V is triple isomorphic to a closed subtriple of an ℓ_∞-sum of discrete factors.*

Proof. By (3.72), the second dual V^{**} is an ℓ_∞-sum

$$V^{**} = V_a \oplus V_a^\square$$

of two weak* closed triple ideals in which V_a is an ℓ_∞-sum of discrete factors. Let $P : V^{**} \longrightarrow V_a$ be the canonical contractive projection and $\widehat{} : V \longrightarrow V^{**}$ the canonical embedding. Both maps are triple homomorphisms and therefore $P(\widehat{V})$ is a closed subtriple of V_a. If $P(\widehat{v}) = 0$, then $\widehat{v} \in V_a^\square$. For each extreme point f in the closed unit ball of V^*, there exists a unique minimal tripotent $u_f \in V^{**}$ such that $f(u_f) = 1$, by [37, Lemma 3.3.15], which implies $u_f \in V_a$ and

$$f(\widehat{v}) = f\{u_f, \widehat{v}, u_f\} = 0$$

(see [37, (3.10)]). Hence $\widehat{v} = 0$ and $P \circ \widehat{} : V \longrightarrow P(\widehat{V})$ is a triple isomorphism. $\square$

By a result of Horn and Neher [88], we can decompose further the non-atomic part of a JBW*-triple V in (3.72) (via a triple isomorphism) into an ℓ_∞-sum as follows.

$$V = \left(\bigoplus_u J(u) \right) \oplus V_1 \oplus \mathcal{R} \oplus H(\mathcal{A}, \beta) \tag{3.73}$$

where V_1 is a (non-atomic) discrete JBW*-triple, $\mathcal{R}$ is a weak* closed right ideal of a continuous von Neumann algebra and

$$H(\mathcal{A}, \beta) = \{a \in \mathcal{A} : \beta(a) = a\}$$

for some continuous von Neumann algebra $\mathcal{A}$, with a linear involutive *-antiautomorphism $\beta : \mathcal{A} \longrightarrow \mathcal{A}$. The second and third summand inherit the Jordan triple product from the underlying von Neumann algebra. Of course, any summand in (3.73) may vanish.

We recall that a von Neumann algebra $\mathcal{A}$ is said to be *continuous* if it does not contain a non-zero projection p such that the algebra $p\mathcal{A}p$ is commutative in which case, viewed as a JBW*-algebra, $\mathcal{A}$ does not contain any non-zero abelian tripotent and is therefore non-discrete, by the following lemma.

Lemma 3.8.14. *Let e be an abelian tripotent in a von Neumann algebra $\mathcal{A}$. Then the algebra $ee^*\mathcal{A}ee^*$ is commutative.*

Proof. Write $p = ee^*$ and $q = e^*e$, which are projections in $\mathcal{A}$. We have $qe^* = e^*$ and the Peirce 2-space $\mathcal{A}_2(e) = p\mathcal{A}q$ is an abelian JBW*-triple (cf. Example 2.4.4). It can be seen that $p\mathcal{A}q$ is triple isomorphic to the JBW*-triple $p\mathcal{A}qe^* = p\mathcal{A}e^*$, which is therefore abelian. Since $p\mathcal{A}p$ is a JBW*-subtriple of $p\mathcal{A}e^*$, it is also an abelian JBW*-triple and hence a commutative algebra, by Lemma 3.8.6. $\square$

In the decomposition (3.73), one can classify the first two summands which are discrete. This has been carried out by Horn in [86, (1.7)], which is stated below.

Theorem 3.8.15. *A JBW*-triple is discrete if and only if it is triple isomorphic to an ℓ_∞-sum*

$$\bigoplus_\alpha C(\Omega_\alpha, C_\alpha)$$

where Ω_α is a hyperstonean space, C_α is a discrete factor and $C(\Omega_\alpha, C_\alpha)$ is the complex Banach space of C_α-valued continuous functions on Ω_α, which is a JBW*-triple in the pointwise triple product

$$\{f, g, h\}(\omega) = \{f(\omega), g(\omega), h(\omega)\} \qquad (f, g, h \in C(\Omega_\alpha, C_\alpha), \omega \in \Omega_\alpha).$$

We recall that a *stonean space* is a compact (Hausdorff) extremely disconnected topological space, and a stonean space Ω is *hyperstonean* if and only if $C(\Omega)$ is a von Neumann algebra (cf. [154, p. 46]).

To classify finite dimensional bounded symmetric domains, it suffices to classify the irreducible ones. For infinite dimensional irreducible bounded symmetric domains in dual Banach spaces, we have the following classification theorem.

Theorem 3.8.16. *A bounded symmetric domain D in a dual Banach space is irreducible if and only if it is biholomorphic to the open unit ball of a JBW*-factor.*

Proof. Let D be realised as the open unit ball of a JBW*-triple V. If V admits a non-trivial weak* closed triple ideal J. Then V decomposes into an ℓ_∞-direct sum

$$V = J \oplus_\infty J^\square$$

of two non-trivial JBW*-triples J and $J^\square$. It follows that D is biholomorphic to the product of the open unit balls of J and $J^\square$, contradicting irreducibility of D.

Conversely, let D be the open unit ball of a JBW*-factor. If D is biholomorphic to the product $D_1 \times D_2$ of two bounded symmetric domains D_1 and D_2, which are realised as the open unit balls of the JBW*-triples V_1 and V_2 respectively, then $D \approx D_1 \times D_2$ is the open unit ball of the ℓ_∞-sum $V_1 \oplus_\infty V_2$, which is not a factor since $V_1 \oplus \{0\}$ is a non-trivial weak*-closed triple ideal. This contradicts the assumption that D is the open unit ball of a factor.   □

Hence JBW*-factors can be used to classify irreducible bounded symmetric domains in dual Banach spaces, including all finite dimensional irreducible ones.

We see from (3.73) and Theorem 3.8.15 that a JBW*-factor must be one of the following forms:

$$\text{(i)} \ \ C(\Omega_\alpha, C_\alpha) \quad \text{(ii)} \ \ \mathcal{R} \quad \text{(iii)} \ \ H(\mathcal{A}, \beta)$$

where the hyperstonean space Ω_α reduces to a singleton since for each $\omega \in \Omega_\alpha$, the subspace $J = \{f \in C(\Omega_\alpha, C_\alpha) : f(\omega) = 0\}$ is a weak* closed triple ideal in $C(\Omega_\alpha, C_\alpha)$. Hence $C(\Omega_\alpha, C_\alpha)$ identifies with the discrete factor C_α. Examples of continuous von Neumann algebras can be constructed by Murray and von Neumann's group measure space construction. Let G be an infinite countable discrete group with identity e, in which all conjugacy classes

$$\{hgh^{-1} : h \in G\} \quad (g \in G\backslash\{e\})$$

except $\{e\}$, are infinite. For instance, G can be the free group of two or more generators. Let $L_2(G)$ be the Hilbert space of square integrable functions on G with respect to the left Haar measure and let $\rho : G \longrightarrow L(L_2(G))$ be the left regular representation defined by

$$\rho(x)f(y) = f(x^{-1}y) \qquad (x, y \in G, f \in L(L_2(G))).$$

The group von Neumann algebra $\mathcal{M}(G)$ is defined to be the von Neumann algebra generated by $\rho(G)$ in the von Neumann algebra $L(L_2(G))$

of bounded linear operators on $L_2(G)$. Pick a projection $p \in \mathcal{M}(G)$. Then $\mathcal{R} := p\mathcal{M}(G)$ is an example for (ii) above.

A JBW*-factor must be discrete if it contains a minimal tripotent, by Lemma 3.8.11. In particular, Cartan factors are discrete. Hence all finite-rank JBW*-triples are discrete since they are finite ℓ_∞-sums of Cartan factors, by Theorem 3.3.5.

To conclude the section, we classify the discrete JBW*-factors. It turns out that they are exactly the Cartan factors and in view of Theorem 3.8.16 and Corollary 3.8.18 below, we have extended Cartan's classification to irreducible bounded symmetric domains in dual Banach spaces. The following result has been proved by Horn in [86, (1.8)].

Theorem 3.8.17. *A JBW*-factor is discrete if and only if it is a Cartan factor.*

Proof. The proof for the necessity is lengthy. We sketch the main steps, following the *grid approach* in [52].

Let V be a discrete JBW*-factor. Then there is an abelian tripotent $e \in V$ such that $V = J(e)$. Since V is a factor, e must be a minimal tripotent, for otherwise we can find two non-zero mutually orthogonal tripotents u and v such that $e = u + v$ and $J(u)$ is a proper weak* closed triple ideal in V, which is impossible.

The next step is to show that the rank of the Peirce 1-space $V_1(e)$ is at most 2. Indeed, given a non-zero tripotent $v \in V_1(e)$, if it is not already a minimal tripotent in $V_1(e)$, then v is the sum

$$v = v_1 + v_2$$

of two mutually orthogonal tripotents $v_1, v_2 \in V_1(e)$. We need only show that v_1 and v_2 are minimal tripotents.

Take v_1 say. We see that e lies outside the Peirce 2-space $V_2(v_1)$ of v_1 for otherwise $V_2(e) \subset V_2(v_1)$ which would imply v_2 is orthogonal to

e — impossible! But e is not in $V_0(v_1)$ either since this would imply e is orthogonal to $v_1 = 2\{e, e, v_1\}$.

We show $e \in V_1(v_1)$. Consider the Peirce decomposition

$$e = P_2(v_1)e + P_1(v_1)e + P_0(v_1)e$$

and apply to it the Peirce 2-projection of e, we obtain

$$e = P_2(e)e = P_2(e)P_2(v_1)e + P_2(e)P_1(v_1)e + P_2(e)P_0(v_1)e.$$

By the Peirce multiplication rules, we have

$$P_2(v_1)e = \{v_1, \{v_1, e, v_1\}, v_1\} \in V_2(e)$$

and hence

$$P_2(v_1)e = P_2(e)P_2(v_1)e = \lambda e$$

for some $\lambda \in \mathbb{C}$, as e is a minimal tripotent. But $P_2(v_1) : V \longrightarrow V$ is a contractive projection, we can only have $\lambda = 0$ or $\lambda = 1$. Since we have already establish $e \notin V_2(v_1)$, we must have $\lambda = 0$ and thus

$$e = P_2(e)P_1(v_1)e + P_2(e)P_0(v_1)e.$$

Further, we have $P_0(v_1)e = e - 2\{v_1, v_1, e\} + P_2(v_1)e = e - 2\{v_1, v_1, e\}$, where $\{v_1, v_1, e\} \in V_2(e)$ implies

$$P_0(v_1)e = P_2(e)P_0(v_1)e = e - 2\{v_1, v_1, e\} = \mu e \qquad (\mu \in \mathbb{C})$$

by minimality of e. Again, we have $\mu \in \{0, 1\}$ since $P_0(v_1)$ is a contractive projection. If $\mu = 1$, the preceding identity implies $\{v_1, v_1, e\} = 0$, contradicting $e \notin V_0(v_1)$. It follows that $\mu = 0$ and $\{v_1, v_1, e\} = \frac{1}{2}e$, that is, $e \in V_1(v_1)$.

By the Peirce multiplication rules again, we have

$$\{e, v_1, e\} = \{v_1, e, v_1\} = 0.$$

Hence

$$\{e + v_1, e + v_1, e + v_1\} = 2(e + v_1)$$

and $u := \frac{1}{\sqrt{2}}(e + v_1)$ is a tripotent. By a remark following (3.29), the Peirce symmetry $B(u, 2u) : V \longrightarrow V$ is a triple isomorphism and it follows from

$$B(u, 2u)e = v_1$$

that v_1 is a minimal tripotent. Likewise v_2 is a minimal tripotent. This proves

$$\operatorname{rank} V_1(e) \leq 2.$$

We now classify V case by case, depending on the rank of $V_1(e)$.

Case 0. $V_1(e) = \{0\}$ in which case, we also have $V_0(e) = \{0\}$ for otherwise, $V_2(e) = \mathbb{C}e$ would be a proper weak* closed triple ideal in $V = V_2(e) \oplus V_0(e)$, which is impossible. Hence $V \approx \mathbb{C}$.

Case 1. $\operatorname{rank} V_1(e) = 1$. Let v be a minimal tripotent in $V_1(e)$. Then there are two possibilities:

(a) $e \in V_1(v)$: it can be shown in this case that V is a Hilbert space, which is a Type I Cartan factor (cf. Example 2.4.9).

(b) $e \in V_2(v)$: it can be shown that V is a Type III Cartan factor.

Case 2. $\operatorname{rank} V_1(e) = 2$. Let v_1 and v_2 be two mutually orthogonal minimal tripotents in $V_1(e)$. Then we have $e \in V_1(v_1) \cap V_1(v_2)$. Let $\tilde{e} = \{v_1, e, v_2\}$. Then $\tilde{e}$ is a minimal tripotent orthogonal to e and there are six possibilities:

(c) $V_1(e + \tilde{e}) = \{0\}$: one can show that V is a spin factor.

(d) $V_1(e + \tilde{e}) \neq \{0\}$: in this case, $\dim V_2(e + \tilde{e}) = 4, 6, 8$ or 10 and the following possibilities occur:

(e) $\dim V_2(e + \tilde{e}) = 4$: one shows that V is a Type I Cartan factor.

(f) $\dim V_2(e + \tilde{e}) = 6$: one shows that V is a Type II Cartan factor.

(g) $\dim V_2(e + \tilde{e}) = 8$: one shows that V is a Type V Cartan factor.

(h) $\dim V_2(e + \tilde{e}) = 10$: one shows that V is a Type VI Cartan factor.

We refer to [52] for a complete proof of the above cases. $\qquad \square$

Finally, we arrive at a Jordan formulation of Cartan's classification.

Corollary 3.8.18. *A finite dimensional irreducible bounded symmetric domain is biholomorphic to the open unit ball of a Cartan factor.*

Proof. Let D be an irreducible bounded symmetric domain in $\mathbb{C}^n$. By Theorem 3.8.16, D is biholomorphic to the open unit ball of a JBW*-factor V, which is finite dimensional. By Theorem 2.2.34, V contains a minimal tripotent u. As noted before, V must be discrete since V is a JBW*-factor and we must have $V = J(u)$. By Theorem 3.8.17, V is a Cartan factor. $\qquad \square$

Corollary 3.8.19. *Let V be a JB*-triple. Then we have*

$$\|\{x, y, z\}\| \leq \|x\| \|y\| \|z\| \qquad (x, y, z \in V).$$

Proof. This is an immediate consequence of the preceding result and Corollary 3.8.13 since the inequality holds in all Cartan factors. $\qquad \square$

Notes. Corollary 3.8.13 has been proved in [65]. The group von Neumann algebra $\mathcal{M}(G)$ constructed before Theorem 3.8.16 is a well-known example of a type II_1 von Neumann algebra with a trivial centre.

Chapter 4

Function theory

4.1 The class $\mathcal{S}$

The beautiful theory of injective holomorphic functions, alias *univalent functions*, is central in complex function theory of one variable. By virtue of the Riemann mapping theorem, the study of these functions on simply connected domains can be transported to the open unit disc $\mathbb{D}$, where it is customary to formulate the theory, for instance, the Bieberbach conjecture [15]. Given a univalent function $h : \mathbb{D} \longrightarrow \mathbb{C}$, we must have $h'(0) \neq 0$ and the univalent function $f := (h - h(0))/h'(0)$ satisfies $f(0) = 0 = f'(0) - 1$, with a Taylor series expansion

$$f(z) = z + a_2 z^2 + \cdots + a_n z^n + \cdots \qquad (z \in \mathbb{D})$$

which is more convenient to handle. For this reason, it suffices to focus one's attention on the class of univalent functions f on $\mathbb{D}$ satisfying $f(0) = 0$ and $f'(0) = 1$. Such a function is called a (*normalised*) *shlicht function*, or a *normalised univalent function*, and the class of these functions is usually denoted by $\mathcal{S}$. The Bieberbach conjecture, proved by de Branges [22], says that $|a_n| \leq n$ for each $f(z) = z + a_2 z^2 + \cdots \in \mathcal{S}$.

A generalisation of the unit disc $\mathbb{D}$ to higher dimensions is the Euclidean balls. However, to confine the study of univalent functions in the setting of Euclidean balls in higher dimensions is of limited generality, which would exclude many interesting domains (e.g. polydiscs) since the Riemann mapping theorem fails in this case. In view of Cartan's classification of bounded symmetric domains and its infinite dimensional extension, bounded symmetric domains, which are biholomorphic to the open unit balls of JB*-triples, appear to be a more appropriate setting for extending the function theory of $\mathbb{D}$.

Let D be a bounded symmetric domain realised as the open unit ball of a JB*-triple V. A holomorphic map $f : D \longrightarrow V$ is called *normalised* if $f(0) = 0$ and the derivative $f'(0) : V \longrightarrow V$ is the identity map.

By the open mapping theorem, the image $f(\mathbb{D})$ of each schlicht function $f \in \mathcal{S}$ is a domain in $\mathbb{C}$ and $f : \mathbb{D} \longrightarrow \mathbb{C}$ is a holomorphic embedding, as defined in Definition 3.7.1. Generalising the class $\mathcal{S}$, we will denote by $\mathcal{S}(D)$ the class of normalised holomorphic embeddings of D in V. In particular, $\mathcal{S} = \mathcal{S}(\mathbb{D})$.

For each $f = z + a_2 z^2 + \cdots \in \mathcal{S}$, a well-known consequence of the inequality $|a_2| \leq 2$ is the Koebe distortion theorem:

$$\frac{1 - |z|}{(1 + |z|)^3} \leq |f'(z)| \leq \frac{1 + |z|}{(1 - |z|)^3} \qquad (f \in \mathcal{S})$$

which implies a growth estimate of f:

$$\frac{|z|}{(1 + |z|)^2} \leq |f(z)| \leq \frac{|z|}{(1 - |z|)^2} \qquad (f \in \mathcal{S}).$$

Although the Koebe distortion theorem is not true for all maps $f \in \mathcal{S}(B)$ on higher dimensional Euclidean balls B, there are numbers of growth and distortion results in literature for *convex* maps in $\mathcal{S}(B)$ and for some other domains (see, for example, [68]). A function $f \in \mathcal{S}(D)$ is called *convex* if its image $f(D)$ is a convex domain.

We shall only present two distortion theorems in this section for bounded symmetric domains to sample the Jordan connections. Further discussion of distortion results related to the Bloch constant will be given in the next section. First, the following growth and distortion results for convex maps have been proved in [75] and [76], respectively.

Lemma 4.1.1. *Let B be the open unit ball of a complex Banach space V. For each convex map $f \in \mathcal{S}(B)$, we have*

$$\frac{\|x\|}{1 + \|x\|} \leq \|f(x)\| \leq \frac{\|x\|}{1 - \|x\|} \qquad (x \in B). \tag{4.1}$$

Lemma 4.1.2. *Let B be the open unit ball of a complex Banach space V. For each convex map $f \in \mathcal{S}(B)$, we have*

$$\frac{1 - \|x\|}{1 + \|x\|} \mathcal{C}_B(x, y) \leq \|f'(x)y\| \leq \frac{1 + \|x\|}{1 - \|x\|} \mathcal{C}_B(x, y) \tag{4.2}$$

for each $x \in B$ and $y \in V$, where $\mathcal{C}_B(x, y)$ is the Carathéodory differential metric on B.

We prove the following distortion result for convex maps on bounded symmetric domains.

Theorem 4.1.3. *Let D be a bounded symmetric domain realised as the open unit ball of a JB*-triple V. For each convex map $f \in \mathcal{S}(D)$, we have*

(i) $\dfrac{1}{(1 + \|x\|)^2} \leq \|f'(x)\| \leq \dfrac{1}{(1 - \|x\|)^2} \qquad (x \in D).$

(ii) $\dfrac{(1 - \|x\|)\|y\|}{(1 + \|x\|)\|B(x,x)^{1/2}\|} \leq \|f'(x)y\| \leq \dfrac{\|y\|}{(1 - \|x\|)^2} \qquad (x \in D, y \in V)$

where $B(x, x)$ is the Bergmann operator on V.

Proof. For the lower bound in (i), we apply (4.2) which gives

$$\|f'(x)y\| \geq \frac{1 - \|x\|}{1 + \|x\|} \mathcal{C}_D(x, y) = \frac{1 - \|x\|}{1 + \|x\|} \|B(x,x)^{-1/2}y\| \quad (x \in D, y \in V).$$

Taking supremum over $y \in D$, we get

$$\|f'(x)\| \geq \frac{1 - \|x\|}{1 + \|x\|} \|B(x,x)^{-1/2}\| = \frac{1 - \|x\|}{(1 + \|x\|)(1 - \|x\|^2)} = \frac{1}{(1 + \|x\|)^2}.$$

The upper bound in (i) is a consequence of the right inequality in (ii), which follows immediately from (4.2):

$$\|f'(x)y\| \leq \frac{1 + \|x\|}{1 - \|x\|} C_D(x,y) = \frac{1 + \|x\|}{1 - \|x\|} \|B(x,x)^{-1/2}y\| \leq \frac{\|y\|}{(1 - \|x\|)^2}$$

by Lemmas 3.2.28 and 3.5.11.

For the left inequality in (ii), a simple application of (4.2) and Lemma 3.5.11 again gives

$$\|f'(x)y\| \geq \frac{1 - \|x\|}{1 + \|x\|} C_D(x,y) = \frac{1 - \|x\|}{1 + \|x\|} \|g'_{-x}(x)(y)\|,$$

where g_{-x} is the Möbius transformation on D. Since

$$\|y\| = \|B(x,x)^{1/2}g'_{-x}(x)(y)\| \leq \|B(x,x)^{1/2}\|\|g'_{-x}(x)(y)\|,$$

we have

$$\|f'(x)y\| \geq \frac{(1 - \|x\|)\|y\|}{(1 + \|x\|)\|B(x,x)^{1/2}\|}.$$

$\square$

Given an element x in the open unit ball D of a JB*-triple V, although the Bergman operator $B(x,x) : V \longrightarrow V$ need not be a hermitian operator [100, Example 4.5], its square root $B(x,x)^{1/2}$ is the exponential of a hermitian operator on V [37, (3.3)] and it follows from [20] that the norm of $B(x,x)^{1/2}$ equals its spectral radius. By Lemma 3.2.10, the spectrum of $B(x,x)^{1/2}$ is contained in $[0, 1]$ and therefore we have

$$\|B(x,x)\| \leq \|B(x,x)^{1/2}\|^2 \leq 1$$

and the lower bound in Theorem 4.1.3 (ii) is greater than $\frac{(1-\|x\|)\|y\|}{1+\|x\|}$. If D is a Hilbert ball and $x \in D$, we have

$$\|B(x,x)^{1/2}\|^2 = \|B(x,x)\| = \begin{cases} (1 - \|x\|^2)^2 & \text{if } D = \mathbb{D} \\ 1 - \|x\|^2 & \text{if } D \neq \mathbb{D} \end{cases} \quad (4.3)$$

which follows from Example 3.4.9 (cf. [37, Lemma 3.2.8]).

We now discuss two-point distortion of convex maps on bounded symmetric domains. We will use the estimates in (4.1) and arguments similar to those in the proof of [69, Theorem 7] to deduce the following result.

Theorem 4.1.4. *Let D be a bounded symmetric domain realised as the open unit ball of a JB*-triple V. For each convex map $f \in \mathcal{S}(D)$ and two distinct points $a, b \in D$, we have*

(i) $\|f(a) - f(b)\| \geq \dfrac{\sinh c_D(a,b)}{\exp c_D(a,b)} \max \left\{ \dfrac{1 - \|a\|^2}{\|f'(a)^{-1}\|}, \dfrac{1 - \|b\|^2}{\|f'(b)^{-1}\|} \right\}.$

(ii) $\|f(a) - f(b)\| \leq$
$\quad \sinh c_D(a,b) e^{c_D(a,b)} \min \left\{ \|f'(a)\| \|B(a,a)^{1/2}\|, \|f'(b)\| \|B(b,b)^{1/2}\| \right\}$

where c_D is the Carathéodory distance.

Proof. Using analogous arguments of the proof in [69, Theorem 7], one can show that

$$\|f(a) - f(b)\|$$
$$\geq \frac{\sinh c_D(a,b)}{\exp c_D(a,b)} \max\{\|(f'(a)g_a'(0))^{-1}\|^{-1}, \|(f'(b)g_b'(0))^{-1}\|^{-1}\}$$

where g_a and g_b are Möbius transformations. Since

$$\|g_x'(0)^{-1}\| = \|B(x,x)^{-1/2}\| = \frac{1}{1 - \|x\|^2}$$

by Lemmas 3.2.25 and 3.2.28, we have

$$\|(f'(x)g_x'(0))^{-1}\| \leq \|g_x'(0)^{-1}\| \|f'(x)^{-1}\| \leq \frac{\|f'(x)^{-1}\|}{1 - \|x\|^2}$$

for $x \in D$. This proves (i).

For the upper bound of $\|f(a) - f(b)\|$, define

$$F(x) = g_a'(0)^{-1} f'(a)^{-1} (f(g_a(x)) - f(a)) \qquad (x \in D).$$

Then F is a convex map in $\mathcal{S}(D)$ and hence (4.1) implies

$$\|F(x)\| \leq \frac{\|x\|}{1 - \|x\|} = \sinh c_D(x, 0) \exp c_D(x, 0) \qquad (x \in D)$$

where $2c_D(x, 0) = \log \frac{1 + \|x\|}{1 - \|x\|}$. Substituting $g_{-a}(b)$ for x gives

$$
\begin{aligned}
\|f(b) - f(a)\| &= \|f'(a) g_a'(0) F(g_{-a}(b))\| \leq \|f'(a) g_a'(0)\| \|F(g_{-a}(b))\| \\
&\leq \|f'(a)\| \|g_a'(0)\| \sinh c_D(g_{-a}(b), 0) \exp c_D(g_{-a}(b), 0) \\
&= \|f'(a)\| \|B(a, a)^{1/2}\| \sinh c_D(a, b) \exp c_D(a, b).
\end{aligned}
$$

Changing the roles of a and b yields the upper bound in (ii).   □

We will introduce Bloch maps in the following section and prove a distortion theorem for these functions in finite dimensions.

Notes.   A detailed treatment of univalent function theory and generalisations in several complex variables has been given in the book [70]. For a schlicht function $f(z) = z + a_2 z^2 + \cdots \in \mathcal{S}$, the inequality $|a_2| \leq 2$ was first proved by Bieberbach [15]. The inequality $|a_n| \leq n$ for $n = 3, 4, 5, 6$ were already proven by 1972, prior to de Branges's proof for all n. Theorems 4.1.3 and 4.1.4 are taken from [41]. The two-point distortion theorem in [105, Remark] for the unit disc $\mathbb{D}$ is a special case of Theorem 4.1.4.

4.2   Bloch constant and Bloch maps

Let $H(\mathbb{D})$ be the class of complex-valued holomorphic functions on the unit disc $\mathbb{D}$ and let $f \in H(\mathbb{D})$ with $f'(0) = 1$. The celebrated Bloch's theorem states that f maps a subdomain in $\mathbb{D}$ biholomorphically onto a disc with radius $r(f)$ greater than some positive universal constant. The 'best possible' constant $\mathfrak{b}$ for all such functions, that is,

$$\mathfrak{b} = \inf\{r(f) : f \in H(\mathbb{D}) \text{ and } f'(0) = 1\},$$

is called the Bloch constant.

As a direct consequence of his extension of the Schwarz lemma, Ahlfors [6] gave an elementary proof of Bloch's theorem with an explicit lower bound $\mathfrak{b} \geq \sqrt{3}/4$. Using the fact that

$$\mathfrak{b} = \inf\{r(f) : f \in H(\mathbb{D}), f'(0) - 1 = f(0) = 0, \sup_{z\in\mathbb{D}}(1 - |z|^2)|f'(z)| \leq 1\}$$
(4.4)

(cf. [117]) and proving a distortion theorem in [18] for functions f in (4.4), which says

$$\mathrm{Re}\, f'(z) \geq \frac{1 - \sqrt{3}|z|}{\left(1 - \frac{|z|}{\sqrt{3}}\right)^3} \qquad \left(|z| \leq \frac{1}{\sqrt{3}}\right),$$
(4.5)

Bonk was able to improve the lower bound for the Bloch constant to $\mathfrak{b} > \frac{\sqrt{3}}{4} + 10^{-14}$.

A holomorphic function $f : \mathbb{D} \longrightarrow \mathbb{C}$ satisfying

$$|f|_{\mathcal{B}} := \sup_{z\in\mathbb{D}}(1 - |z|^2)|f'(z)| < \infty$$

is known as a *Bloch function*, where $|f|_{\mathcal{B}}$ is called the *Bloch semi-norm*. We have therefore seen that one can derive an estimate of the Bloch constant via some distortion result for normalised Bloch functions f on $\mathbb{D}$ satisfying $|f|_{\mathcal{B}} \leq 1$ (equivalently, $|f|_{\mathcal{B}} = 1$ as f is normalised). This will be the theme of our discussion for higher dimensional extension throughout this section, where the following convention is adopted.

Convention. In the rest of this chapter, $\mathbb{C}^n$ is always equipped with the Euclidean norm

$$\|z\|_2 = \sqrt{|z_1|^2 + \cdots + |z_n|^2} \qquad (z = (z_1, \ldots, z_n) \in \mathbb{C}^n)$$

unless stated otherwise. The open Euclidean ball centred at $a \in \mathbb{C}^n$ with radius $r > 0$ will be denoted by

$$\mathbb{B}_n(a, r) := \{z \in \mathbb{C}^n : \|z - a\|_2 < r\}.$$

The Bloch semi-norm $|f|_\mathcal{B}$ on $\mathbb{D}$ can be expressed as

$$|f|_\mathcal{B} = \sup\{|(f \circ g)'(0)| : g \in \operatorname{Aut}\mathbb{D}\}.$$

Indeed, for each $g \in \operatorname{Aut}\mathbb{D}$, we have $|g'(0)| = 1 - |g(0)|^2$ which implies $|(f \circ g)'(0)| = |f'(g(0))|(1 - |g(0)|^2)$. On the other hand, for each $z \in \mathbb{D}$, and for the Möbius transformation g_{-z} induced by $-z$, we have

$$(1 - |z|^2)|f'(z)| = (1 - |z|^2)|(f \circ g_z)'(0)||g'_{-z}(z)| = |(f \circ g_z)'(0)|.$$

The above observation leads to a natural generalisation of the concept of a Bloch function to higher dimensions.

Definition 4.2.1. Let D be a bounded symmetric domain realised as the open unit ball of a JB*-triple. The *Bloch semi-norm* of a holomorphic map $f : D \longrightarrow \mathbb{C}^n$ is defined by

$$|f|_\mathcal{B} = \sup\{\|(f \circ g)'(0)\| : g \in \operatorname{Aut}D\}$$

where $\mathbb{C}^n$ is equipped with the Euclidean norm. We call f a *Bloch map* if $|f|_\mathcal{B} < \infty$. A Bloch map $f : D \longrightarrow \mathbb{C}$ is often called a *Bloch function*.

Remark 4.2.2. Let D be a bounded symmetric domain in a Banach space $(V, |\cdot|)$, and let $\|\cdot\|_{sp}$ be an equivalent norm so that $(V, \|\cdot\|_{sp})$ is a JB*-triple and D identifies with the open unit ball $\{v \in V : \|v\|_{sp} < 1\}$ (cf. Theorem 3.2.18). The Bloch semi-norm $|f|_\mathcal{B}$ is defined in terms of the norm $\|\cdot\|_{sp}$. However, if one uses the norm of $(f \circ g)'(0) : (V, |\cdot|) \longrightarrow \mathbb{C}^n$, given by

$$|(f \circ g)'(0)|_* = \sup\{\|(f \circ g)'(0)(v)\|_2 : |v| \leq 1\},$$

to define another semi-norm

$$|f|_\mathcal{B}^* := \sup\{|(f \circ g)'(0)|_* : g \in \operatorname{Aut} D\},$$

then we have $\alpha|f|_\mathcal{B}^* \leq |f|_\mathcal{B} \leq \beta|f|_\mathcal{B}^*$ for some $0 < \alpha \leq \beta$ and there is no difference between defining a Bloch map by the condition $|f|_\mathcal{B}^* < \infty$, or by $|f|_\mathcal{B} < \infty$ although the two semi-norms are generally different.

Evidently, the Bloch semi-norm of $f : D \longrightarrow \mathbb{C}^n$ is invariant under the automorphism group $\operatorname{Aut} D$ and as before, the inequality

$$
\begin{aligned}
(1 - \|z\|^2)\|f'(z)\| &\leq (1 - \|z\|^2)\|(f \circ g_z)'(0)\|\|g'_{-z}(z)\| \\
&= (1 - \|z\|^2)\|(f \circ g_z)'(0)\|\|B(z, z)^{-1/2}\| \quad (z \in D) \\
&= \|(f \circ g_z)'(0)\|
\end{aligned}
$$

implies

$$\sup\{(1 - \|z\|^2)\|f'(z)\| : z \in D\} \leq |f|_\mathcal{B} \tag{4.6}$$

although the two sides above need not be equal in higher dimensions, even in the case of a bidisc. We will give an example later.

Example 4.2.3. Let $f : D \longrightarrow \mathbb{C}^n$ be a bounded holomorphic map on the open unit ball D of a JB*-triple V and denote the supremum norm of f by

$$\|f\|_\infty := \sup\{\|f(z)\|_2 : z \in D\}.$$

Then f is a Bloch map and $|f|_\mathcal{B} \leq \|f\|_\infty$. Indeed, let $\|f\|_\infty < 1$. Then for each $g \in \operatorname{Aut} D$, the holomorphic map $f \circ g$ maps D into the Euclidean ball $\mathbb{B}_n \subset \mathbb{C}^n$. Hence the Schwarz-Pick lemma (cf. Proposition

3.5.19) gives $\|(f \circ g)'(0)\| \leq \|B(f(g(0)), f(g(0)))^{1/2}\| \leq 1$. From this, one deduces $|f|_{\mathcal{B}} \leq \|f\|_\infty$.

On the unit disc $\mathbb{D}$, the holomorphic function $\psi(\zeta) = \tanh^{-1}(\zeta)$ is a simple example of a unbounded Bloch function with Bloch semi-norm $|\psi|_{\mathcal{B}} = 1$. Given $a \in D \backslash \{0\}$, let $\phi_a \in V^*$ be the support functional at a, that is, $\phi_a(a) = \|a\|$ and $\|\phi_a\| = 1$. The holomorphic function $\psi \circ \phi_a : D \longrightarrow \mathbb{C}$ is clearly unbounded on D, but it is a Bloch function with Bloch semi-norm $|\psi \circ \phi_a|_{\mathcal{B}} = |\psi|_{\mathcal{B}}$ since

$$\|(\psi \circ \phi_a)'(0)\| = \|\psi'(\phi_a(0))\phi_a'(0)\| = \|\psi'(0)\phi_a\| = 1$$

and for each $g \in \operatorname{Aut} D$, we have

$$\begin{aligned}
\|(\psi \circ \phi_a \circ g)'(0)\| &\leq |\psi'(\phi_a(g(0)))| \|(\phi_a \circ g)'(0)\| \\
&\leq \frac{|\psi|_{\mathcal{B}}}{1 - |\phi_a(g(0))|^2}(1 - |\phi_a(g(0))|^2) = |\psi|_{\mathcal{B}}
\end{aligned}$$

by (4.6) and Proposition 3.5.19.

Bloch's theorem fails in dimension 2 [180]. Nevertheless, one can define the Bloch constant for various families of Bloch maps in higher dimensions.

We first extend Bonk's distortion theorem to locally biholomorphic maps on finite dimensional bounded symmetric domains and will return to the Bloch constant and Block maps afterwards.

Definition 4.2.4. Let D be a domain in a complex Banach space V, and W a Banach space. A holomorphic map $f : D \longrightarrow W$ is called *locally biholomorphic* if the derivative $f'(a) : V \longrightarrow W$ has a continuous inverse for every $a \in V$. We denote by $H_{loc}(D, W)$ the class of locally biholomorphic maps $f : D \longrightarrow W$.

We begin with extensions of Bonk's distortion theorem. We observe that Bonk's theorem concerns normalised Bloch functions f with

$|f|_\mathcal{B} = 1$. In higher dimensional Euclidean balls $\mathbb{B}_n \subset \mathbb{C}^n$ and polydiscs $\mathbb{D}^n$, replacing $(f \circ g)'(0)$ in the definition of $|f|_\mathcal{B}$ by its Jacobian, the Bloch semi-norm decreases to

$$\|f\|_0 := \sup\{|\det(f \circ g)'(0)|^{1/n} : g \in \operatorname{Aut} D\} \le |f|_\mathcal{B} \quad (D = \mathbb{B}_n \text{ or } \mathbb{D}^n)$$

where $|\det f'(0)|^{1/n} \le \|f\|_0$ and $\|f\|_0 = |f|_\mathcal{B}$ for $n = 1$. Using this reduction of the Bloch semi-norm, one can show the following distortion results for $\mathbb{B}_n$ and $\mathbb{D}^n$.

Theorem 4.2.5. *Let* $f : D \longrightarrow \mathbb{C}^n$ *be a locally biholomorphic map satisfying* $\det f'(0) = 1 = \|f\|_0$*, where* $D = \mathbb{B}_n$ *or* $\mathbb{D}^n$*. Then*

$$|\det f'(z)| \ge \operatorname{Re} \det f'(z) \ge \frac{\exp\left(\dfrac{-(n+1)\|z\|}{1 - \|z\|}\right)}{(1 - \|z\|)^{n+1}} \quad (z \in \mathbb{B}_n)$$

and

$$|\det f'(z)| \ge \operatorname{Re} \det f'(z) \ge \frac{\exp\left(\dfrac{-2n\|z\|}{1 - \|z\|}\right)}{(1 - \|z\|)^{2n}} \quad (z \in \mathbb{D}^n).$$

Both inequalities are sharp.

Proof. The first inequality is proved in [121, Theorem 7], and the second one is proved in [174, Theorem 3.2]. $\qquad\square$

The preceding distortion results have also been used to estimate suitably defined Bloch constants on $\mathbb{B}_n$ and $\mathbb{D}^n$, which will be discussed later. For now, two natural questions arise:

Question 4.2.6. Can we explain the difference of the exponents in the distortion bounds in Theorem 4.2.5?

Question 4.2.7. Can we extend Theorem 4.2.5 to other bounded symmetric domains in $\mathbb{C}^n$?

Indeed, we are going to show a unified result, thereby extend Theorem 4.2.5 to all finite dimensional bounded symmetric domains, and answer both questions affirmatively.

Let D be a finite dimensional bounded symmetric domain realised as the open unit ball of a JB*-triple V of dimension $\dim V = n$. Denote by r the rank of D, and by p the genus of D, as defined in Definition 3.3.9. We introduce a constant

$$c(D) := \frac{pr}{2} \tag{4.7}$$

which satisfies $(n+1)/2 \le c(D) \le n$ by (3.33) and will play the role of the constant term in the exponents in Theorem 4.2.5. In fact, we have $c(B_n) = (n+1)/2$ for the Euclidean ball $B_n \subset \mathbb{C}^n$ as B_n has rank 1, and $c(\mathbb{D}^n) = 2n/2 = n$ for the polydisc $\mathbb{D}^n \subset \mathbb{C}^n$.

Given a holomorphic map $f : D \longrightarrow \mathbb{C}^n$, we define

$$\|f\|_d = \sup \left\{ (1 - \|z\|^2)^{c(D)/n} |\det f'(z)|^{1/n} : z \in D \right\}$$

which is not a norm, but may be called the *quasinorm* of f. Note that

$$|\det f'(0)|^{1/n} \le \|f\|_d, |f|_{\mathcal{B}}.$$

For $D = B_n$, we have

$$\|f\|_d = \|f\|_0$$

(cf. [121, p. 356]), and for $D = \mathbb{D}^n$, we have

$$\begin{aligned} \|f\|_d &\le \|f\|_0 \\ &= \sup_{z=(z_1,\ldots,z_n)\in\mathbb{D}^n} \{(1 - |z_1|^2)^{1/2} \cdots (1 - |z_n|^2)^{1/2} |\det f'(z)|^{1/n}\} \end{aligned}$$

(cf. [174, p. 654]). We will use the quasinorm $\|f\|_d$, instead of $\|f\|_0$, in the generalisation of Theorem 4.2.5. First, we need two lemmas, one of which is the following version of Julia's lemma.

Lemma 4.2.8. (Julia's lemma) *Let* $\rho, \psi : \mathbb{D} \cup \{1\} \longrightarrow \mathbb{C}$ *be holomorphic functions such that* $\rho(\mathbb{D}) \subset \mathbb{D}$, $\rho(1) = 1$, $\psi(1) = 0$ *and* $\psi(\mathbb{D}) \subset \{\zeta \in \mathbb{C} : \operatorname{Re} \zeta > 0\}$. *Then we have* $\rho'(1) > 0$ *and for each* $r > 0$,

$$\psi(\overline{H}(1, 1/r)) \subset \{\zeta \in \mathbb{C} : |\zeta - dr| \leq dr\}$$

where $d = -\psi'(1) > 0$ *and* $\overline{H}(1, 1/r)$ *is the closure of the horodisc*

$$H(1, 1/r) = \left\{ \zeta \in \mathbb{C} : \frac{|1 - \zeta|^2}{1 - |z|^2} < r \right\} = \frac{1}{1+r} + \frac{r}{1+r} \mathbb{D}.$$

Lemma 4.2.9. *Let* $\rho : \mathbb{D} \cup \{1\} \longrightarrow \mathbb{C}$ *be a holomorphic function satisfying* $\rho(\mathbb{D}) \subset \mathbb{D} \setminus \{0\}$ *and* $\rho(1) = 1$. *Then* $\rho'(1) > 0$ *and*

$$|\rho(x)| \geq \exp\left\{ -2\rho'(1) \frac{1-x}{1+x} \right\}$$

for all $x \in (-1, 1)$.

Proof. By Julia's lemma, we have $\rho'(1) > 0$. Let $\psi = -\log \rho$. Then we have

$$\operatorname{Re} \psi(z) = -\log |\rho(z)| > 0$$

for all $z \in \mathbb{D}$ and $\psi(1) = 0$. Moreover, $\psi'(1) = -\log \rho'(1) < 0$.

Given $-1 < x < 1$, let $r = \frac{1-x}{1+x}$ and $d = -\psi'(1)$. Then $x \in \overline{H}(1, 1/r)$ and by Julia's lemma again, we have $|\psi(x) - dr| \leq dr$ and

$$\operatorname{Re} \psi(x) \leq |\psi(x)| \leq 2dr = -2\psi'(1) \frac{1-x}{1+x}$$

which gives

$$|\rho(x)| \geq \exp\left\{ -2\rho'(1) \frac{1-x}{1+x} \right\}.$$

$\square$

Here is a unified distortion result which includes Theorem 4.2.5 as a special case.

Theorem 4.2.10. *Let D be a bounded symmetric domain realised as the open unit ball of a JB^*-triple V and $\dim V = n$. Let $f : D \longrightarrow \mathbb{C}^n$ be locally biholomorphic, satisfying $\det f'(0) = 1 = \|f\|_d$. Then we have*

$$|\det f'(z)| \geq \frac{1}{(1 - \|z\|)^{2c(D)}} \exp\left\{\frac{-2c(D)\|z\|}{1 - \|z\|}\right\} \qquad (z \in D). \quad (4.8)$$

The estimate in (4.8) is sharp.

Proof. Write $c = c(D)$ to simplify notation. Let $z \in D$ and $u = z/\|z\|$. Define a holomorphic function $\rho : \mathbb{D} \cup \{1\} \longrightarrow \mathbb{C}$ by

$$\rho(\zeta) = \left(\frac{1+\zeta}{2}\right)^{2c} \det f'\left(\frac{1-\zeta}{2}u\right).$$

Evidently, $\rho(1) = 1$. Since f is locally biholomorphic, we have $\rho(\mathbb{D}) \subset \mathbb{D}\backslash\{0\}$.

Observe from the definition of the quasinorm $\|f\|_d = 1$ that

$$|\det f'(w)| \leq \frac{\|f\|_d^n}{(1 - \|w\|^2)^c} = (1 + \|w\|^2 + \|w\|^4 + \cdots)^c \qquad (w \in D)$$

where $\det f'(0) = 1$. Therefore $\rho'(1) = c$ and

$$0 < |\rho(\zeta)| = \left|\frac{1+\zeta}{2}\right|^{2c} \left|\det f'\left(\frac{1-\zeta}{2}u\right)\right| \leq \left|\frac{1+\zeta}{2}\right|^{2c} \left|\frac{1}{1 - \left|\frac{1-\zeta}{2}\right|^2}\right|^c < 1$$

for $\zeta \in \mathbb{D}$. By Lemma 4.2.9, we have

$$|\rho(x)| \geq \exp\left\{-2c\frac{1-x}{1+x}\right\}$$

for all $x \in (-1, 1)$. This inequality and (4.8) are identical for $x = 1 - 2\|z\|$.

To show that (4.8) is sharp, we fix a boundary point $u \in \partial D$ and pick a linear functional $\ell \in V^*$ such that $\ell(u) = 1 = \|\ell\|$. Define a holomorphic map $F : D \longrightarrow \mathbb{C}^n$ by

$$F(z) = \left(\int_0^{\ell(z)} \psi(t)dt\right) u + z - \ell(z)u$$

where $\psi : \mathbb{D} \longrightarrow \mathbb{C}$ is the holomorphic function

$$\psi(\zeta) = \frac{1}{(1-\zeta)^{2c}} \exp\left\{\frac{-2c\zeta}{1-\zeta}\right\}.$$

We have $F(0) = 0$ and $\det F'(z) = \psi(\ell(z))$. In particular, $\det F'(0) = \psi(0) = 1$.

Let $z \in D$ and $\zeta = \ell(z)$. Then we have

$$
\begin{aligned}
(1 - \|z\|^2)^c |\det F'(z)| &\leq (1 - |\ell(z)|^2)^c |\psi(\ell(z))| \\
&= \left(\frac{1 - |\zeta|^2}{|1 - \zeta|^2}\right)^c \left|\exp\left(\frac{-2c\zeta}{1-\zeta}\right)\right| \\
&= \left(\frac{1 - |\zeta|^2}{|1 - \zeta|^2} \exp\left(1 - \operatorname{Re}\left(1 + \frac{2\zeta}{1-\zeta}\right)\right)\right)^c \\
&= \left(\frac{1 - |\zeta|^2}{|1 - \zeta|^2} \exp\left(1 - \frac{1 - |\zeta|^2}{|1 - \zeta|^2}\right)\right)^c \leq 1
\end{aligned}
$$

where the last inequality follows from $te^{1-t} \leq 1$ for $t > 0$. Therefore $\|F\|_d \leq 1$ and hence $\|F\|_d = 1 = \det F'(0)$.

Since $\det F'(\pm\|z\|u) = \psi(\pm\|z\|)$ for all $z \in D$, we see that F attains the equality in (4.8). $\qquad\square$

Now we resume the discussion of Bloch maps. As mentioned at the beginning of this section, given a holomorphic map $f : D \longrightarrow \mathbb{C}^n$, the Bloch semi-norm $|f|_\mathcal{B}$, unlike the one-dimensional case, need not be equal to

$$\sup\{(1 - \|z\|^2)\|f'(z)\| : z \in D\} \leq |f|_\mathcal{B}.$$

In spite of this, one can still ask a weaker question whether both semi-norms are finite at the same time so that, if the answer is positive, one can also use the *'simpler'* condition $\sup\{(1-\|z\|^2)\|f'(z)\| : z \in D\} < \infty$ as an equivalent definition for a Bloch function. Unfortunately, the answer is negative even in the case of a bidisc, but *positive* for a Hilbert ball!

Example 4.2.11. Let $D = \mathbb{D} \times \mathbb{D}$ be the bidisc and let $f : D \longrightarrow \mathbb{C}$ be defined by

$$f(z_1, z_2) = (1 - z_2) \log \frac{1}{1 - z_1}, \qquad (z_1, z_2) \in D.$$

Then we have $\sup_{z \in D}(1 - \|z\|^2)\|f'(z)\| < \infty$, but $|f|_{\mathcal{B}} = \infty$.

Indeed, for $(x_1, x_2) \in \mathbb{C}^2$,

$$f'(z_1, z_2)(x_1, x_2) = \frac{(1 - z_2)x_1}{1 - z_1} - x_2 \log \frac{1}{1 - z_1}.$$

Observe that

$$
\begin{aligned}
\|f'(z_1, z_2)\| &\leq \left| \frac{1 - z_2}{1 - z_1} \right| + \left| \log \frac{1}{1 - z_1} \right| \\
&\leq \frac{|1 - z_2|}{|1 - z_1|} + \frac{2}{|1 - z_1|} + \log 2 + \pi
\end{aligned}
$$

where $1/|1 - z_1| \geq 1/2$ implies

$$2/|1 - z_1| \geq \log(2/|1 - z_1|) \geq |\log(1/|1 - z_1|)| - \log 2.$$

Pick $z = (z_1, z_2) \in D$, where $|z_1| > |z_2|$ or $|z_1| \leq |z_2|$. In the former case, we have

$$
\begin{aligned}
(1 - \|z\|^2)\|f'(z)\| &\leq (1 - |z_1|^2)\left(\frac{|1 - z_2|}{|1 - z_1|} + \frac{2}{|1 - z_1|} + \log 2 + \pi \right) \\
&\leq (1 + |z_1|)(|1 - z_2| + 2 + (\log 2 + \pi)(1 - |z_1|)).
\end{aligned}
$$

In the case of $|z_1| \leq |z_2|$, we have $|1 - z_1| \geq 1 - |z_1| \geq 1 - |z_2|$ and hence

$$
\begin{aligned}
(1 - \|z\|^2)\|f'(z)\| &= (1 - |z_2|^2)\|f'(z)\| \\
&\leq (1 + |z_2|)(|1 - z_2| + 2 + (\log 2 + \pi)(1 - |z_2|)).
\end{aligned}
$$

We deduce from these inequalities that

$$\sup_{z \in D}(1 - \|z\|^2)\|f'(z)\| < \infty.$$

On the other hand, for $z = (z_1, 0) \in D$, we have

$$\left| f'(z_1, 0) \left(\frac{1 - |z_1|^2}{2}, 1 \right) \right| = \left| \frac{1 - |z_1|^2}{2(1 - z_1)} - \log \frac{1}{1 - z_1} \right|.$$

Let $g_z : D \longrightarrow D$ be the Möbius transformation induced by z. We have

$$\begin{aligned}
|f'(z)(x_1, x_2)| &\leq \|(f \circ g_z)'(0)\| \|g'_{-z}(z)(x_1, x_2)\| \\
&\leq |f|_{\mathcal{B}} \|B(z, z)^{-1/2}(x_1, x_2)\|.
\end{aligned}$$

For $z = (z_1, 0)$ and $(x_1, x_2) = \left(\frac{1 - |z_1|^2}{2}, 1 \right)$, we have

$$\|B(z, z)^{-1/2}(x_1, x_2)\| = \|B(z, z)^{-1/2}((1 - |z_1|^2)/2, 1)\| = \|(1/2, 1)\| = 1.$$

Hence

$$\begin{aligned}
|f|_{\mathcal{B}} &\geq \left| \frac{1 - |z_1|^2}{2(1 - z_1)} - \log \frac{1}{1 - z_1} \right| \\
&\geq \left| \log \frac{1}{1 - z_1} \right| - \frac{1 - |z_1|^2}{2|1 - z_1|} \\
&\geq \left| \log \frac{1}{1 - z_1} \right| - \frac{1 + |z_1|}{2} \geq \left| \log \frac{1}{1 - z_1} \right| - 1
\end{aligned}$$

where the right-hand side is unbounded on D.

Example 4.2.12. Let D be the open unit ball of a Hilbert space H. Let $f : D \longrightarrow \mathbb{C}$ be a holomorphic function satisfying

$$\|f\|_D := \sup\{(1 - \|z\|^2)\|f'(z)\| : z \in D\} < \infty. \tag{4.9}$$

Then the Bloch semi-norm of f satisfies

$$|f|_{\mathcal{B}} \leq (2 + 2\sqrt{2})(\sup_{z \in D}(1 - \|z\|^2)\|f'(z)\|) \tag{4.10}$$

and hence f is a Bloch function. If $H = \mathbb{C}$, we actually have $|f|_{\mathcal{B}} = \|f\|_D$, already noted before. To show (4.10) for $\dim H \geq 2$, we first consider the case of $\dim H = 2$, where

$$D = \mathbb{B}_2 = \{(z_1, z_2) \in \mathbb{C}^2 : |z_1|^2 + |z_2|^2 < 1\}.$$

For $f : \mathbb{B}_2 \longrightarrow \mathbb{C}$ satisfying (4.9), we derive the inequality

$$\left| \frac{\partial f}{\partial z_2}(x_1, 0) \right| \sqrt{1 - |x_1|^2} \leq (1 + 2\sqrt{2}) \|f\|_{\mathbb{B}_2} \qquad ((x_1, 0) \in \mathbb{B}_2). \quad (4.11)$$

Given $(x_1, 0) \in \mathbb{B}_2$, let $r = \frac{1}{\sqrt{2}} \sqrt{1 - |x_1|^2}$ so that $|x_1|^2 + r^2 < 1$. By Cauchy formula, we have

$$\frac{\partial f}{\partial z_2}\left(\frac{\partial f}{\partial z_1} \right)(x_1, 0) = \frac{1}{2\pi i} \int_{|w|=r} \frac{\frac{\partial f}{\partial z_1}(x_1, w)}{w^2} dw.$$

Hence

$$\left| \frac{\partial f}{\partial z_2}\left(\frac{\partial f}{\partial z_1} \right)(x_1, 0) \right| \leq \frac{1}{2\pi} \frac{2\pi r \|f\|_{\mathbb{B}_2}}{r^2(1 - |x_1|^2 - r^2)} = \frac{2\sqrt{2}\|f\|_{\mathbb{B}_2}}{(1 - |x_1|^2)^{3/2}}.$$

Observe that

$$\frac{\partial f}{\partial z_2}(x_1, 0) - \frac{\partial f}{\partial z_2}(0, 0) = \int_0^1 \frac{\partial^2 f}{\partial z_2 \partial z_1}(tx_1, 0) x_1 dt$$

which gives

$$\begin{aligned}
\left| \frac{\partial f}{\partial z_2}(x_1, 0) \right| &\leq \int_0^1 \frac{2\sqrt{2}\|f\|_{\mathbb{B}_2}}{(1 - t^2|x_1|^2)^{3/2}} dt + \left| \frac{\partial f}{\partial z_2}(0, 0) \right| \\
&\leq \frac{2\sqrt{2}\|f\|_{\mathbb{B}_2}}{\sqrt{1 - |x_1|^2}} + \frac{\|f\|_{\mathbb{B}_2}}{\sqrt{1 - |x_1|^2}}
\end{aligned}$$

and therefore

$$\left| \frac{\partial f}{\partial z_2}(x_1, 0) \right| \sqrt{1 - |x_1|^2} \leq (1 + 2\sqrt{2}) \|f\|_{\mathbb{B}_2}.$$

Returning to the case of an arbitrary Hilbert ball D and $f : D \longrightarrow \mathbb{C}$ in (4.9), we note that

$$|f|_{\mathcal{B}} = \sup\{\|(f \circ g_a)'(0)\| : a \in D\}$$

where g_a is the Möbius transformation induced by a and we have used the fact that each automorphism of D is of the form $\varphi \circ g_a$ for some linear isometry φ of D and $a \in D$, by Theorem 3.2.32.

Let $a \in D \backslash \{0\}$. Then $u = a/\|a\|$ is an extreme point of the closed ball $\overline{D}$ and hence a maximal tripotent in H by Theorem 3.4.8. It follows from Example 3.4.9 that H is the orthogonal sum $H_1(u) \oplus H_2(u)$ of the Peirce spaces and

$$g_a'(0)v = B(a,a)^{1/2}(v) = \begin{cases} \sqrt{1 - \|a\|^2}\, v & (v \in H_1(u)) \\ (1 - \|a\|^2)v & (v \in H_2(u)). \end{cases}$$

Let $v \in H_1(u)$ with $\|v\| \le 1$. Define a holomorphic function $F :$ $\mathbb{B}_2 \longrightarrow \mathbb{C}$ by

$$F(z_1, z_2) = f(z_1 u + z_2 v) \quad \text{for} \quad (z_1, z_2) \in \mathbb{B}_2.$$

Then we have $\|F'(z_1, z_2)\| \le \|f'(z_1 u + z_2 v)\|$ and F satisfies (4.9) since $\|F\|_{\mathbb{B}_2} \le \|f\|_D < \infty$. It follows from (4.11) that

$$\begin{aligned} |f'(a)v|\sqrt{1 - \|a\|^2} &= \left| \frac{\partial F}{\partial z_2}(\|a\|, 0) \right| \sqrt{1 - \|a\|^2} \\ &\le (1 + 2\sqrt{2})\|F\|_{\mathbb{B}_2} \le (1 + 2\sqrt{2})\|f\|_D. \end{aligned}$$

Finally, for $v = v_1 \oplus v_2 \in H_1(u) \oplus H_2(u)$ with $\|v\| \le 1$, we have

$$\begin{aligned} |(f \circ g_a)'(0)v| &= |f'(a)g_a'(0)v| = |f'(a)(\sqrt{1 - \|a\|^2}\, v_1 + (1 - \|a\|^2)v_2)| \\ &\le (1 + 2\sqrt{2})\|f\|_D + \|f\|_D = (2 + 2\sqrt{2})\|f\|_D \end{aligned}$$

which proves (4.10).

The previous two examples are somewhat unsettling. In view of the Hilbert ball case, can one not simply use the condition

$$\sup\{(1 - \|z\|^2)\|f'(z)\| : z \in D\} < \infty$$

for the definition of a Bloch function in higher dimensions?

We give a justification that the condition $|f|_{\mathcal{B}} < \infty$ is in fact a more appropriate choice. Indeed, for the unit disc $\mathbb{D}$, another equivalent condition for $f : \mathbb{D} \longrightarrow \mathbb{C}$ to be a Bloch function is that the family

$$F_f = \{f \circ g - f(g(0)) : g \in \text{Aut}\,(\mathbb{D})\}$$

is a normal family (cf. Definition 1.1.10). The following lemma should justify the choice of $|f|_{\mathcal{B}} < \infty$.

Lemma 4.2.13. *Let D be a bounded symmetric domain in $\mathbb{C}^n$, realised as the open unit ball of norm on $\mathbb{C}^n$ equivalent to the Euclidean norm, and $f : D \longrightarrow \mathbb{C}^d$ a holomorphic function. The following two conditions are equivalent.*

(i) $|f|_{\mathcal{B}} < \infty$.

(ii) $F_f = \{f \circ g - f(g(0)) : g \in \mathrm{Aut}\,(D)\}$ *is a normal family.*

Proof. (i)$\Rightarrow$(ii). By Lemma 1.1.11, it suffices to show that F_f is uniformly bounded on compact subsets of D, and by Theorem 1.1.1, suffice it to show

$$\sup_{z \in K}\{\|(f \circ g)'(z)\| : g \in \mathrm{Aut}\, D\} < \infty$$

for each compact subset K of D. This follows from

$$\|(f \circ g)'(z)\| \;=\; \|(f \circ (g \circ g_z))'(0)\|\|g'_{-z}(z)\| \leq |f|_{\mathcal{B}}\|\|B(z,z)^{-1/2}\|$$

$$=\; \frac{|f|_{\mathcal{B}}}{1 - \|z\|^2}.$$

(ii)$\Rightarrow$(i). This follows from the Cauchy inequality since

$$(f \circ g)'(0) = (f \circ g - f(g(0)))'(0) \qquad (g \in \mathrm{Aut}\, D).$$

$\square$

We see now that for the holomorphic function $f : \mathbb{D} \times \mathbb{D} \longrightarrow \mathbb{C}$ in Example 4.2.11, the family F_f is not normal although $\sup_{z \in D}(1 - \|z\|^2)\|f'(z)\| < \infty$.

Corollary 4.2.14. *Let D be a bounded symmetric domain in $\mathbb{C}^n$, realised as the open unit ball of an equivalent norm in $\mathbb{C}^n$, and let (f_k) be a* sequence of $\mathbb{C}^d$-valued Bloch maps, with uniformly bounded Bloch semi-norm. Then there is a subsequence of $(f_k - f_k(0))$ converging locally uniformly to a Bloch map on D.

Proof. Let $|f_k|_\mathcal{B} \leq C$ for all k and some $C > 0$. Analogous to the proof of the preceding lemma, one shows that, on a compact subset K of D,

$$\sup_{z \in K}\{\|(f_k - f_k(0))'(z)\| \leq \frac{|f_k|_\mathcal{B}}{1 - \|z\|^2} \leq \frac{C}{1 - \|z\|^2}$$

for all k. Hence Theorem 1.1.1 implies that $\{f_k - f_k(0)\}_k$ is uniformly bounded on compact sets in D and by Lemma 1.1.11, it is a normal family and the conclusion follows. $\square$

Now we introduce the Bloch constant for a family of Bloch maps and will use the distortion result in Theorem 4.2.10 to estimate its lower bound.

We first introduce two more constants besides $c(D)$, attached to a finite dimensional bounded symmetric domain D in $\mathbb{C}^n$, which are related to the squeezing constant $\hat{\sigma}_D$ discussed in Section 3.7. Since D identifies with the open unit ball with respect to an equivalent norm $\| \cdot \|$ on the Euclidean space $(\mathbb{C}^n, \| \cdot \|_2)$, it is sandwiched between two open Euclidean balls. We define

$$s(D) := \inf\{\|z\|_2 : \|z\| = 1\}, \qquad t(D) := \sup\{\|z\|_2 : \|z\| = 1\}. \quad (4.12)$$

It can be seen that

$$\mathbb{B}_n(0, s) \subset D \subset \mathbb{B}_n(0, t)$$

where $s = s(D) > 0$ is the radius of the largest Euclidean ball $\mathbb{B}_n(0, s)$ contained in D, and $t = t(D)$ is the radius of the smallest Euclidean ball $\mathbb{B}_n(0, t)$ containing D. It follows that $0 < \frac{s}{t} \leq \sigma_D(0) = \hat{\sigma}_D$. If D happens to be the Euclidean ball $\mathbb{B}_n(0, 1)$, we have of course $s = t =$

$1 = \hat{\sigma}_D$. If $D = \mathbb{D}^n \subset \mathbb{C}^n$, then $s(D) = 1$ and $t(D) = \sqrt{n}$, in which case $\hat{\sigma}_D = s(D)/t(D)$. In fact, it follows from [115, Lemma 2] that $s(D) \geq 1$ for all classical domains D.

Let D be a bounded symmetric domain realised as the open unit ball of a JB*-triple V with $\dim V = n$ (where $V = (\mathbb{C}^n, \|\cdot\|)$ with $\|\cdot\|$ equivalent to the Euclidean norm).

Definition 4.2.15. Let $f : D \longrightarrow \mathbb{C}^n$ be a holomorphic map. By a *schlicht ball* of f, we mean a Euclidean ball of the form $\mathbb{B}_n(f(a), r) \subset \mathbb{C}^n$ of radius $r > 0$, which is the biholomorphic image $f(G)$ of an open set $G \subset D$ under f. If $n = 1$, a schlicht ball in $f(D)$ is usually called a *schlicht disc*, in which case one can find a holomorphic map $h : \mathbb{D} \longrightarrow D$ such that $f \circ h : \mathbb{D} \longrightarrow \mathbb{B}_1(f(a), r)$ is biholomorphic and

$$(f \circ h)(\zeta) = f(a) + r\zeta \qquad (\zeta \in \mathbb{D}).$$

For each point $a \in D$, let $r(a, f)$ denote the radius of the largest schlicht ball of a holomorphic map $f : D \longrightarrow \mathbb{C}^n$ centered at $f(a)$. We have

$$r(a, f) = \sup\{r > 0 : \mathbb{B}_n(f(a), r) \subset f(D),\ f^{-1} \in H_{emb}(\mathbb{B}_n(f(a), r), D)\}.$$

Let

$$r(f) = \sup\{r(a, f) : a \in D\}.$$

Lemma 4.2.16. *Let D be a bounded symmetric domain in $\mathbb{C}^n$ and let $f : D \longrightarrow \mathbb{C}^n$ be locally biholomorphic, mapping an open subset $G \subset D$ biholomorphically onto a schlicht ball $\mathbb{B}^n(f(a), r(a, f))$ for some $a \in D$. Then $\partial G \cap \partial D \neq \emptyset$. In particular,*

$$r(a, f) = \|f(a) - b\|_2$$

where $b = \lim_k f(z_k) \in \partial \mathbb{B}_n(f(a), r(a, f)) \cap \partial f(D)$ for some sequence (z_k) in G, converging to a boundary point in $\partial G \cap \partial D$.

Proof. Assume $\partial G \cap \partial D = \emptyset$. We deduce a contradiction. In this case, there is a sequence (G_k) of open sets in D satisfying $G_k \supset \overline{G}_{k+1}$ and

$$\bigcap_{n=1}^{\infty} G_k = \overline{G}.$$

By the definition of schlicht balls, one can find two distinct points $z_k, w_k \in G_k$ such that $f(z_k) = f(w_k)$. We may assume without loss of generality that (z_k) converges to some $z \in \partial G$ and (w_k) converges to some $w \in \partial G$. It follows that

$$f(z) = f(w) \in \partial \mathbb{B}^n(f(a), r(a, f)).$$

If $z = w$, then f is not locally biholomorphic, which is impossible. Hence $z \neq w$, but the equality $f(z) = f(w)$ would imply that f is not injective on G, which is also impossible. This proves the first assertion, which implies the second one readily. $\square$

Definition 4.2.17. Let D be a bounded symmetric domain in $\mathbb{C}^n$, realised as the open unit ball of an equivalent norm in $\mathbb{C}^n$, and let $H_{loc}^B(D, \mathbb{C}^n)$ denote the subclass of $H_{loc}(D, \mathbb{C}^n)$ consisting of locally biholomorphic Bloch maps on D. The *Bloch constant* $\mathfrak{b}(D)$ for the class $H_{loc}^B(D, \mathbb{C}^n)$ is defined to be

$$\mathfrak{b}(D) = \inf\{r(f) : f \in H_{loc}^B(D, \mathbb{C}^n), |f|_\mathcal{B} = 1 = \det f'(0)\}.$$

We prove two lemmas before giving a lower estimate of $\mathfrak{b}(D)$.

Lemma 4.2.18. *Let D be a bounded symmetric domain in $\mathbb{C}^n$, identified as the open unit ball of a JB*-triple $V = (\mathbb{C}^n, \|\cdot\|_{sp})$ in a norm $\|\cdot\|_{sp}$ equivalent to the Euclidean norm $\|\cdot\|_2$ in $\mathbb{C}^n$. Given a (non-zero) bounded linear operator $T : V \longrightarrow \mathbb{C}^n$, we have*

$$\|T(v)\|_2 \geq \frac{s(D) |\det T|}{\|T\|^{n-1}} \qquad (\|v\|_{sp} = 1) \qquad (4.13)$$

where $s(D)$ is defined in (4.12).

Proof. We can consider T as a bounded linear operator from the Euclidean space $\mathbb{C}^n$ to itself since $\|\cdot\|_{sp}$ is equivalent to the Euclidean norm. Let T^* be the adjoint operator. Then T^*T is a positive operator with eigenvalues

$$0 \le \lambda_1 \le \cdots \le \lambda_n = \|T^*T\|.$$

By the minimax principle, we have

$$\lambda_1 = \inf\left\{\frac{\langle T^*T(u), u\rangle}{\|u\|_2^2} : u \in V\backslash\{0\}\right\} \le \inf\left\{\frac{\langle T^*T(u), u\rangle}{\|u\|_2^2} : \|u\|_{sp} = 1\right\}.$$

By the definition of $s(D)$ in (4.12), we have $\|u\|_2 \ge s(D)$ whenever $\|u\|_{sp} = 1$. In particular, for $\|v\|_{sp} = 1$, we have

$$
\begin{aligned}
|\det T|^2 &= \det(T^*T) = \lambda_1 \cdots \lambda_n \le \lambda_1 \|T^*T\|^{n-1} \\
&\le \frac{\langle T^*T(v), v\rangle}{\|v\|_2^2}\|T^*T\|^{n-1} \le \frac{\|T(v)\|_2^2 \|T\|^{2(n-1)}}{s(D)^2}
\end{aligned}
$$

which proves the desired inequality. $\square$

For a locally biholomorphic Bloch map f, we have the following lower estimate for the radius of the largest schlicht ball of f centered at $f(0)$.

Theorem 4.2.19. *Let D be a bounded symmetric domain in $\mathbb{C}^n$. Given $f \in H_{loc}^B(D, \mathbb{C}^n)$ with $\|f\|_d = 1 = \det f'(0)$, we have*

$$r(0, f) \ge s(D) \int_0^1 \frac{(1-t^2)^{n-1}}{(1-t)^{2c(D)}} \exp\left\{\frac{-2c(D)t}{1-t}\right\} dt \ge \frac{s(D)}{2c(D)} \ge \frac{1}{2c(D)}$$

where $s(D)$ is defined in (4.12), and $c(D)$ in (4.7).

Proof. Write $c = c(D)$. As before, D identifies with the open unit ball of a norm equivalent to the Euclidean norm of $\mathbb{C}^n$.

By Lemma 4.2.16, $r(0, f)$ is the Euclidean distance from $f(0)$ to a boundary point of $f(D)$. Let $\Gamma : [0, 1] \longrightarrow \mathbb{C}^n$ be the line segment from

$f(0)$ to a point in $\partial f(D)$, which is of Euclidean length $\ell(\Gamma) = r(0, f)$. Define $\gamma : [0, 1) \longrightarrow D$ by

$$\gamma(s) = f^{-1}(\Gamma(s))$$

where $\gamma(0) = 0$ and $\lim_{s \to 1} \|\gamma(s)\| = 1$.

By (4.13), we have

$$r(0, f) = \ell(\Gamma) = \int_0^1 \|(f \circ \gamma)'(s)\|_2 \, ds$$

$$= \int_0^1 \|f'(\gamma(s))\gamma'(s)\|_2 ds = \int_0^1 \left\| f'(\gamma(s)) \left(\frac{\gamma'(s)}{\|\gamma'(s)\|} \right) \right\|_2 \|\gamma'(s)\| ds$$

$$\geq s(D) \int_0^1 \frac{|\det f'(\gamma(s))|}{\|f'(\gamma(s))\|^{n-1}} d\|\gamma(s)\|$$

where, by Theorem 4.2.10 (i) and (4.6),

$$\int_0^1 \frac{|\det f'(\gamma(s))|}{\|f'(\gamma(s))\|^{n-1}} d\|\gamma(s)\|$$

$$\geq \int_0^1 \frac{(1 - \|\gamma(s)\|^2)^{n-1}}{(1 - \|\gamma(s)\|)^{2c}} \exp \left\{ \frac{-2c\|\gamma(s)\|}{1 - \|\gamma(s)\|} \right\} d\|\gamma(s)\|$$

$$= \int_0^1 \frac{(1 - t^2)^{n-1}}{(1 - t)^{2c}} \exp \left\{ \frac{-2ct}{1 - t} \right\} dt.$$

Finally, $c(D) \geq (n + 1)/2$ by (3.33) and hence we have

$$r(0, f) \geq s(D) \int_0^1 \frac{1}{(1 - t)^2} \exp \left\{ \frac{-2c(D)t}{1 - t} \right\} dt \geq \frac{s(D)}{2c(D)}.$$

This completes the proof. $\qquad \square$

Before making use of Theorem 4.2.19 to estimate the Bloch constant $\mathfrak{b}(D)$, we derive two lemmas first.

Lemma 4.2.20. *Let D be a bounded symmetric domain in $\mathbb{C}^n$. For each $f \in H^B_{loc}(D, \mathbb{C}^n)$ with $\det f'(0) = 1$, there is a Bloch function $h \in H^B_{loc}(D, \mathbb{C}^n)$ such that $|h|_{\mathcal{B}} \leq |f|_{\mathcal{B}}$, $\det h'(0) = 1$, $r(h) \leq r(f)$ and*

$$d(h) := \sup\{|\det(h \circ \varphi)'(0)| : \varphi \in \text{Aut } D\} = 1.$$

Proof. Let $d(f) = \sup\{|\det(f \circ \varphi)'(0)| : \varphi \in \operatorname{Aut} D\}$. We have $d(f) \geq 1$. If $d(f) = 1$, there is nothing to prove. Otherwise, we can pick a sequence $(\varphi_k) \in \operatorname{Aut} D$ such that

$$d_k := |\det(f \circ \varphi_k)'(0)| > 1$$

and $\lim_k d_k = d(f)$. Define a Bloch function $h_k : D \longrightarrow \mathbb{C}^n$ by

$$h_k = \frac{f \circ \varphi_k - f \circ \varphi_k(0)}{d_k^{1/n}}.$$

Then we have $h_k(0) = 0$, $\det h_k'(0) = 1$ and $|h_k|_{\mathcal{B}} \leq |f|_{\mathcal{B}}$. Since $d_k > 1$, we have $r(h_k) \leq r(f)$. By Lemma 4.2.13, $\{h_k\}$ is a normal family and hence it has a subsequence converging locally uniformly to h, which satisfies $|h|_{\mathcal{B}} \leq |f|_{\mathcal{B}}$, $\det h'(0) = 1$, $r(h) \leq r(f)$ and $d(h) = 1$. $\qquad\square$

Lemma 4.2.21. *Let D be a bounded symmetric domain in $\mathbb{C}^n$. Then one can find a Bloch map $f \in H_{loc}^B(D, \mathbb{C}^n)$ satisfying $|f|_{\mathcal{B}} = 1 = d(f) = \det f'(0)$ and $r(f) = \mathfrak{b}(D)$.*

Proof. For $k = 1, 2, \ldots$, we can find a Bloch function $f_k \in H_{loc}^B(D, \mathbb{C}^n)$ satisfying

$$|f_k|_{\mathcal{B}} = 1 = \det f_k'(0) \quad \text{and} \quad r(f_k) \leq \mathfrak{b}(D) + \frac{1}{k}.$$

By the preceding lemma, there is a sequence (h_k) in $H_{loc}^B(D, \mathbb{C}^n)$ such that $|h_k|_{\mathcal{B}} \leq |f_k|_{\mathcal{B}}$, $\det h_k'(0) = 1 = d(h_k)$ and $r(h_k) \leq r(f_k) \leq \mathfrak{b}(D) + \frac{1}{k}$.

By Corollary 4.2.14, $(h_k - h_k(0))$ contains a subsequence converging locally uniformly to a Bloch function $h \in H_{loc}^B(D, \mathbb{C}^n)$, which satisfies $\det h'(0) = 1 = d(h) = |h|_{\mathcal{B}}$ and $r(h) = \mathfrak{b}(D)$. $\qquad\square$

Corollary 4.2.22. *Let D be a bounded symmetric domain in $\mathbb{C}^n$. Then the Bloch constant $\mathfrak{b}(D)$ satisfies*

$$\mathfrak{b}(D) \geq s(D) \int_0^1 \frac{(1-t^2)^{n-1}}{(1-t)^{2c(D)}} \exp\left\{\frac{-2c(D)t}{1-t}\right\} dt.$$

Proof. By Lemma 4.2.21, we can find a Bloch function $f \in H_{loc}^B(D, \mathbb{C}^n)$ satisfying $|f|_{\mathcal{B}} = 1 = d(f) = \det f'(0)$ and $r(f) = \mathfrak{b}(D)$. For each $z \in D$, let g_z be the Möbius transformation induced by z. Since

$$(1 - \|z\|^2)^{c(D)/n} |\det f'(z)|^{1/n} \leq |\det f'(z)|^{1/n} = |\det(f \circ g_z)'(0)|^{1/n}$$
$$\leq d(f)^{1/n} = 1,$$

we have $1 = |\det f'(0)|^{1/n} \leq \|f\|_d \leq 1$ and $\|f\|_d = 1$.

Now $\mathfrak{b}(D) = r(f) \geq r(0, f)$ and Theorem 4.2.19 applies. □

Notes. The lower bound for the Bloch constant has been improved in [35] to $\mathfrak{b} > \frac{\sqrt{3}}{4} + 2 \times 10^{-4}$. Chern [36] has generalised Bloch's theorem to holomorphic maps between two Hermitian manifolds of the same dimension. The proofs of Lemma 4.2.9 as well as Theorem 4.2.5 and Theorem 4.2.10 follow more or less the same idea in [123] for the disc $\mathbb{D}$, where the classical Julia's lemma plays an important role. Theorem 4.2.10 is a special case of a more general result shown in [42, Theorem 4.1]. The inequality for the Bloch semi-norm in Example 4.10 is an extension of a similar inequality for the Euclidean balls derived in [162, Theorem 4.7].

The notion of a $\mathbb{C}^n$-valued Bloch map on a finite dimensional bounded symmetric domain, under the name of *normal mapping of finite order*, was first introduced by Hahn [73]. An equivalent, but perhaps less technical, definition of a complex-valued Bloch function on a bounded homogeneous domain was later given by Timoney in [162]. In [121], $\mathbb{C}^n$-valued Bloch maps on Euclidean balls are studied. In [17], complex-valued Bloch functions on Hilbert balls were introduced and studied. The definitions for Bloch maps in these papers are all equivalent to the one we introduce in this section.

Theorem 4.2.19 is a particular case of [42, Theorem 5.6] for all

finite dimensional bounded symmetric domains, which reduces to the
results of [121, Theorem 8] and [19, Corollary 3] for Euclidean balls,
and to [174, Theorem 3.4] for polydiscs. Our proof is similar to the
one given in [121]. A lower bound for the Bloch constant for Type I
domains has also been shown in [61]. Lemma 4.2.21 is analogous to [61,
Theorem 1.3].

4.3   Banach spaces of Bloch functions

In this section, we focus our attention on complex-valued Bloch functions on bounded symmetric domains of any dimension. These functions
form a Banach space in the quotient Bloch semi-norm, where functions
of null Bloch semi-norm are identified. We generalise a number of finite
dimensional results on Bloch functions to infinite dimensional bounded
symmetric domains.

In what follows, $H(D, B)$ denotes the class of holomorphic maps
between two domains D and B in complex Banach spaces. We begin
by showing several equivalent criteria of Bloch functions.

Theorem 4.3.1. *Let D be a bounded symmetric domain realized as the
open unit ball of a JB*-triple V and let $f : D \longrightarrow \mathbb{C}$ be a holomorphic
function. The following conditions are equivalent:*

(i) *f is a Bloch function.*

(ii) $\sup\{Q_f(z) : z \in D\} < \infty$ *where*

$$Q_f(z) = \sup\left\{\frac{|f'(z)v|}{\mathcal{K}_D(z, v)} : v \in V \setminus \{0\}\right\}$$

and $\mathcal{K}_D(\cdot, \cdot)$ is the Kobayashi differential metric.

(iii) *The radii of the schlicht discs in $f(D)$ are bounded above.*

(iv) f is uniformly continuous as a function from the metric space (D, k_D) to the Euclidean plane $\mathbb{C}$, where k_D is the Kobayashi distance.

(v) The family $F_f(0) := \{f \circ g - (f \circ g)(0) : g \in \operatorname{Aut} D\}$ is bounded on rD for $0 < r < 1$.

(vi) The family $\{f \circ h : h \in H(\mathbb{D}, D)\}$ consists of Bloch functions on $\mathbb{D}$ with uniformly bounded Bloch semi-norm.

(vii) The family $\{f \circ h - (f \circ h)(0) : h \in H(\mathbb{D}, D)\}$ is normal.

If $\dim V = n < \infty$, these conditions are equivalent to

(viii) $\quad \beta_f := \sup_{z \in D} \left\{ \sup \left\{ \frac{|f'(z)v|}{h_z(v, v)^{1/2}} : v \in V \setminus \{0\} \right\} \right\} < \infty$

where $h_z(v, v)$ is the Bergman metric.

Proof. The equivalence of (ii) and (viii) is immediate from Example 3.5.13, which establishes the inequalities

$$\mathcal{K}_D(p, v) \leq h_p(v, v)^{1/2} \leq \sqrt{2n}\mathcal{K}_D(p, v).$$

(i) $\Leftrightarrow$ (ii). We show, in fact, $|f|_{\mathcal{B}} = \sup\{Q_f(z) : z \in D\} < \infty$. Let $z \in D \setminus \{0\}$ and let $g_z \in \operatorname{Aut} D$ be the Möbius transformation induced by z. Then we have

$$|f'(z)v| \leq \|(f \circ g_z)'(0)\| \|g'_{-z}(z)v\| \leq \kappa(z, v)|f|_{\mathcal{B}}.$$

We also have

$$|f'(0)v| \leq \|v\| |f|_{\mathcal{B}} = \kappa(0, v)|f|_{\mathcal{B}}.$$

This gives $\sup\{Q_f(z) : z \in D\} \leq |f|_{\mathcal{B}}$.

On the other hand, given $v \in V \setminus \{0\}$ and $g \in \operatorname{Aut} D$, we have

$$\left| (f \circ g)'(0) \left(\frac{v}{\|v\|} \right) \right| = \frac{|f'(g(0))g'(0)v|}{\kappa(g(0), g'(0)v)} \leq Q_f(g(0))$$

and hence $\|f\|_{\mathcal{B}} \leq \sup\{Q_f(z) : z \in D\}$.

(i) $\Rightarrow$ (iv). Let $z_1, z_2 \in D$ and let $\gamma : [0,1] \to D$ be an arbitrary piecewise C^1 smooth curve with $\gamma(0) = z_1$ and $\gamma(1) = z_2$. Then

$$
\begin{aligned}
|f(z_1) - f(z_2)| &\leq \int_0^1 |f'(\gamma(t))\gamma'(t)|dt \\
&\leq \int_0^1 Q_f(\gamma(t))\mathcal{K}_D(\gamma(t), \gamma'(t))dt \\
&\leq |f|_{\mathcal{B}} \int_0^1 \mathcal{K}_D(\gamma(t), \gamma'(t))dt.
\end{aligned}
$$

This gives $|f(z_1) - f(z_2)| \leq |f|_{\mathcal{B}} k_D(z_1, z_2)$ and proves uniform continuity of the function $f : (D, k_D) \longrightarrow \mathbb{C}$.

(iv) $\Rightarrow$ (v). Let $r \in (0,1)$. We show that $F_f(0)$ is bounded on rD, where D is the open unit ball of V. By assumption, there exists $\delta_0 > 0$ such that $|f(z_1) - f(z_2)| \leq 1$ whenever $z_1, z_2 \in D$ satisfy $k_D(z_1, z_2) < \delta_0$. For each $s \in (0, r)$, we can find a partition $0 = t_1 < t_2 < \cdots < t_n = s$ such that $\rho(t_j, t_{j-1}) < \delta_0$, where ρ is the Poincaré distance on $\mathbb{D}$, and n does not depend on s. Let $z \in rD \setminus \{0\}$ and $s = \|z\|$. Let $w = z/\|z\| \in \partial D$ and define $\phi : \mathbb{D} \to D$ by $\phi(\zeta) = \zeta w$. For every $g \in \operatorname{Aut} D$, we have

$$
\begin{aligned}
|f(g(z)) - f(g(0))| &\leq \sum_{j=1}^{n-1} |f(g(\phi(t_j))) - f(g(\phi(t_{j+1})))| \\
&\leq n - 1
\end{aligned}
$$

since $k_D(g(\phi(t_j)), g(\phi(t_{j+1}))) = k_D(\phi(t_j), \phi(t_{j+1})) \leq \rho(t_j, t_{j+1}) < \delta_0$ for $1 \leq j \leq n - 1$.

(v) $\Rightarrow$ (i). Fix $r \in (0,1)$. Then there is a constant $M > 0$ such that

$$
|f(g(z)) - f(g(0))| \leq M
$$

for $\|z\| < r$ and $g \in \operatorname{Aut} D$. By the Schwarz lemma, we have

$$
\|(f \circ g)'(0)\| \leq \frac{M}{r}, \qquad (g \in \operatorname{Aut} D)
$$

which gives $|f|_\mathcal{B} \leq M/r$.

(i) $\Rightarrow$ (iii). Let $\Delta = \{\zeta \in \mathbb{C} : |\zeta - \zeta_0| < r\}$ be a schlicht disc in $f(D)$. Then there is a holomorphic map $h : \mathbb{D} \longrightarrow D$ such that

$$(f \circ h)(\zeta) = \zeta_0 + r\zeta \qquad (\zeta \in \mathbb{D}).$$

Let $z = h(0)$. Since $g_{-z} \circ h : U \longrightarrow D$ is a holomorphic map and $g_{-z} \circ h(0) = 0$, we have $\|(g_{-z} \circ h)'(0)\| \leq 1$ by Schwarz lemma. Hence

$$
\begin{aligned}
r &= (f \circ h)'(0) = |f'(h(0))h'(0)| = |f'(z)g_z'(0)(g_{-z} \circ h)'(0)| \\
&\leq \|f'(z)g_z'(0)\| \leq |f|_\mathcal{B}.
\end{aligned}
$$

(iii) $\Rightarrow$ (i). Let the radii of the schlicht discs in $f(D)$ be bounded above by $R > 0$. Let $y \in \partial D$ be arbitrarily fixed and define $h : \zeta \in \mathbb{D} \mapsto \zeta y \in D$. For each $g \in \operatorname{Aut} D$, the map $f \circ g \circ h : \mathbb{D} \longrightarrow \mathbb{D}$ is holomorphic. If $(f \circ g \circ h)'(0) \neq 0$, then Bloch's theorem implies the existence of a schlicht disc in $(f \circ g \circ h)(\mathbb{D})$ of radius

$$\mathfrak{b}|(f \circ g \circ h)'(0)| = \mathfrak{b}|(f \circ g)'(0)y|$$

where $\mathfrak{b}$ is the Bloch constant. Therefore we have $\mathfrak{b}|(f \circ g)'(0)y| \leq R$. We also have this inequality in the case $(f \circ g \circ h)'(0) = 0$. It follows that $|f|_\mathcal{B} \leq R/\mathfrak{b}$.

(iii) $\Rightarrow$ (vi). Let the radii of the schlicht discs in $f(D)$ be bounded above by $R > 0$. Let $h \in H(\mathbb{D}, D)$. Then $f \circ h$ is a holomorphic function on D. If $(f \circ h)'(0) \neq 0$, then Bloch's theorem implies that $(f \circ h)(\mathbb{D})$ contains a schlicht disc of radius $\mathfrak{b}|(f \circ h)'(0)|$ where $\mathfrak{b}$ is the Bloch constant. Therefore, we have $\mathfrak{b}|(f \circ h)'(0)| \leq R$. We also have this inequality if $(f \circ h)'(0) = 0$. It follows that $|f \circ h|_\mathcal{B} \leq R/\mathfrak{b}$.

(vi) $\Rightarrow$ (iii). Let $\Delta = \{\zeta \in \mathbb{C} : |\zeta - \zeta_0| < r\}$ be a schlicht disc in $f(D)$. There is a holomorphic map $h : \mathbb{D} \longrightarrow D$ such that

$$(f \circ h)(\zeta) = \zeta_0 + r\zeta \qquad (\zeta \in \mathbb{D}).$$

This gives
$$r = |(f \circ h)'(0)| \le |f \circ h|_{\mathcal{B}}.$$

(vi) $\Rightarrow$ (vii). By Montel's theorem, it suffices show that the family $\{f \circ h - (f \circ h)(0) : h \in H(\mathbb{D}, D)\}$ is uniformly bounded on $\mathbb{D}(0, r) = \{z \in \mathbb{C} : |z| < r\}$ for $0 < r < 1$. Indeed, applying (i) $\Rightarrow$ (iv) to the Bloch functions $f \circ h$, one finds a constant $M > 0$ such that

$$|f \circ h(z) - f \circ h(0)| \le M\rho(z, 0) = M \tanh^{-1} |z| \le M \tanh^{-1} r$$

for $|z| < r$.

(vii) $\Rightarrow$ (vi). This can verified by a similar argument for (v) $\Rightarrow$ (i). $\qquad\square$

An examination of the proof of condition (iv) in Theorem 4.3.1 reveals that a holomorphic function $f : (D, k_D) \longrightarrow \mathbb{C}$ is a Bloch function if and only if it is a Lipschitz function. Actually, the Lipschitz constant in this case is given by the Bloch seminorm of f.

Proposition 4.3.2. *Let f be a Bloch function on a bounded symmetric domain D. Then we have*

$$|f|_{\mathcal{B}} = \sup \left\{ \frac{|f(z) - f(w)|}{k_D(z, w)} : z, w \in D, z \ne w \right\}.$$

Proof. Write
$$\operatorname{lip}(f) = \sup_{z \ne w} \frac{|f(z) - f(w)|}{k_D(z, w)}.$$

We have shown in Theorem 4.3.1 that $|f(z) - f(w)| \le |f|_{\mathcal{B}} k_D(z, w)$ for all $z, w \in D$. This implies that $\operatorname{lip}(f) \le |f|_{\mathcal{B}}$. On the other hand, given $g \in \operatorname{Aut} D$ and $w \in D$, we have for every $t \in (0, 1)$,

$$
\begin{aligned}
|f(g(tw)) - f(g(0))| &\le \operatorname{lip}(f) k_D(g(tw), g(0)) \\
&= \operatorname{lip}(f) k_D(tw, 0) \\
&= \operatorname{lip}(f) \tanh^{-1}(t\|w\|).
\end{aligned}
$$

It follows that

$$|(f \circ g)'(0)w| \leq \mathrm{lip}(f)\|w\| \qquad (w \in D)$$

and hence $|f|_\mathcal{B} \leq \mathrm{lip}(f)$.   □

For a finite dimensional bounded symmetric domain D, it has been observed in [11] that

$$h(z,w) \geq \sup\{|f(z) - f(w)| : f \in H(D,\mathbb{C}), \beta_f \leq 1\} \qquad (z,w \in D)$$

where $h(\cdot,\cdot)$ is the Bergman distance on D. In view of the fact that this inequality becomes an equality for the unit disc $\mathbb{D}$, a natural question has been raised in [11] whether the equality also holds in higher dimensions. We show below that the equality actually holds for all dimensions as soon as the Bergman distance is replaced by the Kobayashi distance.

Proposition 4.3.3. *Let D be a bounded symmetric domain. Then we have*

$$k_D(z,w) = \sup\{|f(z) - f(w)| : f \in H(D,\mathbb{C}), |f|_\mathcal{B} \leq 1\} \qquad (4.14)$$

for $z,w \in D$.

Proof. By Proposition 4.3.2, we have

$$\sup\{|f(z) - f(w)| : f \in H(D,\mathbb{C}), |f|_\mathcal{B} \leq 1\} \leq k_D(z,w).$$

To prove the reverse inequality, fix any $a \in D \setminus \{0\}$ and let $f = \psi \circ \phi_a$ be the Bloch function on D defined in Example 4.2.3. Then

$$\begin{aligned} k_D(a,0) &= \tanh^{-1}\|a\| = |f(a) - f(0)| \\ &\leq \sup\{|f(a) - f(0)| : f \in H(D,\mathbb{C}), |f|_\mathcal{B} \leq 1\}. \end{aligned}$$

By composing with an automorphism of D, we have

$$k_D(z,w) \leq \sup\{|f(z) - f(w)| : f \in H(D,\mathbb{C}), |f|_\mathcal{B} \leq 1\}$$

for any $z,w \in D$. This completes the proof.   □

Let D be a bounded symmetric domain realised as the open unit ball of a JB*-triple V. Let $\mathcal{B}(D)$ denote the complex vector space of all $\mathbb{C}$-valued Bloch functions on D. Usually, one can turn the Bloch semi-norm $|\cdot|_{\mathcal{B}}$ into a norm on a quotient of $\mathcal{B}(D)$. Instead, we define a norm on $\mathcal{B}(D)$ by

$$\|f\|_{\mathcal{B}} := |f|_{\mathcal{B}} + |f(0)| \qquad (f \in \mathcal{B}(D)).$$

This is indeed a norm by (4.6). We call it the *Bloch norm* on the space $\mathcal{B}(D)$.

Theorem 4.3.4. *Let D be a bounded symmetric domain realised as the open unit ball of a JB*-triple V. Then $\mathcal{B}(D)$ is a Banach space in the Bloch norm $\|\cdot\|_{\mathcal{B}}$.*

Proof. Let (f_k) be a Cauchy sequence in $\mathcal{B}(D)$. By (4.6), we have

$$\|f_k'(z) - f_p'(z)\| \le \frac{\|f_k - f_p\|_{\mathcal{B}}}{1 - \|z\|^2}$$

for all $z \in D$ and $k, p = 1, 2 \ldots$. On the other hand,

$$|f_k(z) - f_p(z)| \le |f_k(0) - f_p(0)| + \int_0^1 \|f_k'(tz) - f_p'(tz)\| \|z\| dt$$

$$\le |f_k(0) - f_p(0)| + \frac{\|f_k - f_p\|_{\mathcal{B}}}{1 - \|z\|^2}$$

for all k, p and $z \in D$. It follows that (f_k) is a Cauchy sequence in $H(D, \mathbb{C})$, in the locally uniform topology. Hence (f_k) converges locally uniformly to some function $f \in H(D, \mathbb{C})$.

To complete the proof, we show $\|f_k - f\|_{\mathcal{B}} \to 0$ as $k \to \infty$. Let $\varepsilon > 0$. There exists $k_0 \in \mathbb{N}$ such that

$$\|f_k - f_p\|_{\mathcal{B}} < \varepsilon$$

for $k, p \ge k_0$, which gives

$$|f_k(0) - f_p(0)| + \|(f_k'(g(0)) - f_p'(g(0)))g'(0)\| < \varepsilon \qquad (g \in \operatorname{Aut} D, \ k, p \ge k_0).$$

For each $p \geq k_0$ and $g \in \operatorname{Aut} D$, the locally uniform convergence of the sequence (f_k) to f implies that $f_k(0) \to f(0)$ and

$$\|(f_k'(g(0)) - f_p'(g(0)))g'(0)\| \to \|(f'(g(0)) - f_p'(g(0)))g'(0)\|$$

as $k \to \infty$, and hence

$$|f(0) - f_p(0)| + \|(f'(g(0)) - f_p'(g(0)))g'(0)\| \leq \varepsilon$$

for $p \geq k_0$ and $g \in \operatorname{Aut} D$. Consequently, $\|f_p - f\|_{\mathcal{B}} \leq \varepsilon$ for $p \geq k_0$, and $f = (f - f_p) + f_p \in \mathcal{B}(D)$ as well as $\lim_{p \to \infty} \|f_p - f\|_{\mathcal{B}} = 0$. $\qquad \square$

Definition 4.3.5. Given a bounded symmetric domain D, the Banach space $(\mathcal{B}(D), \|\cdot\|_{\mathcal{B}})$ is called the *Bloch space* of D.

Remark 4.3.6. In literature, the Bloch norm for a finite dimensional bounded symmetric domain D (realised as the open unit ball of a JB*-triple) is often defined by the semi-norm β_f given in Theorem 4.3.1 (viii), and the Bloch norm β is defined by

$$\beta(f) := \beta_f + |f(0)|$$

for a Bloch function f on D. With this norm, $\mathcal{B}(D)$ also forms a Banach space. It should be pointed out that the two Banach spaces $(\mathcal{B}(D), \|\cdot\|_{\mathcal{B}})$ and $(\mathcal{B}(D), \beta)$ are not identical, but linearly homeomorphic. Throughout the chapter, we only consider the Bloch space $(\mathcal{B}(D), \|\cdot\|_{\mathcal{B}})$.

Although the functions in $\mathcal{B}(D)$ need not be bounded, it is of interest to note that on the disc $\mathbb{D}$, a Bloch function $f(z) = \sum_{n=0}^{\infty} a_n z^n$ has bounded Taylor coefficients a_n. It has been shown in [12, Lemma 2.1] that $|a_n| \leq 2|f|_{\mathcal{B}}$. In fact, this bound can be improved slightly in all dimensions.

Theorem 4.3.7. *Let D be the open unit ball of a JB*-triple V. Given $f \in \mathcal{B}(D)$ with a power series representation*

$$f(z) = \sum_{n=0}^{\infty} p_n(z) \qquad (z \in D)$$

by homogeneous polynomials p_n of degree n, we have

$$\|p_1\| \le |f|_{\mathcal{B}} \quad \text{and} \quad \|p_n\| \le \sqrt{e}|f|_{\mathcal{B}} \quad (n \ge 2).$$

Proof. For each $v \in V$ with $\|v\| = 1$, we define a Bloch function $\phi_v : \mathbb{D} \longrightarrow \mathbb{C}$ by

$$\phi_v(\zeta) = f(\zeta v) = \sum_{n=0}^{\infty} p_n(v)\zeta^n \quad (\zeta \in \mathbb{D}).$$

The first inequality follows from the definition of the semi-norm $|\phi_v|_{\mathcal{B}}$.

Given a Bloch function $\psi(\zeta) = \sum_{n=0}^{\infty} a_n \zeta^n \in \mathcal{B}(\mathbb{D})$ and $r \in (0,1)$, we have

$$a_n r^n = \frac{1}{2\pi} \int_{-\pi}^{\pi} \psi(re^{i\theta}) e^{-in\theta} d\theta.$$

Hence

$$n|a_n|r^{n-1} \le \sup_{|\zeta|=r} |\psi'(\zeta)|$$

for $n \ge 2$, which implies

$$|a_n| \le \frac{1}{nr^{n-1}(1-r^2)} |\psi|_{\mathcal{B}}.$$

Taking $r^2 = 1 - 1/n$, we obtain

$$|a_n| \le \left(1 + \frac{1}{n-1}\right)^{(n-1)/2} |\psi|_{\mathcal{B}} \le \sqrt{e}|\psi|_{\mathcal{B}}$$

for $n \ge 2$. Therefore we have shown $|p_n(v)| \le \sqrt{e}\,|\phi_v|_{\mathcal{B}}$ for $n \ge 2$ and

$v \in V$ with $\|v\| = 1$. It follows that

$$
\begin{aligned}
\|p_n\| &= \sup_{\|v\|=1} |p_n(v)| \leq \sqrt{e} |\phi_v|_{\mathcal{B}} \\
&= \sqrt{e} \, | \sup_{\zeta \in \mathbb{D}} (1 - |\zeta|^2) |\phi_v'(\zeta)| \\
&= \sqrt{e} \, | \sup_{\zeta \in \mathbb{D}} (1 - |\zeta v|^2) |f'(\zeta v) v| \leq \sqrt{e} |f|_{\mathcal{B}}
\end{aligned}
$$

for $n \geq 2$. $\qquad\square$

Corollary 4.3.8. *The Bloch space $\mathcal{B}(\mathbb{D})$ contains an isomorphic copy of the Banach space ℓ_∞ of bounded complex sequences.*

Proof. Let $(a_n) \in \ell_\infty$. Then the function $f(z) = a_0 + \sum_{n=1}^\infty a_n z^{2^n}$ is holomorphic on $\mathbb{D}$ and

$$
\begin{aligned}
|zf'(z)| &\leq \|(a_n)\|_\infty (2|z|^2 + 2^2|z|^4 + \cdots) \\
&= \|(a_n)\|_\infty (1 - |z|)(2|z|^2 + 2|z|^3 + (2 + 2^2)|z|^4 \\
&\qquad + (2 + 2^2)|z|^5 + \cdots + (2 + 2^2 + 2^3)|z|^8 + \cdots) \\
&= \|(a_n)\|_\infty (1 - |z|) \sum_{n=2}^\infty \left(\sum_{\{k : 2^k \leq n\}} 2^k \right) |z|^n \\
&\leq \|(a_n)\|_\infty (1 - |z|) \sum_{n=2}^\infty 2n|z|^n \leq \|(a_n)\|_\infty (1 - |z|) \frac{|z|}{(1 - |z|)^2}.
\end{aligned}
$$

Hence we have $(1 - |z|^2)|f'(z)| \leq 2\|(a_n)\|_\infty$ and f is a Bloch function on $\mathbb{D}$.

One can therefore define a continuous linear monomorphism

$$
(a_n) \in \ell_\infty \mapsto f(z) = a_0 + \sum_{n=0}^\infty a_n z^{2^n} \in \mathcal{B}(\mathbb{D})
$$

which is actually a homeomorphism onto a closed subspace of $\mathcal{B}(\mathbb{D})$, by Theorem 4.3.7. $\qquad\square$

We infer from the above corollary that the Bloch space $\mathcal{B}(\mathbb{D})$ is not reflexive. In fact, it has been shown in [12] that $\mathcal{B}(\mathbb{D})$ is the second dual space of its subspace

$$\mathcal{B}_0(\mathbb{D}) := \{f \in \mathcal{B}(\mathbb{D}) : \lim_{|z| \to 1} (1 - |z|^2)|f'(z)| = 0\}.$$

The Banach space $\mathcal{B}_0(\mathbb{D})$ is separable and is the $|\cdot|_\mathcal{B}$-closure of the polynomials (restricted to $\mathbb{D}$). One can extend the definition of this space to higher dimensions.

Polynomials, when restricted to a bounded symmetric domain D, are Bloch functions, by Example 4.2.3.

Definition 4.3.9. Let D be a bounded symmetric domain. The *little Bloch space* $\mathcal{B}_0(D)$ is defined to be the $\|\cdot\|_\mathcal{B}$-closure of the polynomials on D.

Lemma 4.3.10. *Let $f \in \mathcal{B}_0(D)$. Then we have $\lim_{\|z\| \to 1} (1 - \|z\|^2)\|f'(z)\| = 0$.*

Proof. Let $\varepsilon > 0$ and $\|f - p\|_\mathcal{B} < \varepsilon$ for some polynomial $p \in \mathcal{B}(D)$. Then we have

$$
\begin{aligned}
(1 - \|z\|^2)\|f'(z)\| &\leq (1 - \|z\|^2)\|(f - p)'(z)\| + (1 - \|z\|^2)\|p'(z)\| \\
&\leq \|f - p\|_\mathcal{B} + (1 - \|z\|^2)\|p'(z)\| \\
&< \varepsilon + (1 - \|z\|^2)\|p'(z)\|.
\end{aligned}
$$

Since $\|p'(\cdot)\|$ is bounded on D, we see that $(1 - \|z\|^2)\|f'(z)\| \to 0$ as $\|z\| \to 1$. $\qquad\square$

The converse of the previous lemma is true if D is a Hilbert ball and we actually have the following result.

Theorem 4.3.11. *Let D be a Hilbert ball and $f \in \mathcal{B}(D)$. The following conditions are equivalent.*

(i) $f \in \mathcal{B}_0(D)$.

(ii) $Q_f(z) \to 0$ *uniformly as* $z \to \partial D$, *that is, for each* $\varepsilon > 0$, *there exists* $\delta \in (0,1)$ *such that* $Q_f(z) < \varepsilon$ *whenever* $\delta < \|z\| < 1$.

(iii) $\lim\limits_{\|z\| \to 1} (1 - \|z\|^2)\|f'(z)\| = 0$.

(iv) $\lim\limits_{\|z\| \to 1} \sup\limits_{w \in D \setminus \{z\}} (1 - \|z\|^2)^{1/2}(1 - \|w\|^2)^{1/2} \dfrac{|f(z) - f(w)|}{\|z - w\|} = 0$.

Proof. (i) $\Rightarrow$ (ii). Let $f \in \mathcal{B}_0(D)$. Then

$$\frac{|f'(z)x|}{\kappa(z,x)} = \frac{|f'(z)x|}{\|B(z,z)^{-1/2}x\|} \leq \frac{\|f'(z)\|\|x\|\|B(z,z)^{1/2}\|}{\|x\|}$$

$$= \|f'(z)\|\|B(z,z)^{1/2}\|$$

which implies $Q_f(z) \leq \|f'(z)\|\sqrt{1 - \|z\|^2} \to 0$ as $\|z\| \to 1$ since the null convergence is true for all polynomials.

(ii) $\Rightarrow$ (iii). This follows from the inequality

$$Q_f(z) \geq \|f'(z)\|(1 - \|z\|^2)$$

since

$$\frac{|f'(z)x|}{\kappa(z,x)} = \frac{|f'(z)x|}{\|B(z,z)^{-1/2}x\|} \geq \frac{|f'(z)x|(1 - \|z\|^2)}{\|x\|}$$

for $x \neq 0$.

(iii) $\Rightarrow$ (i). Let $f \in \mathcal{B}(D)$ satisfy condition (iii). Let $\varepsilon > 0$. There exists $r \in (1/\sqrt{2}, 1)$ such that $(1 - \|z\|^2)\|f'(z)\| < \varepsilon/2$ for $r \leq \|z\| < 1$.

Let $f_k(z) = f((1 - 1/k)z)$ for $k = 1, 2, \ldots$. Then f_k is holomorphic on $\overline{D}$ and hence there is a polynomial p_k such that

$$\sup_{z \in \overline{D}} |f_k(z) - p_k(z)| \leq \frac{1}{k}$$

for each $k \geq 1$. By Example 4.2.3, $f_k - p_k \in \mathcal{B}(D)$ and

$$|f_k - p_k|_{\mathcal{B}} \leq \frac{1}{k}.$$

By the maximum principle, we have $(1 - \|z\|^2)\|f_k'(z)\| < \varepsilon/2$ for $r \leq \|z\| < 1$ and for all $k \geq 1$. Since the sequence (f_k) converges to f uniformly on each closed ball in D, there exists $K \geq 1$ such that

$$\|f'(z) - f_k'(z)\| < \varepsilon$$

for $\|z\| \leq r$ and $k \geq K$. This implies $(1 - \|z\|^2)\|f'(z) - f_k'(z)\| < \varepsilon$ for $z \in D$ and $k \geq K$.

By (4.10), we have

$$|f - f_k|_\mathcal{B} \leq \frac{2 + \sqrt{2}}{k}$$

for $k \geq K$. It follows that

$$|f - p_k|_\mathcal{B} \leq |f - f_k|_\mathcal{B} + |f_k - p_k|_\mathcal{B} \leq \frac{3 + \sqrt{2}}{k}$$

for $k \geq K$, proving that $f \in \mathcal{B}_0(D)$.

(i) $\Rightarrow$ (iv). Let $f \in \mathcal{B}_0(D)$. For $z, w \in D$, we have

$$|f(z) - f(w)| = \left| \int_0^1 f'(tz + (1-t)w)(z - w)dt \right|$$

$$\leq \int_0^1 \|f'(tz + (1-t)w)\| \|z - w\| dt \leq \int_0^1 \frac{|f|_\mathcal{B} \|z - w\|}{1 - \|tz + (1-t)w\|^2} dt$$

$$\leq |f|_\mathcal{B} \|z - w\| \int_0^1 \frac{dt}{1 - \|tz + (1-t)w\|}$$

by (4.6). Observe that

$$1 - \|tz + (1-t)w\| \geq 1 - t\|z\| - (1-t)\|w\|$$

$$= t(1 - \|z\|) + (1-t)(1 - \|w\|) \geq 2\sqrt{t(1-t)(1 - \|z\|)(1 - \|w\|)}.$$

Since $\int \frac{dt}{\sqrt{t(1-t)}} = \cos^{-1}(1 - 2t)$, we deduce

$$\frac{|f(z) - f(w)|}{\|z - w\|} \leq \frac{\pi|f|_\mathcal{B}}{2\sqrt{(1 - \|z\|)(1 - \|w\|)}}$$

and

$$(1 - \|z\|^2)^{1/2}(1 - \|w\|^2)^{1/2}\frac{|f(z) - f(w)|}{\|z - w\|} \leq \frac{\pi|f|_\mathcal{B}}{2}. \qquad (4.15)$$

For each $t \in (0, 1)$, define $f_t(z) = f(tz)$ for $z \in D$. Applying (4.15) twice, we get

$$(1 - \|z\|^2)^{1/2}(1 - \|w\|^2)^{1/2}\frac{|(f - f_t)(z) - (f - t_t)(w)|}{\|z - w\|} \leq \frac{\pi|f - f_t|_\mathcal{B}}{2}$$

$$(4.16)$$

and

$$(1 - \|z\|^2)^{1/2}(1 - \|w\|^2)^{1/2}\frac{|f_t(z) - f_t(w)|}{\|tz - tw\|}$$

$$= \frac{t\sqrt{1 - \|z\|^2}\sqrt{1 - \|w\|^2}}{\sqrt{1 - \|tz\|^2}\sqrt{1 - \|tw\|^2}}(\sqrt{1 - \|tz\|^2}\sqrt{1 - \|tw\|^2})\frac{|f_t(z) - f_t(w)|}{\|tz - tw\|}$$

$$\leq \frac{\pi t}{2(1 - t)}(1 - \|z\|^2)^{1/2}|f|_\mathcal{B}. \qquad (4.17)$$

We have seen in the proof of (i) $\Rightarrow$ (iii) above that $|f - f_t|_\mathcal{B} \to 0$ as $t \to 1$. If $t_0 \in (0, 1)$ makes $|f - f_{t_0}|_\mathcal{B}$ sufficiently small, then

$$\lim_{\|z\| \to 1} \frac{\pi t_0}{2(1 - t_0)}(1 - \|z\|^2)^{1/2}|f|_\mathcal{B} = 0.$$

We therefore conclude from (4.16) and (4.17) that

$$\lim_{\|z\| \to 1} \sup_{w \in D \setminus \{z\}} (1 - \|z\|^2)^{1/2}(1 - \|w\|^2)^{1/2}\frac{|f(z) - f(w)|}{\|z - w\|} = 0$$

by the triangle inequality.

(iv) $\Rightarrow$ (iii). For each $\|w\| = 1$ and $z \in D$, we have $z + tw \in D$ for small $t > 0$ and

$$(1 - \|z\|^2)|f'(z)w| = \lim_{t \downarrow 0} \sqrt{1 - \|z\|^2}\sqrt{1 - \|z + tw\|^2}\frac{|f(z) - f(z + tw)|}{t}$$

$$\leq \sup_{w \in D \setminus \{z\}} \sqrt{1 - \|z\|^2}\sqrt{1 - \|w\|^2}\frac{|f(z) - f(w)|}{\|z - w\|}.$$

It follows that

$$\lim_{\|z\|\to 1}(1-\|z\|^2)\|f'(z)\|$$

$$\leq \lim_{\|z\|\to 1}\sup_{w\in D\setminus\{z\}}(1-\|z\|^2)^{1/2}(1-\|w\|^2)^{1/2}\frac{|f(z)-f(w)|}{\|z-w\|}=0.$$

$$\square$$

Example 4.3.12. Let $g = (g_1, \ldots, g_n)$ be a Möbius transformation of the Euclidean ball $\mathbb{B}_n$. Since each component g_j of g extends holomorphically to a neighbourhood of $\overline{\mathbb{B}}_n$, it belongs to $\mathcal{B}_0(\mathbb{B}_n)$ by Theorem 4.3.11.

Notes. In Theorem 4.3.1, condition (viii) was originally used in [162] to define a Bloch function on a finite dimensional bounded homogeneous domain, where the constant β_f is defined to be the Bloch semi-norm of f, and was then shown to be equivalent to the other conditions in Theorem 4.3.1, in finite dimensions. Condition (ii) of Theorem 4.3.1 (in terms of Carathéodory differential metric) has also been used in [53] to define Bloch functions on bounded symmetric domains, which are shown to form a Banach space. A definition of Bloch functions on finite dimensional strongly pseudoconvex domains has been introduced by Krantz and Ma [113]. On a finite dimensional bounded homogeneous domain D, it has been shown in [11, Theorem 3.1] that

$$\beta_f = \sup_{z\neq w}\frac{|f(z)-f(w)|}{h(z,w)}$$

for a Bloch function f on D, where $z, w \in D$ and $h(\cdot, \cdot)$ is the Bergman distance. It has been declared in the proof of [17, Proposition 2.7] that the second inequality in Theorem 4.3.7 is true for a Hilbert ball. Corollary 4.3.8 has been proved in [12].

Theorem 4.3.11 has been proved in [43] and [44]. The proof of
(i) ⇔ (iii) in this theorem is analogous to the one for $\mathcal{B}_0(\mathbb{D})$ given in
[12]. Using similar arguments, the equivalence of (i) and (ii) in this
theorem has also been proved in [175, Theorem 3]. The expression of
the left-hand side in Theorem 4.3.11 (iv) originates from the paper [85],
where it is shown that the Bloch functions on $\mathbb{D}$ are exactly the func-
tions f for which $\sup_{w \in \mathbb{D} \setminus \{z\}} (1 - \|z\|^2)^{1/2} (1 - \|w\|^2)^{1/2} \frac{|f(z) - f(w)|}{\|z - w\|} < \infty$.
This result has been extended to the Euclidean balls by Ren and Tu
[147], and to the Hilbert balls by Chu, Hamada, Honda and Kohr [41].
For Euclidean balls, the equivalence (i) ⇔ (iv) in Theorem 4.3.11 has
also been proved in [147]. The dimension-free inequality (4.15) in the
proof of (i) ⇒ (iv) for Hilbert balls is a modification of a similar one,
albeit dimension-dependent, derived in [147]. For a finite dimensional
bounded symmetric domain which is not a Euclidean ball, the characterizations of the little Bloch space in Theorem 4.3.11 are no longer
valid. It has been shown in [163, Proposition 4.1] that a Bloch function
f on such a domain must be constant if it satisfies condition (ii) in
Theorem 4.3.11.

4.4   Composition operators

The topic of composition operators is an important one in the study
of function spaces and there are many applications. We discuss briefly
the composition operators between Bloch spaces of bounded symmetric domains. The question of boundedness, compactness and isometric
conditions are our main focus.

*Throughout this section, we will denote by D and B two bounded
symmetric domains realised as the open unit balls of two JB*-triples V
and W respectively.*

Given $\varphi \in H(D, B)$, one can form a *composition operator*

$$C_\varphi : f \in \mathcal{B}(B) \mapsto f \circ \varphi \in H(D, \mathbb{C})$$

which is clearly linear. We call φ the *symbol* of C_φ. The first natural question is whether $f \circ \varphi$ is a Bloch function on D, and if so, whether C_φ is a bounded operator between $\mathcal{B}(B)$ and $\mathcal{B}(D)$. The answer is positive.

Theorem 4.4.1. *Let D and B be bounded symmetric domains and let $\varphi \in H(D, B)$. Then $C_\varphi(f) \in \mathcal{B}(D)$ for each $f \in \mathcal{B}(B)$, and $C_\varphi : \mathcal{B}(B) \longrightarrow \mathcal{B}(D)$ is a bounded linear operator satisfying*

$$\max\left\{1, \tanh^{-1} \|\varphi(0)\|\right\} \leq \|C_\varphi\| \leq \max\left\{1, \tanh^{-1} \|\varphi(0)\| + K_\varphi\right\}$$

where

$$K_\varphi = \sup_{z \in D} \sup_{v \neq 0} \frac{\mathcal{K}_B(\varphi(z), \varphi'(z)v)}{\mathcal{K}_D(z, v)} \leq 1.$$

In particular, $\|C_\varphi\| = 1$ whenever $\varphi(0) = 0$.

Proof. Let $f \in \mathcal{B}(B)$. By Theorem 4.3.1,

$$|C_\varphi(f)|_\mathcal{B} = |f \circ \varphi|_\mathcal{B} = \sup\{Q_{f \circ \varphi}(z) : z \in D\}.$$

For $z \in D$ and v in the ambient JB*-triple V, we have

$$\frac{|(f \circ \varphi)'(z)v|}{\mathcal{K}_D(z, v)} \leq \frac{|f'(\varphi(z))\varphi'(z)v|}{\mathcal{K}_B(\varphi(z), \varphi'(z)v)} \leq Q_f(\varphi(z))$$

for $\varphi'(z)v \neq 0$. If $\varphi'(z)v = 0$, then $(f \circ \varphi)'(z)v = f'(\varphi(z))\varphi'(z)v = 0$. Hence

$$Q_{f \circ \varphi}(z) \leq Q_f(\varphi(z)) \leq |f|_\mathcal{B} \qquad (4.18)$$

which implies $|C_\varphi(f)|_\mathcal{B} \leq |f|_\mathcal{B}$ and $C_\varphi(f) \in \mathcal{B}(D)$. Observe that

$$|f(\varphi(0))| \leq |f(0)| + \int_0^1 |f'(t\varphi(0))\varphi(0)| dt$$

$$\leq |f(0)| + |f|_\mathcal{B} \int_0^1 \frac{\|\varphi(0)\|}{1 - t^2\|\varphi(0)\|^2} dt$$

by (4.6). This gives

$$\|C_\varphi(f)\|_\mathcal{B} \le \|f\|_\mathcal{B} \left(1 + \int_0^1 \frac{\|\varphi(0)\|}{1 - t^2\|\varphi(0)\|^2} dt\right)$$

and therefore C_φ is a bounded operator.

To get the desired upper bound for $\|C_\varphi\|$, let us look at $Q_{f\circ\varphi}(z)$ again. We have

$$\frac{|(f\circ\varphi)'(z)v|}{\mathcal{K}_D(z,v)} = \frac{|f'(\varphi(z))\varphi'(z)v|}{\mathcal{K}_D(\varphi(z),\varphi'(z)v)} \frac{\mathcal{K}_D(\varphi(z),\varphi'(z)v)}{\mathcal{K}_D(z,v)}$$

$$\le Q_f(z)\frac{\mathcal{K}_D(\varphi(z),\varphi'(z)v)}{\mathcal{K}_D(z,v)}$$

which gives

$$Q_{f\circ\varphi}(z) = \sup_{v\ne 0}\frac{|(f\circ\varphi)'(z)v|}{\mathcal{K}_D(z,v)} \le |f|_\mathcal{B}\sup_{v\ne 0}\frac{\mathcal{K}_D(\varphi(z),\varphi'(z)v)}{\mathcal{K}_D(z,v)}$$

and therefore

$$|f\circ\varphi|_\mathcal{B} \le \sup_{z\in D}\sup_{v\ne 0}\frac{\mathcal{K}_D(\varphi(z),\varphi'(z)v)}{\mathcal{K}_D(z,v)}|f|_\mathcal{B} = K_\varphi|f|_\mathcal{B}.$$

By Theorem 4.3.1, we have

$$|f(\varphi(0))| \le |f(0)| + |f(\varphi(0)) - f(0)| \le \|f\|_\mathcal{B} - |f|_\mathcal{B} + k_D(\varphi(0),0)|f|_\mathcal{B}$$

and hence

$$\|f\circ\varphi\|_\mathcal{B} \le \|f\|_\mathcal{B} - |f|_\mathcal{B} + k_D(\varphi(0),0)|f|_\mathcal{B} + |f\circ\varphi|_\mathcal{B}$$

$$\le \|f\|_\mathcal{B} + (-1 + \tanh^{-1}\|\varphi(0)\| + K_\varphi)|f|_\mathcal{B}.$$

If $\tanh^{-1}\|\varphi(0)\| + K_\varphi - 1 \le 0$, we have

$$\|f\circ\varphi\|_\mathcal{B} \le \|f\|_\mathcal{B}.$$

On the other hand, if $\tanh^{-1}\|\varphi(0)\| + K_\varphi - 1 > 0$, then

$$\|f\circ\varphi\|_\mathcal{B} \le (\tanh^{-1}\|\varphi(0)\| + K_\varphi)\|f\|_\mathcal{B}$$

and we have obtained the desired upper bound.

Let $\mathbf{1}$ be the constant function on B with value 1. Then $1 = \|C_\varphi(\mathbf{1})\|_\mathcal{B} = \|\mathbf{1}\|_\mathcal{B}$. This implies $\|C_\varphi\| \geq 1$.

For the lower bound, we note that $\varphi(0) = 0$ implies $\|C_\varphi\| \leq \max\{1, K_\varphi\} = 1$.

Consider the case $\varphi(0) \neq 0$. Let $\ell_{\varphi(0)}$ be the support functional at $\varphi(0)$, that is, $\ell_{\varphi(0)}(\varphi(0)) = \|\varphi(0)\|$ and $\|\ell_{\varphi(0)}\| = 1$. By Example 4.2.3, the function $\tanh^{-1} \circ \ell_{\varphi(0)}$ is a Bloch function on B with $\|\tanh^{-1} \circ \ell_{\varphi(0)}\|_\mathcal{B} = 1$. Therefore

$$\|C_\varphi\| \geq \|\tanh^{-1} \circ \ell_{\varphi(0)} \circ \varphi\|_\mathcal{B} \geq \tanh^{-1}\|\varphi(0)\|.$$

This establishes the lower bound for $\|C_\varphi\|$.   □

Corollary 4.4.2. *In the preceding theorem, if $C_\varphi : \mathcal{B}(B) \longrightarrow \mathcal{B}(D)$ is an isometry, then we have $\varphi(0) = 0$ and $K_\varphi = 1$.*

Proof. Let $a = \varphi(0)$ and g_{-a} the Möbius transformation induced by $-a$. We show that $\psi(a) = 0$ for each continuous functional $\psi \in W^*$. Then $\psi \circ g_{-a}$ is a Bloch function on B and we have

$$\begin{aligned}
|\psi(-a)| + |\psi \circ g_{-a}|_\mathcal{B} &= \|\psi \circ g_{-a}\|_\mathcal{B} \\
&= \|C_\varphi(\psi \circ g_{-a})\|_\mathcal{B} \\
&= |\psi \circ g_{-a} \circ \varphi|_\mathcal{B} \\
&\leq |\psi \circ g_{-a}|_\mathcal{B}
\end{aligned}$$

where the last inequality has been shown in the preceding theorem, which implies $\psi(-a) = 0$.

For the second assertion, we have shown in the preceding theorem that

$$|C_\varphi(f)|_\mathcal{B} \leq K_\varphi |f|_\mathcal{B}$$

for all $f \in \mathcal{B}(B)$. Hence we must have $K_\varphi = 1$ as $\varphi(0) = 0$ and C_φ is an isometry.   □

Example 4.4.3. The converse of Corollary 4.4.2 is false. Let D be the open unit ball of an ℓ_∞-sum $V_1 \oplus V_2$ of two JB*-triples and let $\varphi : D \longrightarrow D$ be the projection $\varphi(z_1, z_2) = (z_1, 0)$. Then $\varphi(0,0) = (0,0)$ and $K_\varphi = 1$. Let ψ be the support functional at a point $(0, b) \in D \backslash \{(0,0)\}$, that is, $\psi(0, b) = \|b\|$ and $\|\psi\| = 1$. Then ψ is a Bloch function on D with

$$\|\psi\|_{\mathcal{B}} \geq Q_\psi((0,0)) = 1,$$

but $\|\psi \circ \varphi\|_{\mathcal{B}} = 0$.

In the remaining section, we write $\dim B = n$ to mean that B is a finite dimensional bounded symmetric domain in $\mathbb{C}^n$, realised as the open unit ball of a norm $\|\cdot\|$ equivalent to the Euclidean norm $\|\cdot\|_2$. In this case, a holomorphic map $\varphi : D \longrightarrow B$ can be expressed as

$$\varphi(z) = (\varphi_1(z), \ldots, \varphi_n(z)) \qquad (z \in D)$$

where D is the open unit ball of a JB*-triple V and the holomorphic functions

$$\varphi_j : D \to \mathbb{C}$$

will be called the *coordinate components* of φ.

There is a constant $c > 0$ such that for each $y = (y_1, \ldots, y_n) \in \mathbb{C}^n$,

$$\|y\| \leq c\|y\|_2 = c \left(\sum_j |y_j|^2 \right)^{1/2}.$$

A bounded linear map $\psi : V \longrightarrow \mathbb{C}^n$ can be expressed in the form

$$\psi = (\psi_1, \ldots, \psi_n)$$

where each ψ_j is a continuous linear functional on V. In this case, we have

$$\|\psi\| \leq c \left(\sum_j \|\psi_j\|^2 \right)^{1/2}. \tag{4.19}$$

Given $\varphi = (\varphi_1, \ldots, \varphi_n) \in H(D, B)$, it is natural to ask if the composition operator C_φ sends the little Bloch space $\mathcal{B}_0(B)$ into $\mathcal{B}_0(D)$. We show below that this is the case for a Hilbert ball D exactly when the coordinate components φ_j of the symbol φ are in $\mathcal{B}_0(D)_0$.

Proposition 4.4.4. *Let D be a Hilbert ball and* $\dim B = n < \infty$. *Let* $\varphi = (\varphi_1, \ldots, \varphi_n) \in H(D, B)$. *Then C_φ maps $\mathcal{B}_0(B)$ to $\mathcal{B}_0(D)$ if and only if $\varphi_j \in \mathcal{B}_0(D)$ for $j = 1, \ldots, n$.*

Proof. Let $C_\varphi(\mathcal{B}_0(B)) \subset \mathcal{B}_0(D)$. Since the coordinate maps

$$p_j : (y_1, \ldots, y_n) \in B \mapsto y_j \in \mathbb{C}$$

belong to $\mathcal{B}_0(B)$, we have $\varphi_j = p_j \circ \varphi = C_\varphi(p_j) \in \mathcal{B}_0(D)$ for $j = 1, \ldots, n$.

Conversely, let $\varphi_j \in \mathcal{B}_0(D)$ for $j = 1, \ldots, n$. Let $f \in \mathcal{B}_0(B)$ and $\varepsilon > 0$. There exists a polynomial P on B such that $\|f - P\|_{\mathcal{B}} < \varepsilon/2$. By (4.6) and (4.19), we have

$$
\begin{aligned}
(1 - \|z\|^2)\|(f \circ \varphi)'(z)\| &\leq (1 - \|z\|^2)\|(P \circ \varphi)'(z)\| + Q_{(f-P)}(\varphi(z)) \\
&\leq (1 - \|z\|^2)\|P'(\varphi(z))\|\|\varphi'(z)\| + \frac{\varepsilon}{2} \\
&\leq (1 - \|z\|^2)\|P'(\varphi(z))\|c \sum_{k=1}^{n} \|\varphi_k'(z)\| + \frac{\varepsilon}{2}.
\end{aligned}
$$

Since $\varphi_j \in \mathcal{B}_0(D)$ for $j = 1, \ldots, n$ and $\|P'(\cdot)\|$ is bounded on B, Theorem 4.3.11 implies that there exists $\delta > 0$ for which

$$(1 - \|z\|^2)\|P'(\varphi(z))\|c \sum_{k=1}^{n} \|\varphi_k'(z)\| \leq \frac{\varepsilon}{2}$$

whenever $\|z\| > \delta$. This gives

$$(1 - \|z\|^2)\|(f \circ \varphi)'(z)\| < \varepsilon$$

for $\|z\| > \delta$, proving $C_\varphi(f) \in \mathcal{B}_0(D)$. $\qquad\square$

We now study compactness of the composition operators C_φ.

Proposition 4.4.5. *Let D be a bounded symmetric domain realised as the open unit ball of a JB*-triple V. Let $\varphi \in H(D, B)$, where $\dim B = n < \infty$. Then $C_\varphi : \mathcal{B}(B) \to \mathcal{B}(D)$ is a compact operator if for every $\varepsilon > 0$, there exists a $\delta > 0$ such that*

$$\frac{\mathcal{K}_B(\varphi(z), \varphi(z)'v)}{\mathcal{K}_D(z, v)} < \varepsilon \qquad (v \in V \setminus \{0\}) \qquad (4.20)$$

whenever $\|\varphi(z)\| > \delta$.

Proof. Let φ satisfy the given condition. Let (f_m) be a sequence in $\mathcal{B}(B)$ such that $\|f_m\|_{\mathcal{B}} \leq 1$ for $m = 1, 2, \ldots$. We need to show that $(f_m \circ \varphi)$ contains a convergent subsequence in $\mathcal{B}(B)$.

Since

$$\|f'_m(z)\| \leq \frac{\|f_m\|_{\mathcal{B}}}{1 - \|z\|^2} \leq \frac{1}{1 - \|z\|^2}$$

by (4.6), Montel's theorem implies the existence of a subsequence (f_k) of (f_m) converging locally uniformly to a holomorphic function f on B with $|f|_{\mathcal{B}} \leq 1$.

Let $F_k = f_k - f$ for $k = 1, 2, \ldots$. Then $F_k \to 0$ locally uniformly on B as $k \to \infty$ and

$$\|F_k\|_{\mathcal{B}} \leq 2 \qquad (k = 1, 2, \ldots). \qquad (4.21)$$

Let $\varepsilon > 0$. By assumption, there exists $\delta \in (0, 1)$ such that (4.20) holds whenever $\|\varphi(z)\| > \delta$ which, together with (4.21), implies

$$Q_{F_k \circ \varphi}(z) \leq \varepsilon Q_{F_k}(\varphi(z)) \leq 2\varepsilon \quad \text{for } \|\varphi(z)\| > \delta$$

where the first inequality follows from analogous derivation for (4.18).

On the other hand, if $\|\varphi(z)\| \leq \delta$, we have

$$\inf\{\mathcal{K}_B(w, y) : \|w\| \leq \delta, \|y\| = 1\} > 0$$

since $\mathcal{K}_B(w, y) = \|B(w, w)^{-1/2}y\| > 0$ on the compact subset

$$\{w \in \mathbb{C}^n : \|w\| \leq \delta\} \times \{y \in \mathbb{C}^n : \|y\| = 1\}$$

of $\mathbb{C}^n \times \mathbb{C}^n$.

Since $F_k(w) \to 0$ uniformly for $\|w\|_Y \leq (\delta + 1)/2$ as $k \to \infty$, there exists $K \in \mathbb{N}$ such that

$$Q_{F_k \circ \varphi}(z) \leq Q_{F_k}(\varphi(z)) < \varepsilon$$

for $k > K$ and $\|\varphi(z)\| \leq \delta$.

It follows that $\|F_k \circ \varphi\|_{\mathcal{B}} = |F_k(\varphi(0))| + |F_k \circ \varphi|_{\mathcal{B}} \to 0$ as $k \to \infty$, giving $f_k \circ \varphi \to f \circ \varphi$ in $\mathcal{B}(D)$. This proves compactness of C_φ. $\qquad\square$

In what follows, we consider the special case of Hilbert balls.

Proposition 4.4.6. *Let D be a Hilbert ball and $\dim B = n < \infty$. Then the composition operator $C_\varphi : \mathcal{B}(B) \to \mathcal{B}(D)$ is compact if the symbol $\varphi \in H(D, B)$ satisfies the condition that for every $\varepsilon > 0$, there exists $\delta > 0$ such that*

$$\frac{1 - \|z\|^2}{1 - \|\varphi(z)\|^2} \|\varphi'(z)\| < \varepsilon \tag{4.22}$$

whenever $\|\varphi(z)\| > \delta$.

Proof. Let (f_k) be a sequence in $\mathcal{B}(B)$ such that $\|f_k\|_{\mathcal{B}} \leq 1$ for $k = 1, 2, \ldots$. By (4.6) and Montel's theorem, we may assume by choosing a subsequence, that (f_k) converges to a function f locally uniformly on B. We have $\|f\|_{\mathcal{B}} \leq 1$.

Let $F_k = f_k - f$ for $k = 1, 2, \ldots$. Then $F_k \to 0$ locally uniformly on B as $k \to \infty$ and

$$\|F_k\|_{\mathcal{B}} \leq 2 \tag{4.23}$$

for $k = 1, 2, \ldots$.

Let $\varepsilon > 0$. By assumption, there exists a $\delta \in (0, 1)$ such that

$$\frac{1 - \|z\|^2}{1 - \|\varphi(z)\|^2} \|\varphi'(z)\| < \varepsilon \tag{4.24}$$

for $\|\varphi(z)\|_Y > \delta$.

Since $F_k(w) \to 0$ uniformly for $\|w\| \le (\delta + 1)/2$ as $k \to \infty$, there exists a $K \in \mathbb{N}$ such that

$$\|F_k'(w)\| < \varepsilon \qquad (k > K, \|w\| \le \delta). \tag{4.25}$$

If $\|\varphi(z)\| > \delta$, then we deduce from (4.23), (4.24) and (4.6) that

$$(1 - \|z\|^2)\|F_k'(\varphi(z))\varphi'(z)\| \le \varepsilon(1 - \|\varphi(z)\|^2)\|F_k'(\varphi(z))\| \le \varepsilon\|F_k\|_{\mathcal{B}} \le 2\varepsilon.$$

If $\|\varphi(z)\| \le \delta$, then by (4.25) and the Schwarz-Pick lemma, we have

$$(1 - \|z\|^2)\|F_k'(\varphi(z))\varphi'(z)\| \le \|F_k'(\varphi(z))\| \le \varepsilon$$

for $k > K$. Hence $\|F_k \circ \varphi\|_{\mathcal{B}} = |F_k(\varphi(0))| + |F_k \circ \varphi|_{\mathcal{B}} \to 0$ as $k \to \infty$.
This proves compactness of C_φ. $\qquad\square$

Corollary 4.4.7. *Let D be a Hilbert ball and $\dim B = n < \infty$. If $\varphi \in H(D, B)$ satisfies*

$$\lim_{\|z\| \to 1} \frac{1 - \|z\|^2}{1 - \|\varphi(z)\|^2} \|\varphi'(z)\| = 0, \tag{4.26}$$

then $C_\varphi : \mathcal{B}_0(B) \longrightarrow \mathcal{B}_0(D)$ is a compact operator between little Bloch spaces.

Proof. We note that (4.26) is equivalent to (4.22) and the condition

$$\lim_{\|z\| \to 1} (1 - \|z\|^2)\|\varphi'(z)\| = 0. \tag{4.27}$$

Under condition (4.27), $\varphi_j \in \mathcal{B}_0(D)$ for $j = 1, 2, \ldots, n$. Hence $C_\varphi : \mathcal{B}_0(B) \longrightarrow \mathcal{B}_0(D)$ is a bounded operator by Proposition 4.4.4. Since C_φ is compact on $\mathcal{B}(B)$ by Proposition 4.4.6, its restriction to the closed subspace $\mathcal{B}_0(B)$ is also compact. $\qquad\square$

Theorem 4.4.8. *Let D be a Hilbert ball and* $\dim B = n < \infty$. *Let* $\varphi \in H(D, B)$. *Then the composition operator* $C_\varphi : \mathcal{B}_0(B) \longrightarrow \mathcal{B}_0(D)$ *is compact if and only if*

$$\lim_{\|z\| \to 1} (1 - \|z\|^2) \sup\{\|(f \circ \varphi)'(z)\| : f \in \mathcal{B}_0(B), \|f\|_\mathcal{B} \leq 1\} = 0. \quad (4.28)$$

Proof. Let C_φ be compact so that the set

$$\mathcal{E} = \{C_\varphi(f) : f \in \mathcal{B}_0(B), \|f\|_\mathcal{B} \leq 1\}$$

is relatively compact in $\mathcal{B}_0(D)$.

Let $\varepsilon > 0$. Then there exist $f^1, \dots, f^l \in \mathcal{B}_0(B)$ such that $\mathcal{E} \subset \bigcup_{j=1}^l U(f^j, \varepsilon)$, where

$$U(f^j, \varepsilon) = \left\{ F \in \mathcal{B}_0(D) : \|F - f^j \circ \varphi\|_\mathcal{B} < \frac{\varepsilon}{2} \right\} \qquad (j = 1, \dots, l).$$

Since $f^j \circ \varphi \in \mathcal{B}_0(D)$, there exists $r \in (0, 1)$ such that $(1 - \|z\|^2)\|D(f^i \circ \varphi)(z)\| < \frac{\varepsilon}{2}$ for $\|z\| > r$ and $j = 1, \dots, l$ by Theorem 4.3.11. Let $F \in \mathcal{E}$. Then $F \in U(f^{j_0}, \varepsilon)$ for some j_0 and hence

$$(1 - \|z\|^2)\|F'(z)\| \leq \|F - f^{j_0} \circ \varphi\|_\mathcal{B} + (1 - \|z\|^2)\|(f^{j_0} \circ \varphi)'(z)\| < \varepsilon$$

for $\|z\| > r$ which proves (4.28).

Conversely, assume (4.28). Then $f \circ \varphi \in \mathcal{B}_0(D)$ for each $f \in \mathcal{B}_0(B)$ by Theorem 4.3.11. Let (f_k) be a sequence in $\mathcal{B}_0(B)$ such that $\|f_k\|_\mathcal{B} \leq 1$ for $k = 1, 2, \dots$. By Montel's theorem and choosing a subsequence, we may assume that (f_k) converges locally uniformly to a holomorphic function f on B. We have $\|f\|_\mathcal{B} \leq 1$ and by (4.28), there exists an $r_0 \in (0, 1)$ such that

$$(1 - \|z\|^2)\|(f_k \circ \varphi)'(z)\| < \varepsilon \qquad (\|z\| > r_0, k = 1, 2, \dots).$$

It follows that

$$(1 - \|z\|^2)\|(f \circ \varphi)'(z)\| \leq \varepsilon \qquad (\|z\| > r_0).$$

This implies $f \circ \varphi \in \mathcal{B}_0(D)$ and $\|f_k \circ \varphi - f \circ \varphi\|_\mathcal{B} \to 0$ as $k \to \infty$, proving compactness of C_φ. $\qquad\qquad\qquad\qquad\qquad\qquad\qquad\qquad\qquad\square$

Notes. The results in this section are taken from [43] and [44]. The upper bound for the composition operator C_φ in Theorem 4.4.1 is analogous to the one in [11, Theorem 3.2] for finite dimensional bounded homogeneous domains. For Euclidean balls, Theorem 4.4.1 has been shown in [11, Corollary 3.1]. Corollary 4.4.2 has been shown in [11, Theorem 6.1(b)] for composition operators on Bloch spaces of finite dimensional classical bounded symmetric domains and it has been speculated in [11, Remark 6.1] that the result might be true for *all* finite dimensional bounded symmetric domains. Proposition 4.4.4 is an extension of a result in [157, Theorem 2] for the Euclidean balls.

Proposition 4.4.5, together with its converse proved in [43, Proposition 5.3], generalises simultaneously the main theorem in [186] for finite dimensional classical domains, as well as [157, Theorem 3] for finite dimensional bounded symmetric domains and [157, Theorem 4] for the Euclidean balls. For the unit disc $\mathbb{D}$, the result of Proposition 4.4.6 has been shown in [127], and also in [157, Theorem 5] for Euclidean balls. Corollary 4.4.7 has been shown in [127] and [157] for the disc $\mathbb{D}$ and Euclidean balls respectively.

Bibliography

[1] M. Abate, *Horoshperes and iterates of holomorphic maps*, Math. Z. **198** (1988) 225-238.

[2] M. Abate, *Iteration theory of holomorphic maps on taut manifolds* (Cosenza: Mediterranean Press, 1989).

[3] M. Abate and G. Patrizio, *Finsler metrics - a global approach* (Berlin: Springer-Verlag, 1971).

[4] M. Abate and J. Raissy, *Wolff-Denjoy theorems in nonsmooth convex domains*, Ann. Mat. Pura Appl. **193**(2014) 1503-1518.

[5] A, Ash, D. Mumford, M. Rapoport and Y. Tai, *Smooth compactification of locally symmetric varieties*, (Massachusetts: Math. Sci. Press, 1975).

[6] L.V. Ahlfors, *An extension of Schwarz's lemma*, Trans. Amer. Math. Soc. **43** (1938) 359-364.

[7] L.V. Ahlfors, *Complex analysis* (New York: McGraw-Hill, 1966).

[8] A.A. Albert, On a certain algebra of quantum mechanics, Annals of Math. **35** (1934) 65-73.

[9] A.A. Albert, *On Jordan algebras of linear transformations*, Trans. Amer. Math. Soc. **59** (1946) 524-555.

[10] H. Alexander, *Extremal holomorphic embeddings between the ball and polydisc*, Proc. Amer. Math. Soc. **68** (1978) 200–202.

[11] R.F. Allen and F. Colonna, *On the isometric composition operators on the Bloch space in $\mathbb{C}^n$*, J. Math. Anal. Appl. **355** (2009), 675-688.

[12] J.M. Anderson, J. Clunie and Ch. Pommerenke, *On Bloch functions and normal functions*, J. Reine Angew. Math. **270** (1974) 12-37.

[13] D. Bao, S.S. Chern and Z. Shen, *An introduction to Riemann-Finsler geometry* (New York: Springer-Verlag, 2000).

[14] T.J. Barth, *Some counterexamples concerning intrinsic distances*, Proc. Amer. Math. Soc. **66** (1977) 49-53.

[15] L. Bieberbach, *Über der Koeffizienten derjenigen Potenzreihen, welche eine schlichte Addildung des Einheitskreises vermitteln*, S.-B. Preuss, Aka. Wiss. (1916) 940-955.

[16] G. Birkhoff, *Analytic groups*, Trans. Amer. Math. Soc. **43** (1938) 61-101.

[17] O. Blasco, P. Galindo and A. Miralles, *Bloch functions on the unit ball of an infinite dimensional Hilbert space*, J. Funct. Anal. **267** (2014) 1188-1204.

[18] M. Bonk, *On Bloch's constant*, Proc. Amer. Math. Soc. **110** (1990) 889-894.

[19] M. Bonk, D. Minda and H. Yanagihara, *Distortion theorems for locally univalent Bloch functions*, J. Anal. Math. **69** (1996) 73-95.

[20] F.F. Bonsall and J. Duncan, *Numerical range of operators on normed spaces and elements of normed algebras* (Cambridge: Cambridge Univ. Press, 1971).

[21] N. Bourbaki, *Lie groups and Lie algebras, Chapters 1-3*, (Berlin: Springer-Verlag, 1989).

[22] L. de Branges, *A proof of the Bieberbach conjecture*, Acta. Math. **154** (1985) 137-152.

[23] H. Braun and M. Koecher, *Jordan-Algebren* (Berlin: Springer-Verlag, 1966).

[24] R. Braun, W. Kaup and H. Upmeier, *A holomorphic characterization of Jordan C*-algebras*, Math. Z. **161** (1978) 277-290.

[25] H.-J. Bremermann, *Über die Äqivalenz der pseudokonvexen Gebiete und der Holomorphie-gebiete in Raume non n komlexen veränderlichen*, Math. Ann. **128** (1954) 63-91.

[26] M. Budzynska, *The Denjoy-Wolff theorem in $\mathbb{C}^n$*, Nonlinear Anal. **75** (2012) 22-29.

[27] L.J. Bunce and C-H. Chu, *Dual spaces of JB*-triples and the Radon-Nikodym property*, Math. Z. **308** (1991) 327-334.

[28] L.J. Bunce and C-H. Chu, *Compact operations, multipliers and Radon-Nikodym property in JB*-triples*, Pacific J. Math. **153** (1992) 249-265.

[29] R.B. Burckel, *Iterating analytic self-maps on discs*, Amer. Math. Monthly **88** (1981) 396-407.

[30] M. Cabrera and A. Rodriguez-Palacios, *Non-associative normed algebras*, **Vol. 2**, (Cambridge: Cambridge Univ. Press, 2018).

[31] C. Carathéodory, *Über eine spezielle Metrik, die in der Theorie der analytischen Funktionen auftritt,* atti. Pontif. Acad. Sci. Nuovi Lincei **80** (1927) 99-105.

[32] É. Cartan, *Sur les domaines bornés homogènes de l'espace de n variables complexes,* Abh. Math. Semin. Univ. Hamburg **11** (1935) 116–162.

[33] H. Cartan, *Les fonctions de deux variables complexes et le problème de la représentation analytique,* J. Math. Pures et Appl. **10** (1931) 1-114.

[34] H. Cartan, *Sur les groupes de transformations analytiques,* Act. Sci. Ind. **198**, (Paris: Hermann, 1935).

[35] H. Chen and P. Gauthier, *On Bloch's constant,* J. Anal. Math. **69** (1996) 275-291.

[36] S.S. Chern, *On holomorphic mappings of hermitian manifolds of the same dimension,* In Proc. Symp. Pure Math. **XI** (Providence: Amer. Math. Soc. 1968), pp. 157-170.

[37] C-H. Chu, *Jordan structures in geometry and analysis* (Cambridge: Cambridge Univ. Press, 2012).

[38] C-H. Chu, *Iteration of holomorphic maps on Lie balls,* Adv. Math. **264** (2014) 114-154.

[39] C-H. Chu, *Jordan structures in bounded symmetric domains,* In Geometric theory in higher dimension, Springer INdAM Ser. **26** (Cham: Springer Nature, 2017) pp. 43-61.

[40] C-H. Chu, *Infinite dimensional Jordan algebras and symmetric cones,* J. Alg. **491** (2017) 357-371.

[41] C-H. Chu, H. Hamada, T. Honda and G. Kohr, *Distorsion theorems for convex mappings on homogeneous balls*, J. Math. Anal. Appl. **369** (2010) 437-442.

[42] C-H. Chu, H. Hamada, T. Honda and G. Kohr, *Distorsion of locally biholomorphic Bloch mappings on bounded symmetric domains*, J. Math. Anal. Appl. **441** (2016) 830-843.

[43] C-H. Chu, H. Hamada, T. Honda and G. Kohr, *Bloch functions on bounded symmetric domains*, J. Funct. Anal. **272** (2017) 2412-2441.

[44] C-H. Chu, H. Hamada, T. Honda and G. Kohr, *Bloch space of a bounded symmetric domain and composition operators*, Complex Anal. Oper. Theory **13** (2019) 479-492.

[45] C-H. Chu and B. Iochum, *Complementation of Jordan triples in von Neumann algebras*, Proc. Amer. Math. Soc. **108** (1990) 19-24.

[46] C-H. Chu, K-T. Kim and S. Kim, *Infinite dimensional holomorphic homogeneous regular domains*, J. Geom. Anal. **30** (2020) 223-247.

[47] C-H. Chu and P. Mellon, *Iteration of compact holomorphic maps on a Hilbert ball*, Proc. Amer. Math. Soc. **125** (1997) 1771-1777.

[48] C-H. Chu and L. Oliveira, *Tits-Kantor-Koecher Lie algebras of JB*-triples*, J. Alg. **512** (2018) 465-492.

[49] C-H. Chu and J.M. Rigby, *Iteration of self-maps on a product of Hilbert balls*, J. Math. Anal. Appl. **411** (2014) 773-786.

[50] C-H. Chu and J.M. Rigby, *Horoballs and iteration of holomorphic maps on bounded symmetric domains*, Adv. Math. **311** (2017) 338-377.

[51] J-L. Clerc, *Geometry of the Shilov boundary of a bounded symmetric domain*, J. Geom. Symmetry Phys. **13** (2009) 25-74.

[52] T. Dang and Y. Friedman, *Classification of JBW*-triple factors and applications*, Math. Scand. **61** (1987) 292-330.

[53] F. Deng, Q. Guan and L. Zhang, *Some properties of squeezing functions on bounded domains*, Pacific J. Math. **257** (2012) 319–341.

[54] F. Deng, Q. Guan, and L. Zhang, *Properties of squeezing functions and global transformations of bounded domains*, Trans. Amer. Math. Soc. **368** (2016) 2679–2696.

[55] A. Denjoy, *Sur l'itération des fonctions analytiques*, C.R. Acad. Sc. Paris **182** (1926) 255-257.

[56] J. Diestel and J.J. Uhl, Jr. *Vector measures*, Mathematical Surveys **15** (Providence: Amer. Math. Soc. 1977).

[57] J. Dieudonné, *Foundations of modern analysis* (London: Academic Press, 1969).

[58] N. Dunford and J.T. Schwartz, *Linear operators* (New York: Wiley Classics Library Edition, 1988).

[59] C.J. Earle and R.S. Hamilton, *A fixed point theorem for holomorphic mappings*, Proc. Symp. Pure Math. **16** (1969) 61-65.

[60] J. Faraut and A. Koranyi, *Analysis on symmetric cones* (Oxford: Clarendon Press, 1994).

[61] C.H. FitzGerald and S. Gong, *The Bloch theorem in several complex variables*, J. Geom. Anal. **4** (1994) 35-58.

[62] J. E. Fornaess and F. Rong, *Estimate of the squeezing function for a class of bounded domains*, Math. Ann. **371** (2018) 1087-1094.

[63] T. Franzoni and E. Vessentini, *Holomorphic maps and invariant distances*, Math. Studies **40** (Amsterdam: North-Holland, 1980)

[64] Y. Friedman, *Physical applications of homogeneous balls* (Boston: Birkhäuser, 2005).

[65] Y. Friedman and B. Russo, *The Gelfand Naimark theorem for JB*-triples*, Duke Math. J. **53** (1986) 139-148.

[66] C.M. Glennie, *Some identities valid in special Jordan algebras but not valid in all Jordan algebras*, Pacific J. Math. **16** (1966) 47-59.

[67] K. Goebel, *Fixed points and invariant domains of holomophic mappings of the Hilbert ball*, Nonlinear Analysis **6** (1982) 1327-1334.

[68] S. Gong, *Convex and starlike mappings in several complex variables* (Dordrecht: Kluwer Acad. Press 1999)

[69] I. Graham, G. Kohr and J. Pfaltzgraff, *Growth and two-point distortion for biholomorphic mappings of the ball*, Complex Var. Elliptic Equ. **52** (2007) 211-223.

[70] I. Graham and G. Kohr, *Geometric function theory in one and higher dimensions* (New York: Marcel Dekker, 2003).

[71] R.E. Green, K-T. Kim and S.G. Krantz, *The geometry of convex domains* (Boston: Birkhäuser, 2011).

[72] S.J. Greenfield and N.R. Wallach, *Automorphism groups of bounded domains in Banach spaces*, Trans. Amer. Math. Soc. **166** (1972) 45-57.

[73] K.T. Hahn, *Holomorphic mappings of the hyperbolic space into the complex Euclidean space and the Bloch theorem*, Canad. J. Math. **27** (1975) 446-458.

[74] K.T. Hahn, *On completeness of the Bergman metric and its subordinate metric*, Proc. Nat. Acad. Sci. USA **73** (1976) 4294.

[75] H. Hamada and G. Kohr, *Growth and distortion results for convex mappings in infinite dimensional spaces*, Complex Var. Theory Appl. **47** (2002) 291-301.

[76] H. Hamada and G. Kohr, *Φ-like and convex mappings in infinite dimensional spaces*, Rev. Roumaine Math. Pures Appl. **47** (2002) 315-328.

[77] H. Hanche-Olsen and E. Størmer, *Jordan operator algebras* (London: Pitman, 1984).

[78] Harish-Chandra, *Representations of semi-simple Lie groups VI*, Amer. J. Math. **78** (1956) 564-628.

[79] L.A. Harris, *Bounded symmetric domains in infinite dimensional spaces*, In Lecture Notes in Math. **364**, (Berlin: Springer-Verlag 1974), pp. 13-40.

[80] S. Helgason, *Differential geometry, Lie groups and symmetric spaces* (London: Academic Press, 1980).

[81] R. Hermann, *Geometric aspects of potential theory in the symmetric bounded domains*, II, Math. Ann. **151**(1963) 143-149.

[82] M. Hervé, *Quelques propriétés des applications analytiques d'une boule a m dimensions dans elle-meme*, J. Math. Pures et Appl. **42** (1963) 117-147.

[83] M. Hervé, *Iteration des transformations analytiques dans le bicercle-unité*, Ann. Sci. Ecole Norm. Sup. **71** (1954) 1-28.

[84] E. Hille and R.S. Phillips, *Functional analysis and semi-groups*, AMS Colloquium Publ. **31** (Providence: Amer. Math. Soc. 1957).

[85] F. Holland and D. Walsh, *Criteria for membership of Bloch space and its subspace, BMOA*, Math. Ann. **273** (1986) 317-335.

[86] G. Horn, *Classification of JBW*-triples of type* I, Math. Z. **196** (1987) 271-291.

[87] G. Horn, *Charaterization of the predual and ideal structure of a JBW*-triple*, Math. Scand. **61** (1987) 117-133.

[88] G. Horn and E. Neher, *Classification of continuous JBW*-triples*, Trans. Amer. Math. Soc. **306** (1988) 553-578.

[89] H. Horstmann, *Carathéodory Metrik und Regularitätshullen*, Math. Ann. **108** (1933) 208-217.

[90] L.K. Hua, *Harmonic analysis of functions of several complex variables in the classical domains*, Translations of Math. Monographs **6** (Providence: Amer. Math. Soc. 1963).

[91] M. Ito and B.F. Lorenço, *The automorphism group and the non-self-duality of p-cones*, J. Math. Anal. Appl. **471** (2019) 392-410.

[92] N. Jacobson, *Structure and representations of Jordan algebras*, Amer. Math. Soc. Colloq. Publ. **39** (Providence: Amer. Math. Soc. 1968).

[93] R.C. James, *Reflexivity and the supremum of linear functionals*, Ann. of Math. **66** (1957) 159-169.

[94] P. Jordan, J. von Neumann and E. Wigner, *On an algebraic generalisation of the quantum mechanical formalism*, Ann. of Math. **36** (1934) 29–64.

[95] B. Josefson, *A counterexample in the Levi problem*, In Proc. on infinite dim. holomorphy, Lecture Notes Math. **364** (Berline: Springer-Verlag, 1974) pp.168-177.

[96] I.L. Kantor, *Classification of irreducible transitive differential groups*, Dokl. Akad. Nauk SSSR **158** (1964) 1271-1274.

[97] I.L. Kantor, *Transitive differential groups and invariant connections on homogeneous spaces*, Trudy Sem. Vecktor. Tenzor. Anal. **13** (1966) 310-398.

[98] W. Kaup, *Algebraic characterization of symmetric complex Banach manifolds*, Math. Ann. **228** (1977) 39-64.

[99] W. Kaup, *A Riemann mapping theorem for bounded symmetric domains in complex Banach spaces*, Math. Z. **183**(1983) 503-529.

[100] W. Kaup, *Hermitian Jordan triple systems and their automorphisms of bounded symmetric domains*, In Non-associative algebra and its applications, Oviedo 1993, ed. S. González. (Dordrecht: Kluwer Acad. Publ. 1994) pp. 204-214.

[101] W. Kaup, *On a Schwarz lemma for bounded symmetric domains*, Math. Nachr. **197** (1999) 51-60.

[102] W. Kaup and J. Sauter, *Boundary structures in bounded symmetric domains*, Manuscripta Math. **101** (2000) 351-360.

[103] W. Kaup and H. Upmeier, *Jordan algebras and symmetric Siegel domains in Banach spaces*, Math. Z. **157** (1977) 179-200.

[104] K-T. Kim and L. Zhang, *On the uniform sequeezing property of bounded convex domains in* $\mathbb{C}^n$, Pacific J. Math. **282** (2016) 341-358.

[105] S. A. Kim and D. Minda, *Two-point distortion theorems for univalent functions*, Pacific J. Math. **163** (1994) 137-157.

[106] W. Klingenberg, *Riemannian Geometry* (Berlin: Walter der Gruyter, 1982).

[107] S. Kobayashi, *Intrinsic metrics on complex manifolds*, Bull. Amer. Math. Soc. **73** (1967) 347-349.

[108] S. Kobayashi, *Hyperbolic manifolds and holomorphic mappings*, Second Edition, (Singapore: World Scientific, 2005).

[109] M. Koecher, *Jordan algebras and their applications*, Lecture Notes, University of Minnesota, Minneapolis, 1962 (Edited reprint: Lecture Notes in Math. **1710**, Heidelberg: Springer-Verlag, 1999).

[110] M. Koecher, *Imbedding of Jordan algebras into Lie algebras* I, Bull. Amer. J. Math. **89** (1967) 787-816.

[111] M. Koecher, *An elementary approach to bounded symmetric domains*, Lecture Notes (Rice University, 1969).

[112] A. Koranyi and J.A. Wolf, Realization of Hermitian symmetric spaces as generalized half-planes, Ann. of Math. **81** (1965) 265-288.

[113] S. G. Krantz and D. Ma, *Bloch functions on strongly pseudoconvex domains*, Indiana Univ Math J. **37** (1988), 145-163.

[114] A. Kryczka and T. Kuczumow, *The Denjoy-Wolff-type theorem for compact K_{B_H}-nonexpansive maps on a Hilbert ball*, Annales Univ. Marie Curie-Skłodowska Sect. A **51** (1997) 179-183.

[115] Y. Kubota, *A note on holomorphic imbeddings of the classical Cartan domains into the unit ball*, Proc. Amer. Math. Soc. **85**(1982) 65-68.

[116] S. Kwapien, *Isomorphic characterizations of Hilbert spaces by orthogonal series with vector valued coefficients*, Séminaire d'analyse fonctionnelle (Polytechnique) exp. no.8 (1972-1973) 1-7.

[117] E. Landau, *Über die Blochsche Konstante und zwei verwandte Weltkonstanten*, Math. Z. **30** (1929) 608-634.

[118] S. Lang, *Differential and Riemannian manifolds* (Heidelberg: Springer-Verlag, 1995).

[119] K. Liu, X. Sun and S.-T. Yau, *Canonical metrics on the moduli space of Riemann surfaces* I, J. Diff. Geom. **68** (2004) 571-637.

[120] K. Liu, X. Sun and S.-T. Yau, *Canonical metrics on the moduli space of Riemann surfaces* II, J. Diff. Geom. **69** (2005) 163-216.

[121] X.Y. Liu, *Bloch functions of several complex variables*, Pacific J. Math. **152** (1992) 347-363.

[122] K.H. Look, *Schwarz lemma and analytic invariants*, Scientia Sinica **7** (1958) 453-504.

[123] X.Y. Liu and D. Minda, *Distortion theorems for Bloch functions*, Trans. Amer. Math. Soc. **333** (1992) 325-338.

[124] O. Loos, *Jordan pairs*, Lecture Notes in Math. **460** (Heidelberg: Springer-Verlag, 1975).

[125] O. Loos, *Bounded symmetric domains and Jordan pairs*, Mathematical Lectures (Irvine: University of California, Irvine, 1977).

[126] B.D. MacCluer, *Iteration of holomorphic self-maps on the open unit ball of* $\mathbb{C}^n$, Michigan Math. J. **30** (1983) 97-106.

[127] K. Madigan and A. Matheson, *Compact composition operators on the Bloch space*, Trans. Amer. Math. Soc. **347** (1995) 2679-2687.

[128] K. McCrimmon, *A taste of Jordan algebras*, Universitext (Heidelberg: Springer-Verlag, 2004).

[129] P. Mellon, *Holomorphic invariance on bounded symmetric domains*, J. Reine Angew. Math. **523** (2000) 199-223.

[130] K. Meyberg, *Jordan-Tripelsysteme und die Koecher-Konstruktion von Lie-Algebren*, Math. Z. **115** (1970) 58-78.

[131] N. Mok, *Metric rigidity theorems on Hermitian locally symmetric manifolds* (New Jersey: World Scientific Publ. Co. 1989).

[132] N. Mok and I.H. Tsai, *Rigidity of convex realization of irreducible bounded symmetric domains of rank* ≥ 2, J. Reine Angew. Math. **431** (1992) 91-122.

[133] J. Mujica, *Complex analysis in Banach spaces*, Math. Studies **120** (Amsterdam: North Holland, 1986).

[134] R. Narasimhan, *Several complex variables* (Chicago: Uni. Chicago Press, 1971).

[135] T. Nomura, *Grassmann manifold of a JH-algebra*, Annals of Global Analysis and Geometry **12** (1994) 237-260.

[136] F. Norguet, *Sur les domaines d'holomorphie des fonctions uniformes de plusieurs variables complexes (passage du local au global)*, Bull. Soc. Math. France **82** (1954) 139-159.

[137] P. Noverraz, *Pseudo-convexité, convexité polynomiale et domains d'holomorphie en dimension infinite*, Math. Studies **3** (Amsterdam: North-Holland, 1973).

[138] R. D. Nussbaum, *Finsler structures for the part metric and Hilbert's projective metric and applications to ordinary diffeential equations*, Diff. Integ. Eq. **7** (1994) 1649-1707.

[139] K. Oka, *Sur les fonctions analytiques des plusieurs variables IX. Domaines finis sans point critique intérieur*, Japan J. Math. **23** (1953) 97-155.

[140] H. Omori, *Infinite-dimensional Lie groups*, Transl. Math. Monograps **158** (Providence: Amer. Math. Soc. 1997).

[141] A. Pelczynski and C. Bessaga, *Some aspects of the present theory of Banach spaces*, In S. Banach, Travaux sur L'Analyse Fonctionnelle (Warszaw: 1979).

[142] V.P. Potapov, *The multiplicative structure of J-contractive matrix functions*, Amer. Math. Soc. Transl. **15** (1960) 131-243.

[143] I. Pyatetzki-Shapiro, *On a problem posed by É. Cartan*, Dokl. Akad. Nauk. SSSR **124** (1959) 272-273.

[144] I. Pyatetzki-Shapiro, *Automorphic functions and the geometry of classical domains* (New York: Gordon-Breach, 1969).

[145] S. Reich and D. Shoikhet, *The Denjoy-Wolff theorem*, Ann. Univ. Mariae Curie-Skłodowska Sec. A **51** (1997) 219-240.

[146] S. Reich and D. Shoikhet, *Nonlinear semigroups, fixed points, and geometry of domains in Banach spaces* (Singapore: World Scientific, 2005).

[147] G. Ren and C. Tu, *Bloch space in the unit ball of* $\mathbb{C}^n$, Proc. Amer. Math. Soc. **133** (2005) 719-726.

[148] J.M. Rigby, *Holomorphic dynamics on bounded symmetric domains of finite rank*, PhD Thesis (University of London, 2015).

[149] T. Robart, *Sur l'intégrabilité des sous-algèbres de Lie en dimension infinie*, Canad. J. Math. **49** (1997) 820-839.

[150] G.J. Roos, *Exceptional symmetric domains*, In Symmetries in complex analysis, Contemp. Math. **468** (2008) 157-189.

[151] H. Rossi, *Holomorphically convex sets in several complex variables*, Ann. of Math. **74** (1961) 470-493.

[152] H.L. Royden, *Renarks on the Kobayashi metric, Several complex variables*, II. In Lecture Notes in Math. **185** (Berlin: Springer-Verlag 1971), pp. 125-137.

[153] W. Rudin, *Functional analysis* (New York: McGraw-Hill, 1973).

[154] S. Sakai, *C*-algebras and W*-algebras* (Berlin: Springer-Verlag, 1971).

[155] I. Satake, *Algebraic strutures of symmetric domains* (Princeton: Princeton Univ. Press, 1980).

[156] R.D. Schafer, *An introduction to nonassociative algebras* (London: Academic Press, 1966).

[157] J. Shi and L. Luo, *Composition operators on the Bloch space of several complex variables*, Acta Math. Sinica **16** (2000) 85-98.

[158] H. Shima, *A differential geometric characterization of homogeneous self-dual cones*, Tsukuba J. Math. **6** (1982) 79-88.

[159] Y-T. Siu, *Pseudoconvexity and the problem of Levi*, Bull. Amer. Math. Soc. **84** (1978) 481-512.

[160] A. Stachura, *Iterates of holomorphic self-maps of the unit ball in Hilbert space*, Proc. Amer. Math. Soc. **93** (1985) 88-90.

[161] M. Takesaki, *Theory of operator algebras* I (Berlin: Springer-Verlag, 1979).

[162] R.M. Timoney, *Bloch functions in several complex variables*, I, Bull. London Math. Soc. **12** (1980) 241-267.

[163] R.M. Timoney, *Bloch functions in several complex variables*, II, J. Reine Angew. Math. **319** (1980) 1-22.

[164] A.C. Thompson, *On certain contraction mappings in a partially ordered vector space*, Proc. Amer. Math. Soc. **14** (1963) 438-443.

[165] J. Tits, *Algèbres alternatives, algèbres de Jordan et algèbres de Lie exceptionnelles*, I. *Construction*, Nederl. Indag. Math. **28** (1966) 223-237.

[166] T. Tsuji, *A characterization of homogeneous self-dual cones*, Proc. Japan Acad. **57** (1981) 185-187.

[167] H. Upmeier, *Über die Automorphismengruppe von Banach-Mannigfaltigkeiten mit invarianter Metrik*, Math. Ann. **223** (1976) 279-288.

[168] H. Upmeier, *Symmetric Banach manifolds and Jordan C*-algebras*, Math. Studies **104** (Amsterdam: North Holland, 1985).

[169] E. Vessentini, *Invariant metrics on convex cones*, Scuola Norm. Sup. Pisa **3** (1974) 671-696.

[170] E.B. Vinberg, *The theory of convex homogeneous cones*, Trudy Moskov. Mat. Obsc. **12** (1963) 303-358.

[171] E.B. Vinberg, S.G. Gindikin and I.I. Piatetski-Shapiro, *Classification and canonical realization of complex homogeneous bounded domains*, Trudy Moscow Mat. Obšč **12** (1963) 359-388.

[172] J.P. Vigué, *Le groupe des automorphismes analytiques d'un domaine borné d'un espace de Banach complexe. Application aux domaines bornés symétriques*, Ann. Sc. Ec. Norm. Sup. **9** (1976) 203-282.

[173] W. van Est and T. Korthagen, *Nonenlargeable Lie algebras*, Indag. Math. **26** (1964) 15-31.

[174] J.F. Wang and T.S. Liu, *Bloch constant of holomorphic mappings on the unit polydisk of* $\mathbb{C}^n$, Sci. China Ser. A **51** (2008) 652-659.

[175] F.D. Wicker, *Generalized Bloch mappings in complex Hilbert space*, Canad. J. Math. **29** (1977) 299-306.

[176] J. Wolf, *Fine structure of Hermitian symmetric spaces*, In Symmetric Spaces (Marcel Dekker, 1972), pp. 271-357.

[177] J. Wolff, *Sur une génŕalisation d'un theéorème de Schwarz*, C.R. Acad. Sc. Paris **182** (1926) 918-920.

[178] J. Wolff, *Sur l'itération des fonctions bornées*, C.R. Acad. Sc. Paris **182** (1926) 42-43 and 200-201.

[179] J.D.M. Wright, *Jordan C^*-algebras*, Michigan Math. J. **24** (1977) 291-302.

[180] H. Wu, *Normal families of holomorphic mappings*, Acta Math. 119 (1967) 193-233.

[181] M. Youngson, *Hermitian operators on Banach Jordan algebras*, Proc. Edinburgh Math. Soc. **22** (1979) 169-180.

[182] S-K. Yeung, *Geometry of domains with the uniform squeezing property*, Adv. Math. **221**(2009) 547-569.

[183] E.I. Zelmanov, *On prime Jordan algebras*, Algebra i Logika **8** (1979) 162-175. English Transl. Algebra and Logic **18** 1979.

[184] E.I. Zelmanov, *On prime Jordan algebras* II, Sibirsk Mat. Zh. **24** (1983) 89-104. English Transl. Siberian Math. J. **24** 1993.

[185] E.I. Zelmanov, *Lie algebras with a finite grading*, Math. USSR Sbornik, **52** (1985) 347-385.

[186] Z. Zhou and J. Shi, *Compactness of composition operators on the Bloch space in classical bounded symmetric domains*, Michigan Math. J. **50** (2002) 381-405.

Index

Printed in the United States
by Baker & Taylor Publisher Services